Sustainable Eco-technologies for Water and Wastewater Treatment

In Focus – Special Book Series

Sustainable Eco-technologies for Water and Wastewater Treatment

Editors

Eldon R. Rene, Li Shu and Veeriah Jegatheesan

IWA PUBLISHING

Published by **IWA Publishing**
Alliance House
12 Caxton Street
London SW1H 0QS, UK
Telephone: +44 (0)20 7654 5500
Fax: +44 (0)20 7654 5555
Email: publications@iwap.co.uk
Web: www.iwapublishing.com

First published 2020

British Library Cataloguing in Publication Data
A CIP catalogue record for this book is available from the British Library

ISBN: 9781789061369 (paperback)
ISBN: 9781789061376 (eBook)
ISBN: 9781789061383 (ePUB)

Contents

Editorial: Sustainable eco-technologies for water and wastewater treatment

One of the major challenges in the world is to provide clean water and sanitation for all. With 3% fresh water reserves in the earth, there are more than 1 billion people who still lack access to clean drinking water. The declining water quality has not only reduced the life expectancy of humans, but it has also contributed to the deleterious negative impacts on aquatic/marine life, flora, fauna and the ecosystem. However, with rapid technological advancements and the availability of advanced scientific instruments, there has been substantial improvement in the design and operation of water and wastewater treatment systems. Recently, these sustainable eco-technologies have been designed and operated to offer the following advantages: (i) a smaller footprint, (ii) less maintenance, (iii) >99% removal of contaminants, (iv) provides the option for resource recovery, (v) less energy consumption, (vi) minimal use of chemicals, and (vii) less investment and operational costs.

Although these eco-technologies have proven to be reliable to treat water and wastewater containing a complex mixture of pollutants, the development and application of innovative hybrid processes and concepts are still needed in order to meet the stringent pollutant control regulations. Besides, due to changing lifestyles, such as the use of numerous health care products, cosmetics and pharmaceutical compounds, the challenges faced by the treatment systems have become increasingly complex. In the last decade, there has been a marked rise in the occurrence of emerging/persisting contaminants and antimicrobial resistant bacteria in the majority of drinking water sources. Due to the development of sophisticated detection techniques and analytical instruments, the detection and monitoring of very low concentrations of these compounds have been made possible by researchers. From the perspective of wastewater treatment, the conventional technologies such as activated sludge process, trickling filters, sequencing batch reactors, ponds and lagoons, among others, are not designed to remove these pollutants. In such cases, stand-alone technologies such as membrane filtration, ultrafiltration (UF), nano-filtration (NF), reverse osmosis (RO), ozone and catalytic oxidation, and membrane bioreactors (MBRs) could only partially remove some these emerging/persisting contaminants from water or wastewater. Thus, researchers have also focused on the development of hybrid eco-technologies that combine the operational advantages of advanced technologies or physico-chemical treatment systems and conventional biotechnologies (e.g. activated sludge process + a membrane filtration unit, or adsorption column + a biotrickling filter) in order to reduce the level of these pollutants to the desired regulatory standards.

This special issue on 'Sustainable eco-technologies for water and wastewater treatment' highlights the technologies used for the removal of pollutants such as dyes, uranium, cyanotoxins, faecal contamination and P/N compounds from water environments. Several case-studies from landfills, bioretention systems, membrane bioreactors and pumping stations are also covered in this special issue. Figure 1 shows a representative infographic of the different ecotechnologies used worldwide for water, wastewater treatment and resource recovery.

Numerous anthropogenic sources such as the metal finishing industry, electroplating processes, mining extraction operations, textile industries, and nuclear power plants, as well as natural sources such as soil erosion, urban runoff, aerosols, and particulates contribute to the deposition of metals or metallic elements in water bodies. The need for environmentally friendly biotechnologies for the removal of organic and inorganic water pollutants has been the focus of many published reports in recent years, with the focus on sustainably produced adsorbents for removal of water pollutants. In this regard, Baltrėnaitė *et al.* (2019) identified the characteristics of accumulated metallic elements ligneous biomass and related it to the adsorptive properties of the biochar derived from such modified biomasses. The authors of this paper investigated the forms and quantities of metallic elements present in the biomass and their transformation to biochar. The results from their study suggest that additional syngenetical present metallic elements increases the adsorptive properties of the resultant biomass-derived biochar. When compared with a reference biochar, i.e. commercially available biochar, the lignin biochar showed a higher amount of ash content and higher electrical conductivity values. The mean concentrations of metallic elements in lignin biochar were higher than in reference biochar: 4 times for Pb, 70 times for Cu, 10 times for Ni, 38

doi: 10.2166/aqua.2019.100

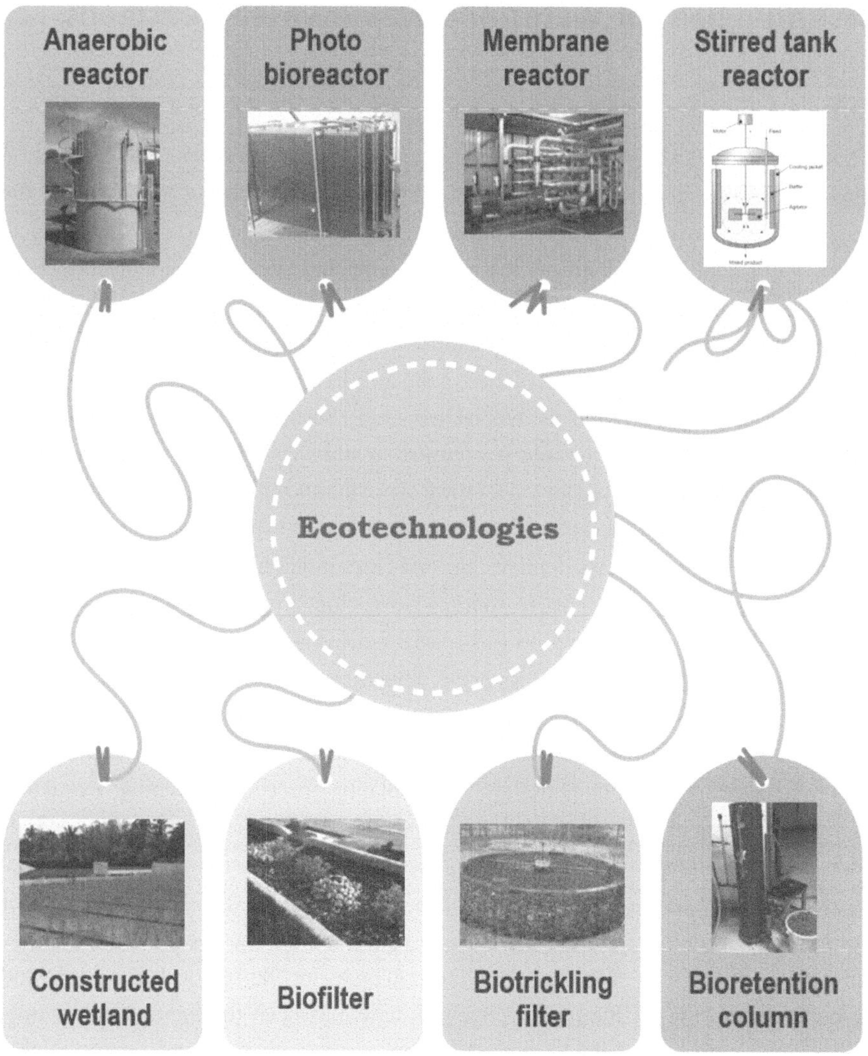

Figure 1 | Representative ecotechnologies used for water, wastewater treatment and resource recovery. Most of these technologies use naturally available resources (e.g. micro-organisms or packing material or plant species) for treating the pollutant.

times for Cr, 1.7 times for Mn, 3.5 times for Zn and 3.8 times for Cd, respectively. In another study, Zeng *et al.* (2019) collected anaerobic granular sludge from a citric acid wastewater treatment plant and enriched the sludge under sulfate-reducing conditions. Thereafter, the authors used the biomass for the removal of high uranium concentrations (20 mg/L) from synthetic wastewater. The authors also identified the dominant microbial communities involved in removing uranium from wastewater. The Bacteroidetes phylum had the largest part in the composition, which accounted for 42.25% with 3,504 reads. The second most dominant phylum was Firmicutes (26.55%) followed by Proteobacteria (20.66%). Accordingly, the sulfate-reducing granular sludge exhibited high (>90%) removal efficiencies during four feeding operations, lasting for a total of 24 days from synthetically prepared 20 mg/L uranium acid mine drainage. The interesting results obtained from this study proved that the microbial community plays an important role in uranium recovery from contaminated sites, which is also beneficial for the control of uranium contamination.

Among the emerging pollutants reported in many water bodies worldwide, microcystins (MCs) are one of the common and highly potent hepatotoxins formed during algal blooms. Accidental exposure to MCs via skin contact, inhalation, or ingestion of toxin-contaminated water sources can lead to breathing problems, nausea, diarrhoea, skin irritation, and even acute liver damage at a high level of exposure. MCs are difficult to physically detect in water due to their colourless, odourless, and tasteless characteristics. According to Ponnusamy *et al.* (2019), the complete removal of cyanotoxin could be a challenging process for conventional water treatment plants as there is a possibility of dissolved extracellular toxins passing through the treatment process. The authors ascertained the MC-LR removal efficiency at varying ozone dosages and initial toxin concentrations and tested the effectiveness of the MC-LR removal using a novel peroxone process. The experimental results showed that ozone and peroxone can be used to effectively remove MC-LR in drinking water; however, the oxidant dosages were governed by the initial toxin concentrations and the organic carbon content present in the water samples. The authors observed that higher total organic content values led to poor removal of MC-LR, presumably due to competition kinetics between the toxin and organic carbon during the oxidation process. From a practical viewpoint, the authors recommended that, as toxin concentration in real cases exists at extremely low concentration, an ozone dose of 0.75 mg/L would be sufficient to degrade MC-LR.

Nanotechnologies have prominent applications in the field of science and technology owing to their size-tuneable properties, providing a promising approach for the degradation of various pollutants. Nickel ferrites due to their relatively dry and more compact physical structure than ferric hydroxide considerably reduce sludge volume in ferrite formation. In a study by Vijay *et al.* (2019), nickel ferrite was used as the model heterogeneous photo-Fenton catalyst. The authors tested the effect of nickel ferrite nanoparticles on the degradation of Irgalite violet dye by Fenton's reaction using oxalic acid as an oxidizing agent in the presence of sunlight. The effect of pH and adsorbent dosage on the rate of dye degradation was evaluated and 99% dye degradation was achieved under the following conditions: catalyst dosage – 0.2 g, dye concentration – 400 mg/L, oxalic acid dose – 2.0 mM, pH – 3.0 and contact time - 60 min. In another similar study, Valappil *et al.* (2019) tested the photocatalytic potential of Co_3O_4 nanoparticles for the treatment of RB220 (dye) at different time, pH, dye concentration (mg/L), and nanoparticle concentration (mg/L). The authors determined the decolourization efficiency of Co_3O_4 nanoparticles and elucidated the mechanism and kinetics of RB220 decolourization. According to the results, a maximum dye concentration of 10 mg/L of RB220 was decolorized, at Co_3O_4 nanoparticle dose of 150 mg/L.

Apart from dyes and heavy metals, nonpoint source (NPS) pollution also leads to different water quality problems. NPSs of contamination may be attributed to urban and agricultural runoff, leakage from septic or sewer systems, and storm water runoff. Martinez *et al.* (2019) developed methods in which certain microorganisms were used to indirectly identify probable sources of anthropogenic faecal pollution in the Bayou Lafourche area, USA. This is known as microbial source tracking (MST). The overall goal of that study was to investigate the correlations between optical brighteners (OBs), faecal coliforms (FC), *E. coli*, and polymerase chain reaction (PCR) data in selected contaminated sites in Upper Bayou Lafourche and to determine if there was any correlation among these parameters to identify the source of faecal contamination. According to the authors, the OB readings, FC, *E. coli*, and *M. smithii* were found not to be good indicators of human faecal contamination. Therefore, the authors had recommended using the two anthropogenic sewage markers HB and HPyV in order to obtain reliable results from specific water contaminated locations. In another study related to water quality, Senevirathna *et al.* (2019) discussed the problems faced due to the primary and secondary water quality of eight water distribution networks in New South Wales (NSW), Australia. Based on the analysis of a large number of drinking water samples (more than 11,000), the authors identified that maintaining microbial water quality and the required free chlorine level (>0.2 mg/L) are challenging issues for regional water distribution networks. Another interesting observation from this study was the fact that water taps installed in public places are rarely used by the regional population and most of them failed to meet drinking water quality guidelines. The authors recommended basic modifications in design and operational procedures to solve regional drinking water systems.

In the same magnitude of environmental and health effects as that of persisting/emerging contaminants, pesticide and insecticide usage has become a necessary evil in developing countries. It has increased several-fold where agriculture is anticipated to be the backbone of the economy. During the past few decades, agrochemicals have been widely used in most agricultural sectors for enhancing crop yield and improving the quality of the product. Consequently, the extensive application of pesticides poses potential risks to the biodiversity of freshwater aquatic environments because of their bioaccumulation and intrinsic toxicity. The insecticide imidacloprid is a chloronicotinyl nitroguanidine and due to its systemic nature, it is ranked as the second most widely used pesticide globally. Muazzam *et al.* (2019) evaluated the toxicity of combined exposures of endosulfan and imidacloprid on zebrafish in terms of oxidative stress and deoxyribonucleic acid (DNA) damage in liver and histological alterations in gills and muscles. Zebrafish were exposed to three different sub-lethal concentrations of endosulfan and imidacloprid along with control selected for each treatment for 21 days. The authors reported a substantial increase in DNA damage after 21 days' exposure to pesticides and significant morphological changes were observed in the gills of the fish.

The presence of high nitrate concentrations in drinking water results in health issues such as blue baby syndrome (methemoglobinemia), stomach cancer, high blood pressure, thyroiditis, cytogenetic malfunction, and several other defects in newborn babies. Conventional treatment processes include reverse osmosis, adsorption, ion exchange, chemical denitrification, photocatalytic reduction and electro-dialysis for the remediation of water containing excessive levels of nitrate. As a low-cost and eco-technology, phytoremediation can be considered as a versatile treatment technology for removing nutrients and trace contaminants from wastewater and surface water sources. As a proof-of-concept, Shyamala *et al.* (2019) demonstrated the application of ornamental plants for the treatment of nitrate containing water by performing laboratory-scale experiments using two ornamental plants: money plant and arrowhead plant. The authors reported that the individual (main) effect of the process variables, i.e. the initial nitrate concentration, growth period, plant density, was found to play a major role in affecting the nitrate removal process. Furthermore, it was reported that an increase in the growth period, i.e. the time, increased the rate of nitrate removal with k values of 0.22 d^{-1} for money plant and 0.35 d^{-1} for arrowhead plant, respectively.

In natural systems, the removal of ammonia nitrogen (NH_3-N) is mainly through adsorption or ion exchange in the soil and filler. Zhang *et al.* (2019) proposed a practical long-term treatment method for treating high nitrate river water in the Taihu basin (China): the traditional emergency method (e.g. water diversion) or short-term method (e.g. sediment dredging). The authors designed six enhanced submerged media, operating at four hydraulic retention times (HRTs) for the bioretention cells to treat the highly nitrogenous river water. According to the authors, the bioretention with activated carbon showed the highest removal of nitrate (93–96%) compared to surfactant-modified activated carbon (SMAC), surfactant-modified zeolite (SMZ), zeolite, fly ash and ceramsite. For long-term performance, the activated carbon, non-modified zeolite and SMZ showed high nitrate removal efficiency due to the formation and stabilization of the denitrification system. The authors recommended the most cost-efficient design as follows: 150 mm activated carbon +150 mm gravel operating under hydraulic loading of 4 L/h. In another bioretention study, Yang *et al.* (2019) determined the proper composition of filter media using locally available materials, which are suitable for the specific/local climate and environmental conditions in the Yangtze River delta region (China). The results of that study showed that the sand with discontinuous gradation contained a certain amount of clay, leading to unsatisfying hydraulic performance (hydraulic conductivity ranged from 423 to 1,054 mm/h, and 1,500 to 29 mm/h). On the contrary, the mixture of locally available sand, which consisted of continuous gradation of coarse sand (40–70%, by mass), fine sand (0–40%, by mass), very fine sand (10–60%, by mass) and nutrient soil (0–3%, by mass), had a hydraulic conductivity ranging from 200 to 400 mm/h and relatively stable structure. Furthermore, during the 70-day flooding test, the hydraulic conductivity changed within the first 20 days due to the migration of particles (mainly $<$0.6 mm) and thereafter the bioretention cell became stable. Thus, the authors recommended that easy availability and a simple production process make these materials more economic and easier to promote their utilization in the construction of sponge cities in China. In another novel bioretention study, Wang *et al.* (2019) operated a bioretention cell (BRC) and an enhanced system combined bioretention cell with microbial fuel cell (BRC-MFC) to treat domestic wastewater. The authors investigated the nitrogen removal characteristics and permeation characteristics of these two systems by adjusting the influent carbon/nitrogen ratio ($C/N = 2.0$–20.0). Based on the results, it was delineated that the nitrification and denitrification performances were mainly influenced by organic matter present in the wastewater and the system combination influenced/promoted better nitrogen removal.

The discharge of excess nutrients into the water bodies affects the self-purification ability of aquatic ecosystems, resulting in algal blooms. Eutrophication leads to the sharp decline of dissolved oxygen in water and the deterioration of water quality which destroy freshwater ecosystems. Therefore, in order to solve the problem of poor P removal efficiency of traditional floating beds under low P concentration, Liu *et al.* (2019) proposed an efficient P removal composite (EPRC) as the growth substrate of the ecological floating island. The main components of the EPRC were fly ash and steel slag. According to the authors, the composite floating bed was able to significantly enhance the P removal efficiency of *Ipomoea aquatica* and promote its growth.

The overload of organic matter pollution in surface water leads to operational challenges during treatment in conventional drinking water treatment plants. These challenges include higher consumption of coagulants and higher disinfection by-product formation potential. There has been limited research on membrane bioreactors (MBR) in treating polluted surface water. Li *et al.* (2019) analyzed and compared the endogenous respiration process in an aerobic membrane bioreactor (aMBR) and UV/O_3-aMBR systems in treating polluted surface water by the respirometry method. The authors compared the microbial activities, including heterotrophic and autotrophic entry time of endogenous respiration, and ability under anti-decay conditions. The UV/O_3-aMBR system was able to make full use of the advanced oxidation process (AOP) and biological process, leading to a higher treatment performance, and has the potential to mitigate total energy demand. According to the authors, the study provided useful information for the understanding and practical application of aMBR and UV/O_3-aMBR processes in treating polluted surface water for drinking purposes in a more environmentally friendly way.

With the worldwide construction of reservoirs and inter-basin water transfer projects, an optimal joint operation of multi-reservoir systems has recently become a research hotspot. In this context, double-reservoir-and-double-pumping-station systems are commonly used for irrigation water supply in hilly regions. Gong *et al.* (2019) proposed an optimization model for a water supply system in order to minimize water shortage. The optimization models include two types of decision variables (water supplies and spills) that increase with the number of reservoirs; increased variables lead to the 'curse of dimensionality' in the solving process. The authors developed an optimization method based on the decomposition-aggregation theory for the joint operation of a double-reservoir-and-double-pumping station system that can provide optimal results following normal operation regulations. The authors integrated a joint operation rule of the system into the optimization model to obtain an appropriate operation scheme of both reservoirs and pumping stations.

Landfill leachate can be defined as an aqueous liquid generated as a consequence of the percolation of precipitation (rain, snow melt) and infiltration of groundwater through wastes, biochemical processes in waste's cells and the inherent water content of wastes themselves. Leachate is a complex wastewater comprising of a large number of different substances or compounds of mainly organic matter, ammoniacal nitrogen, heavy metals and other dissolved solids. In this special issue, two case studies focused on practical issues related to landfill sites.

In the first case study, Narayan *et al.* (2019) proposed a sequencing batch reactor (SBR) for the treatment of landfill leachate in the Lany municipality of Czech Republic. The authors also performed an economic assessment on the proposed design of SBR using cost benefit analysis (CBA) and evaluated the economic performance of the project by estimating cash flow, net present value and internal return rate. According to the authors, the net present value of the project was 19,526 €, with an internal return rate of 21.6%. Furthermore, an assessment of environmental, economic, and community impact on the existing practice (total score 55.1%) and on-site treatment (total score 59.6%) was comparable in that study. In the second case study, Anvarov *et al.* (2019) assessed different strategies for the remediation of groundwater contaminated with the leachate from an old, non-functional landfill located next to Hořkovec open cast mine, in the Czech Republic. The leachate consisted of mainly chlorinated aliphatic compounds and aromatic volatile compounds. The authors presented an analysis of the advantages, disadvantages and limitations of different remediation technologies. From a practical perspective and in view of their suitability for the treatment of the landfill leachate, a combination of different technologies was proposed by the authors. Additionally, the authors proposed financial assessment and an environmental impact assessment of the proposed treatment. Some recommendations from this study include: (i) consideration of the geology and geochemistry of the area while evaluating the suitability of different technologies, and (ii) to perform pilot tests before designing the remediation treatment in order to characterise the pollutants present and the characteristics of the treated area.

Based on the topics covered in this special issue, it is evident that ecotechnologies are becoming more and more important and playing critical role in removing a wide variety of organic and inorganic pollutants from water. This special issue required substantial commitment from the authors and the reviewers to meet the high scientific standards of quality maintained by the *Journal of Water Supply: Research and Technology-AQUA*. We would like to thank them for their efforts. We thank and appreciate the continuous support and invaluable guidance received from Ms Emma Gulseven and her team at IWA Publishing in making this special issue possible.

Guest Editors
Eldon R. Rene
IHE-Delft Institute of Water Education, Delft, The Netherlands

Li Shu
LJS Environment, Melbourne, Australia

Veeriah Jegatheesan
Royal Melbourne Institute of Technology (RMIT) University, Melbourne, Australia

REFERENCES

Anvarov, A., Angel, A. P., Rando, B. F. & Gil, J. L. 2019 Remediation of groundwater contamination from an old, non-functional landfill in Hořkovec open cast mine, Czech Republic. *J. Water Supply Res. Technol. Aqua* **68** (8), 829–841.

Baltrėnaitė, E., Baltrėnas, P. & Huisingh, D. 2019 Technogenic metallic elements in biomass and their effects on biomass product properties. *J. Water Supply Res. Technol. Aqua* **68** (8), 623–644.

Gong, Z., Jiang, X., Cheng, J., Gong, Y., Chen, X. & Cheng, H. 2019 Optimization method for joint operation of a double-reservoir-and-double-pumping-station system: a case study of Nanjing, China. *J. Water Supply Res. Technol. Aqua* **68** (8), 803–815.

Li, L., Song, K. & Visvanathan, C. 2019 Endogenous respiration process analysis between aMBR and UV/O_3-aMBR for polluted surface water treatment. *J. Water Supply Res. Technol. Aqua* **68** (8), 793–802.

Liu, Y., Lv, J. & Singh, R. P. 2019 Removal of low-concentration phosphorus by efficient phosphorus removal composite-based ecological floating beds. *J. Water Supply Res. Technol. Aqua* **68** (8), 782–792.

Martinez, S., Kilgen, M., Corbin, A., Nathaniel, R., Ramachandran, B. & Boopathy, R. 2019 Anthropogenic markers for source tracking of fecal contamination in Bayou Lafourche: a major drinking water source in Southeast Louisiana, USA. *J. Water Supply Res. Technol. Aqua* **68** (8), 687–707.

Muazzam, B., Munawar, K., Khan, I. A., Jahan, S., Iqbal, M., Asi, M. R., Farooqi, A., Nazli, A., Hussain, I. & Zafar, M. I. 2019 Stress response and toxicity studies on zebrafish exposed to endosulfan and imidacloprid present in water. *J. Water Supply Res. Technol. Aqua* **68** (8), 718–730.

Narayan, R. B., Zargham, B. I., Ngambia, A. & Riyanto, A. R. 2019 Economic and environmental impact analysis of ammoniacal nitrogen removal from landfill leachate using sequencing batch reactor: a case study from Czech Republic. *J. Water Supply Res. Technol. Aqua* **68** (8), 816–828.

Ponnusamy, G., Francis, L., Loganathan, K., Ogunbiyi, O. O., Jasim, S. & Saththasivam, J. 2019 Removal of cyanotoxins in drinking water using ozone and ozone-hydrogen peroxide (peroxone). *J. Water Supply Res. Technol. Aqua* **68** (8), 655–665.

Senevirathna, S. T. M. L. D., Goncher, A. M. & Hollier, A. 2019 Assessment of drinking water quality in regional New South Wales, Australia. *J. Water Supply Res. Technol. Aqua* **68** (8), 708–717.

Shyamala, S., Manikandan, N. A., Pakshirajan, K., Tang, V. T., Rene, E. R., Park, H.-S. & Behera, S. K. 2019 Phytoremediation of nitrate contaminated water using ornamental plants. *J. Water Supply Res. Technol. Aqua* **68** (8), 731–743.

Valappil, R. S. K., Vijayanandan, A. S. & Balakrishnan, R. M. 2019 Decolorization of Reactive Blue 220 aqueous solution using fungal synthesized Co_3O_4 nanoparticles. *J. Water Supply Res. Technol. Aqua* **68** (8), 675–686.

Vijay, S., Balakrishnan, R. M., Rene, E. R. & Priyanka, U. 2019 Photocatalytic degradation of Irgalite violet dye using nickel ferrite nanoparticles. *J. Water Supply Res. Technol. Aqua* **68** (8), 666–674.

Wang, Y., Singh, R. P., Zhang, J., Xu, Y. & Fu, D. 2019 Nitrogen removal performance of microbial fuel cell enhanced bioretention system. *J. Water Supply Res. Technol. Aqua* **68** (8), 769–781.

Yang, F., Singh, R. P. & Fu, D. 2019 Experimental study on filter media using locally available materials in bioretention. *J. Water Supply Res. Technol. Aqua* **68** (8), 757–768.

Zhang, J., Singh, R. P., Liu, Y. & Fu, D. 2019 Design and operation of submerged layer in bioretention for enhanced nitrate removal. *J. Water Supply Res. Technol. Aqua* **68** (8), 744–756.

Zeng, T., Zhang, S., Liao, W., Ma, H., Lens, P. N. L. & Xie, S. 2019 Bacterial community analysis of sulfate-reducing granular sludge exposed to high concentrations of uranium. *J. Water Supply Res. Technol. Aqua* **68** (8), 645–654.

Technogenic metallic elements in biomass and their effects on biomass product properties

Edita Baltrėnaitė, Pranas Baltrėnas and Donald Huisingh

ABSTRACT

Availability of clean water and sustainable management of water are among the 17 global 'Sustainable Development Goals,' established by the United Nations. The increasing technogenesis-related pollution causes increases the concentrations of many metallic elements (MEs) throughout the biosphere. The syngenetic bio-incorporation of MEs changes the biomass's chemical and physical composition and impacts the quality of products, e.g. biochar (BC). This has impacts upon BCs potential for adsorption of water pollutants. While direct, engineered modification of BC can improve its adsorption characteristics, the more indirect alterations of BC composition, based on syngenetic elements, also modifies BCs adsorption functionality. The objective of this study was to identify the characteristics of syngenetically accumulated MEs in ligneous biomass on the adsorptive properties of the BCs derived from such modified biomasses. The authors of this paper investigated the syngenetic pathway of MEs into the ligneous biomass and provided an overview of the forms and quantities of MEs in the biomass and their transformation to BC. The early results of the effects of syngenetic MEs on the adsorptive properties of BC from ligneous biomass are presented. The results suggest that additional syngenetical MEs result in increases in the adsorptive properties of the resultant biomass-derived BCs.

Key words | adsorption, biochar, metallic elements, syngenetic elements, technogenesis

Edita Baltrėnaitė (corresponding author)
Pranas Baltrėnas
Institute of Environmental Protection,
Vilnius Gediminas Technical University,
Saulėtekio al. 11, LT-10223,
Vilnius,
Lithuania
E-mail: *baltrenaite@aol.com*

Donald Huisingh
The Institute for a Secure and Sustainable Environment,
University of TN,
Knoxville,
USA

INTRODUCTION

According to the World Health Organization, 2.1 billion people live without safe water at home. Contaminated water can transmit diseases such as diarrhea, cholera, dysentery, typhoid, and polio. Contaminated drinking water is estimated to cause 502,000 diarrheal deaths each year. At least 10% of the world's population has been documented to consume food irrigated by wastewater, which usually contains contaminants such as metallic elements (MEs) (e.g. Cu is a typical metal in wastewater (Shen 2019)), and organic contaminants (WHO 2017).

Both natural sources (such as soil erosion, urban runoff, aerosols, and particulates) and anthropogenic sources (metal finishing and electroplating processes, mining extraction operations, textile industries, nuclear power (Zeng *et al.* 2018; Oluwatuyi *et al.* 2019)) of MEs entering waterbodies are driven by intensive technogenesis and increased consumption of society. The need for environmentally friendly biotechnologies for the removal of organic and inorganic water pollutants was investigated (Rene *et al.* 2018) with the focus on sustainably produced adsorbents for the removal of water pollutants (Begum *et al.* 2018; Feng *et al.* 2018; Rai *et al.* 2018).

With the increase in gross domestic product (GDP) and human population by about 225% and 72% respectively since 1994, the world extraction of MEs increased by over 75%. As a result, since 1,983 the emissions of MEs to the biosphere, including waterways, have increased by a factor of 1.5–3. The stocks and flows of the global cycles of Ag, Al, Cr, Co, Fe, Ni, Pb, and Zn comprise over 98% of the

doi: 10.2166/aqua.2019.082

total mass of MEs mobilized by human activities by the turn of the 21st century (Rauch & Pacyna 2009).

The increase in MEs extraction intensifies the problem development of higher water contamination loads, which, in turn, lead to the increases in MEs biophilicity and bioaccumulation (Lietuvninkas 2012; Baltrėnaitė *et al.* 2016, 2018). The ways such technogenic substances are entering biomass during growth has the features of syngeneity and, therefore, can be called syngenetic. Therefore, MEs are subjected to a new – syngenetic – way of entering biomass.

The growing demand for organic products in the market promotes their use in various fields, including environmental engineering. One such product is biochar (BC) and its use in adsorption systems. Being of natural origin (i.e. produced via pyrolysis from organic biomass), such adsorbents are subject not only to adsorption efficiency requirements, but also to requirements of multi-functionality, cascading usage. For example, biochar, which resembles activated carbon, is produced for the adsorption of pollutants from biowaste (an advantage in waste management), it is relatively cheap, locally available, sustainable, promotes nutrient recycling, improves soil properties (when applied to reduce bioavailability of metals in soil), and stimulates healthy soil microbial diversity (Wu *et al.* 2016).

While artificial modification of BCs with MEs can highly improve its adsorptive characteristics (Wei *et al.* 2018) (also for adsorption of MEs from aqueous solution), syngenetically, incorporated MEs may affect BC composition, which enhances the adsorption of water contaminants by BCs as well as decreases BCs modification costs. For example, BC, made of tomato biomass with naturally increased Mg content, adsorbed approximately two times greater amounts of phosphates than BCs from biomass of tomatoes grown in the 'normal way' (Yao *et al.* 2013). It showed that natural or syngenetic entrance of MEs into biomass can be beneficial for altering properties of biomass products. When considering more engineered biomass products, syngeneity of MEs can be advantageous in designing BCs.

The novelty of the paper is the analysis of MEs syngeneity effects on the adsorptive properties of BCs made from this modified biomass. Such BCs can be used in water engineering applications, e.g. removal of contaminants from water.

The authors of this paper investigated the problems associated with the increasing rate of MEs accumulation in living matter in general, with a special focus upon accumulations of MEs in the biomass of ligneous plants. The authors reviewed the advantages of bioaccumulation of syngenetic MEs having technogenic origins, in the biomass of ligneous plants and its product (biochar, BC) from an environmental (water) engineering perspective.

This paper is divided into three parts.

The first two sections review the ways and quantities of MEs that enter the biomass of the ligneous plants. Ligneous biomass was selected to document MEs syngeneity effect for BC adsorptive properties because ligneous biomass is preferred for BC production as plant parts rich in ligneous substances, such as bark, roots and branches, are viewed as waste instead of being used for the production of higher value products. The second part of the paper focuses on the forms and quantities of MEs present during transformation of ligneous biomass into BC.

The BCs produced from ligneous biomass were selected for this research. They are widely produced from different types of biomass for which syngenetic entrance of MEs is especially high.

The BCs have been widely studied for their potential to adsorb water pollutants. Moreover, the syngeneity affected properties of BCs to be utilised in water decontamination processes.

The third part of the paper ('The influence of syngenetic elements on the pyrolysis') provides the preliminary results of the effects of syngenetic MEs on BCs' properties for adsorption of water pollutants. The focus is on the physico-chemical properties which may increase the adsorptive features of BCs.

This research provides new insights. First, it introduces new – syngenetic – origins of MEs entrance into biomass and extends the systematic understanding of origins of MEs. Second, it describes and presents preliminary research results on the syngenetic MEs' impacts on the characteristics of BCs and how those changes impact the absorptive capacities of BCs.

SYNGENETIC METALLIC ELEMENTS AND THEIR PLACE IN CLASSIFICATION BASED ON ORIGIN

The mechanisms of the accumulation of chemical (and metallic) elements (MEs) into the biomass of the ligneous plants vary extensively. Vassilev *et al.* (2012) presented a classification of the biomass containing MEs according to their origin, place and time of generation. These MEs originated from natural (authigenic and detrital) and anthropogenic processes, taking place during pre-syngenesis, syngenesis, epigenesis and post-epigenesis of biomass according to the main formation processes and locations, as well as the times and mechanisms of the formation processes.

The MEs can be integrated into biomass from the aerogenic and edaphic sources as early as during the plant's growth. Aerogenic sources of MEs are usually the stationary or mobile emission sources from which MEs are transported in the form of aerosolic particles. The particles of aerogenic origin primarily include finely dispersed products of chemical ashing of rock (clay minerals) and organic or mineral soil components. Both of them can adsorb MEs, while in organic soil components (i.e. decomposed plants and humus), these components may originate from plants which concentrated the MEs. Thus, MEs can be incorporated within aerosol particles due to the processes taking place on the Earth's surface, such as mineralization, deflation, and evapotranspiration.

Another way MEs enter into biomass is by edaphic origin such as via fertilizers and other soil amendments. They are assimilated and accumulated in the plants as they grow. This path of accumulation MEs in biomass is referred to as a syngenetic mechanism. This implies that the classification of MEs according to their origin offered by Vassilev *et al.* (2012) should be supplemented by the data on MEs of technogenic (authigenic) origin, while their penetration into biomass takes place when a plant is growing (i.e. it is syngenetic) (Figure 1).

With the increase in the intensity of technogenesis, the penetration of MEs into biomass is increasing due to a higher load of MEs in the biosphere. Therefore, syngenetic entrance of MEs into biomass has a growing importance because of the potential effects on biomass and BCs' characteristics.

AMOUNTS AND PHASES OF SYNGENETIC METALLIC ELEMENTS FOUND IN LIGNEOUS BIOMASS

The presence of MEs in ligneous biomass varies depending on their type, species and the availability of MEs in the soil. The differences in local characteristics of the soil, hydrology,

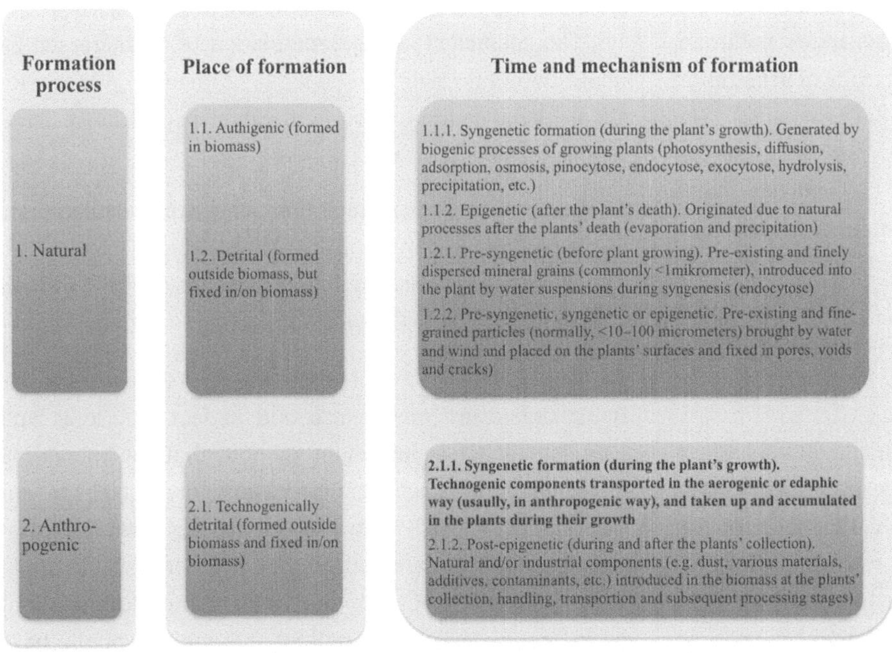

Figure 1 | The origin of metallic elements in ligneous biomass. New – syngenetic – types of metallic elements' entrance into biomass are provided in bold (adapted from Vassilev *et al.* 2012).

Table 1 | Bioaccumulation of chemical elements (according to Lietuvninkas 2012).

Level		Coefficients, K_{ba}				
		100*n	**10*n**	**n**	**0.n**	**0.0n–0.00n**
Accumulation	Energetic	Br, Cl, I, P, S				
	High		B, Ca, K, Mg, Na, Se, Sr, Zn			
Capture	Medium		Ag, As, Ba, Co, Cu, F, Ga, Hg, Mo, Mn, Ni, Pb, Ra, Sn			
	Weak or very weak		Al, Be, Cd, Cr, Cs, Fe, Li, Nb, Rb, Sb, Sc, Si, Ta, Th, Ti, U, V, W, Y, Zr			

Note that the value of n is from 1 to 9.

and the degree of exposure to air pollutants result in various degrees of MEs immobilization, assimilation and bioaccumulation by ligneous plants. Biomass of ligneous plants is 90% of the biomass of terrestrial living matter, while the latter is 99.5% of the mass of living matter (Lietuvninkas 2012). Therefore, bioaccumulation as a phenomenon is important in the case of ligneous plants. The most extensive results and summary on bioaccumulation levels of MEs in living matter in natural conditions are given in Table 1. It is clear that MEs, which are considered potentially toxic (Cr, Cu and Pb), are accumulated at the level 1–0.001*n (where n is a factor with values from 1 to 9) compared to those of the substrate (Table 1).

These limits are corrected by the substrate's pollution, thereby reflecting the increase in the intensity of penetration of MEs into biomass (Baltrėnaitė & Baltrėnas 2019). The changes in the chemical background and the resulting increases in the amounts of MEs in the ligneous plants can be observed in the cases of pollution of edaphic and aerogenic origins. The results revealed that the load of MEs of aerogenic origin was more specific for the accumulation of Ni and Pb emissions, while the load of the edaphic origin was more characteristic for Cu, Cd and Zn emissions (Baltrėnaitė *et al.* 2018).

The global increase in metal concentration in soil and terrestrial biomass from 1990 to 2009 can be observed from data provided in Table 2. The increase in soil Cu, Pb and Cr concentrations by 1.95, 2.7 and 1.44 times, respectively, can be detected. Using data of metal concentration in soil and biomass, the coefficient of bioaccumulation of metals in biomass can be calculated. This coefficient, calculated on the global level, reflects uptake of metals by biomass as affected

Table 2 | Global mean metal (Cu, Zn, Pb, Ni, Cr) concentrations (data in 1990 and 2009) in soils and terrestrial biomass and coefficient of metal bioaccumulation by biomass, mg/kg (Saet *et al.* 1990; Rauch & Pacyna 2009)

	Cu	Zn	Pb	Ni	Cr
Soil (Saet *et al.* 1990)	20	50	10	40	90
Soil (Rauch & Pacyna 2009)	39	48	27	25	130
Terrestrial biomass (ash) (Rauch & Pacyna 2009)	96	720	188	22	18.8
Coefficient of metal bioaccumulation	2.5	15	6.96	0.88	0.14

by long-term metal contamination of soil and indicates the global changes in metal circulation in soil–plant systems.

The coefficient of bioaccumulation of Pb increased by about two times and, consequently, it changed places with Cu and became more important with respect to bioaccumulation in global biomass:

Global biomass bioaccumulation sequence (calculated for data of 2009): $Zn_{15}Pb_{6.96}Cu_{2.5}Ni_{0.88}Cr_{0.14}$.

Global biomass bioaccumulation sequence (in 1990 by Saet *et al.* 1990): $Zn_{19.6}Cu_{9.1}Pb_{3.7}Ni_{1.5}Cr_{1.0}$.

Taking into account that Pb is not essential for living organisms' metabolism and in fact it is toxic and carcinogenic, it is important to note that the increasing bioaccumulation coefficient gives cause for societal concern (Markert *et al.* 2015). Therefore, due to the development of technogenesis, the amount of MEs of anthropogenic origin of releases of Cu, Cr, Pb and Zn into the biosphere will be increasing even more rapidly and will potentially have more negative impacts upon the biosphere and upon human health (Rauch & Pacyna 2009).

Accumulation of MEs in trees due to bioaccumulation, which is a non-equilibrium process, involves the following steps of the element's transport:

a. diffusion of the ME on the surface of the cell;
b. sorption (including adsorption and absorption);
c. binding to the groups on the surface of the cell (passive or active);
d. binding in the organic-metallic form (in this way, metal is excluded from the metabolic processes) (Chojnacka 2010).

The MEs are incorporated into the inorganic matter of the ligneous biomass in any or all of three solid states: crystalline, semi-crystalline and amorphous states (Vassilev *et al.* 2012). The inorganic components in ligneous biomass are typically various salts of carbonates, silicates, phosphates, and oxalates (predominantly, in the bark), in free ionic phase or as cations bound to acidic groups. They are minor constituents of the ligneous materials and, naturally, their total concentrations (measured from the ash) rarely exceed the 1% level (Ivaska & Harju 1999).

The phases of inorganic compounds were studied by Keown *et al.* (2008), Doshi *et al.* (2009), Meyer *et al.* (2012), Vassilev *et al.* (2012, 2014) and Leijenhorst *et al.* (2016) and a summary of the information is presented in Figure 2. The MEs phases in ligneous biomass vary, depending on the element. The most simple species is a free ion dissolved in fluids within the biomass's inner structure. More complex phases include salts in mineral form (e.g. sodium chloride), covalent bonding between inorganic elements and organic structures, solid salt structures embedded on the cell wall or bonded inside the organic matrix, forms of impurities in amorphous or semi-crystalline cellulose, or in sulfate, nitrate or in small (<2 μm) crystal structures.

Meyer *et al.* (2012) suggested that ligneous materials should have more positive associations with respect to divalent MEs ions than to monovalent ions. They found that cellulose can bind more MEs than lignin. Considering ligneous plants as representatives of the living matter, forming the largest biomass, it is evident that accumulation of MEs in ligneous biomass will increase with more intensive technogenic loads of MEs. More divalent MEs are expected to be accumulated in the ligneous biomass than univalent MEs.

EFFECT OF COMPOSITION OF FEEDSTOCK AND CONDITIONS OF PYROLYSIS ON THE ACCUMULATION AND FORMS OF METALLIC ELEMENTS IN BIOCHAR AS AN EXAMPLE OF TYPE OF LIGNEOUS BIOMASS PRODUCT

The growing demand for (natural) organic products in the market fosters new uses/demands. An example of such

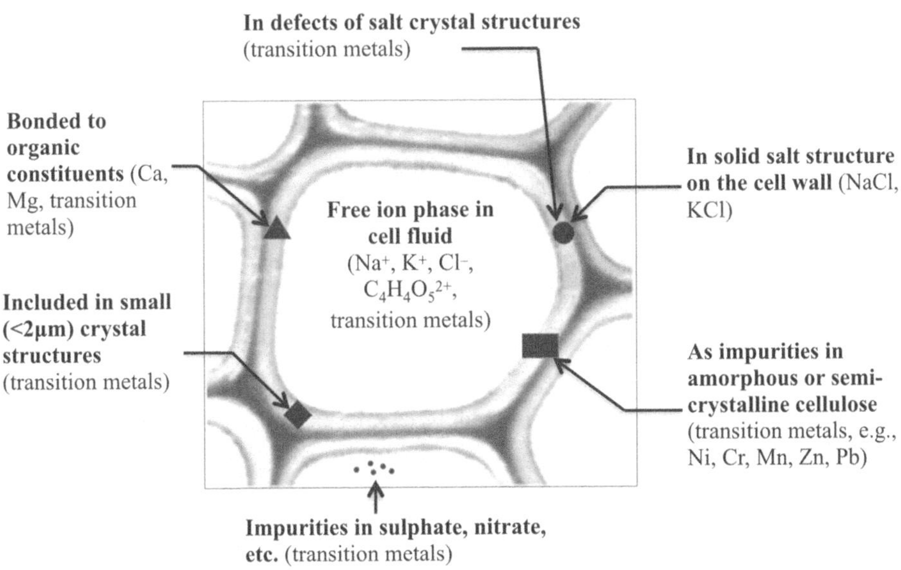

Figure 2 | The phases of the metallic elements in ligneous biomass.

products is BCs which are produced via pyrolysis. It was found that for climate change mitigation, the BC production from lignocellulosic biomass has higher potential for energy production than direct biomass combustion. (Woolf *et al.* 2010).

However, BCs are too valuable to be used to only meliorate soil immediately after its production. They should be used for another useful purpose at least once before being added to the soil. Among the innovative environmental applications, BCs have been used to absorb contaminants in environmental protection engineering systems (Srinivasan & Sarmah 2014).

Also, pyrolysis was proposed as an alternative method to dispose of the MEs' contaminated biomass, such as phytoremediation biomass waste or impregnated wood waste (i.e. wood impregnated with Cu, Cr, and As salts for preservation purposes).

The MEs concentrated in BCs show the potential of adding the practical value for char, which can be used in environmental pollution control technologies.

Subsequent subsections of this article describe the effects of pyrolysis conditions and the composition of feedstock on BCs properties. The role of syngenetic elements is also emphasized in the discussion.

Degradation phases of structural components of biomass

Wood is formed predominantly of lignocellulosic structural components with various infiltrated substances. The content and composition of structural components in ligneous plants affect the amount and quantity of pyrolysis products as well as the concentrations of infiltrated components, such as MEs. Extractive-free softwoods, in general, contain 70–75% holocellulose (the total carbohydrate portion includes true cellulose and hemicellulose), the total carbohydrate portion, including true cellulose and hemicelluloses, and 25–30% lignin. Extractive-free hardwoods contain about 75–82% holocellulose, and 18–25% lignin. The holocellulose of hardwoods usually contains more cellulose than that of softwoods, being considerably richer in pentosans rather than hexosans (as in the case of softwoods) (Stamm 1964). The lignin content depends on the strength of the bound fibers in plants, i.e. woody plant species are richer in lignin due to tightly bound fibers, while herbaceous plants have more loosely bound fibers and, consequently, lower lignin content. Up to 40% of bark is made of lignin. Lignin is the most stable component of the plant's matrix, due to its aromatic nature and tendency to condensation (Azarov *et al.* 1999). Lignin content, composition, and distribution varies among the taxa of trees. In different species of ligneous plants, the lignin content can range from 15 to 36% of the dry weight of the wood (Zobel & van Buijtenen 1989).

Lignin decomposes within the range of 280–500 °C. With the higher content of lignin in softwoods, they tend to give higher BC yields than hardwoods. Lignin from coniferous trees was found to be more stable than lignin of deciduous trees, and the former produces a larger percentage of BC per kg of input materials (Brebu & Vasile 2010).

Cellulose is the main component of ligneous biomass that consists of long polymers of cellulose chains. The aggregation of these chains provides a crystalline structure that is highly inert and inaccessible to some chemicals. Cellulose usually constitutes approximately 50% of dry wood. Cellulose mainly contributes to the production of tar, while lignin is primarily converted into BC and a small amount of water on pyrolysis. However, at a relatively low temperature, cellulose degrades into stable anhydrocellulose which yields high amounts of BC, but at higher temperatures (>350 °C), the cellulose decomposes to produce volatile products (Demirbas 2007).

Unlike cellulose, hemicellulose has a branched structure based on pentose and hexose sugars, while its molecular weight is low. Its content varies from 20 to 40%. Hemicellulose is most sensitive thermally and so decomposes faster, therefore hemicellulose conversion to biochar starts at 200 °C.

Primary pyrolysis involves the degradation of the three major components of wood and converting the biomass into liquid (bio-oil or bio-crude), solid (char/BC) and gaseous fractions (syn-gas). The thermal decomposition patterns of the lignocellulosic structures during the pyrolysis process are summarized in three distinct steps (Table 3).

The products of the primary reactions, e.g. char and volatiles, usually catalyze the secondary reactions. If the purpose is to maximize the char yields, low temperature and low heating rates should be applied (Brewer *et al.* 2009). The target product of slow pyrolysis is BC, but this

Table 3 | Thermal decomposition patterns of the lignocellulosic structures during the pyrolysis process (Demirbas & Arin 2007; Tripathi *et al.* 2016)

Type of process	**Heating and drying**	**Devolatilization**		**Secondary reactions**
Temperature	**100–200 °C**	**200–300 °C**	**200–650 °C**	**>650 °C**
Process description	In this phase, the wood temperature rises to 100 °C, moisture evaporates; the wood particle becomes chemically inert. At around 160 °C, dehydration is accomplished, elimination of volatiles, bond breakage, the formation of free radicals, carbonyl, carboxyl and hydroperoxide groups take place	In the devolatilization of the pyrolysis process, gas (carbon dioxide, carbon monoxide and condensable vapors) and tar species leave the solid matrix of the wood (hemicellulose). At the end of this stage, a porous biochar structure is left. At <300 °C, the governing process is the reduction in the degree of polymerization of cellulose. The thermal degradation of hemicellulose starts in the temperature regime of 200–260 °C. After decomposition of the polymer into soluble fragments, the conversion into monomers takes place which further degrade into volatile products. In comparison to cellulose, decomposition of hemicellulose results in more volatiles, less tar and biochar. At the beginning of the pyrolysis (250–300 °C) process, most of the volatiles are released at a rate 10 times higher than that of the next step	The degradation of cellulose increases and reaches a peak around 320 °C. At that temperature the reaction products are mainly volatiles. The degradation rate of lignin increases at temperatures higher than 320 °C. It is followed by a rapid increase in the carbon content of biochar. In the second step, at >300 °C, the formation of biochar, tar and gaseous products takes place. The major component of tar is levoglucosan that vaporizes and then degrades with the increasing temperature. When lignin is heated to >400–450 °C, only about 50% volatilizes and the rest remains as biochar	Tar produced during the decomposition of the wood can decompose further. Tar decomposition is suggested to be catalyzed by the solid matrix of the fuel, resulting in biochar creation on the pore walls.

is always obtained together with an almost equal percentage of liquid and gaseous products. Slow pyrolysis takes several hours to complete; the process is characterized by slower heating rates (1–20 °C/min), relatively long solid and vapor residence times (5–30 min) and, usually, a lower temperature than that maintained in the fast pyrolysis. If a low temperature (compared to the temperature used in the combustion process, around 1,273 K) is adopted, MEs do not volatilize and are accumulated in the solid pyrolysis fraction (char and ash). At the same time, a reduction in the weight and volume of the ME contaminated biomass is obtained (Lievens *et al.* 2008).

In summary, lignin content is higher in ligneous biomass in comparison to herbaceous plants and among the tree species softwoods in general they contain about 5% higher lignin content than hardwoods. Lignin is the most charring biomass component and biochar is the most abundant component in the products of ligneous biomass pyrolysis.

Cellulose mainly contributes to the production of tar, but at low temperatures it degrades to rather stable compounds, contributing to a high production of char. Therefore, softwoods with some of their morphological parts (e.g. bark) provide the best feedstock for BC production because of the lignin content. The MEs associated with lignin structures have a higher potential to be retained in the char.

Temperature effects

The pyrolysis temperature is of prime importance for the transfer of MEs from the feedstock to BC. The operating temperature of 400–600 °C was found to be adequate to disintegrate the organic structures of biomass, but low enough to avoid volatilization of the inorganics (Leijenhorst *et al.* 2016). It has been shown that alkali earth MEs (Mg, Ca), transition MEs (Cu, Fe, Cd, Ni, Co, Cr, Zn, Mn) and post-transition MEs (Pb, Al) predominantly remain on/in the BC. For the alkali MEs (K, Na), the entrainment of solid BC particles to pyrolysis oil is the primary pathway, although K and Na are also transported primarily by the reactions with volatiles.

The results of research by Lievens *et al.* (2008, 2009); Kan *et al.* (2016) and Kovacs & Szemmelveisz (2017) documented that MEs can be divided into those that are mainly released during decomposition stage (250–475 °C) (i.e. Sc, Cr, Ti, V, Ni from the transition ME group) and those that are mainly released at high temperatures (900–1,300 °C) (i.e. Zn, Cu, Mn from the transition MEs group), thus reflecting their different roles in organic structures. Based upon these findings, the major quantities of alkali MEs (K, Rb, Cs), transition elements (Zn, Cu, Mn) and high-field strength elements (Mo, Ag, Cd) can be expected to be found in the slow pyrolysis BC (~450 °C production temperature) (Table 4).

In summary, the temperature regime of degradation of all three wood components takes place and overlaps between 400 and 450 °C. Low temperature (400–450 °C), low heating rate (1–20 °C/min) and relatively long solid and vapor residence times (>30 min) are the conditions for maximizing the char yield and retaining metals (including Zn, Cu, Mn, Cd) in the char.

Table 4 | The release of chemical elements at two stages of pyrolysis (adapted after Thy *et al.* (2017) and Stals *et al.* (2010))

	The amount of release, %	
Type of element	**Decomposition of the organic matrix (250–475 °C)**	**After decomposition of the organic matrix, high temperature (900–1,300 °C)**
Anions (C, S, Cl)	75–90	
Alkali MEs (K, Rb, Cs)		85–90
Alkali earth elements: (Mg, Sr)	70–90	
Transition elements:		
Sc, Cr, Ti, V, Ni	75–80	
Zn, Cu, Mn		80–90
High-field strength elements:		
Y, Zr, Nb	60–70	
Mo, Ag, Cd		70–90
Rare earth elements: (La)	< 20	
Potentially volatile and toxic elements: (As, Se, Sb, Hg)	60–90	
EPA criteria pollutants:		
Cu, Cd, Tl, Pb		45–90
Cr, Ni	90	

Amounts and forms of metallic elements in BC

A thorough analysis of the composition of BC obtained from the feedstock of natural origin focused on the concentration of MEs in BC was performed by Hilber *et al.* (2017). In the case where the total concentration of the potentially toxic elements is low, the bioavailability and accessibility of toxic elements are low.

Table 5 shows the concentrations of MEs in BC produced from noncontaminated and contaminated feedstock. The sources of the feedstock contamination were wood processing (both mechanical and chemical) and ligneous-based waste materials. It can be stated that, first, BC obtained from non-polluted feedstock did not exceed the limiting values with respect to MEs' content specified by the European Biochar Certificate. A tendency of exceeding a limiting value is characteristic of BC (more exactly, a pyrogenic carbonaceous product) made of ligneous waste or another type of processed wood waste material in the case of all MEs (except Pb). The recommended limit values were exceeded 2–4 times which shows the potential of MEs to increase within the wood waste material.

As presented above, the direct involvement of BC components in natural flows could only increase the intensity of technogenic flows, however, it can be used in a safer way if used as a feedstock for the production of other products, where MEs as the constituent parts have beneficial roles, for example, in the production of adsorbents with specific features influenced by the presence of syngenetic MEs.

Second, BC obtained at the temperature of 400–450 °C demonstrated relatively higher concentrations of MEs (except for Cu) than BC produced at the temperature range of 700–750 °C. This is in agreement with the observations of Thy *et al.* (2017) where the largest amounts of MEs (Cu, Cd, Pb and Zn) evaporate at the temperature of 1,000 °C and above, and only Cr and Ni are characterized by more intense volatilization at a temperature of about 500 °C. Following the findings, in slow pyrolysis BC (~450 °C of the production temperature), the major quantities of MEs considered as potentially toxic (Zn, Cu, Cd) can be expected to be found.

Figure 3 presents the major MEs (Zn, Pb, Hg, Ni, Cu, Cd, Cr, As) phases found in biochar after pyrolysis. The formation of MEs phases are highly dependent on the characteristics of MEs themselves and the behavior of them during thermal treatment. The behavior features of the most relevant MEs are described below.

Zn appears to be less volatile than Pb and Cd. Zn remains stable in mineral form, e.g. $ZnCr_2O_4$, $ZnFe_2O_4$, $ZnAl_2O_4$, Zn_2SiO_4 (Janse *et al.* 2000). Cellulose and hemicellulose degraded at 200 °C producing phenolic compounds and organic acids. Organic acids could react with Zn carbonate. At temperatures of 200–400 °C, cellulose, hemicellulose and other organic materials are decomposed. The volatilization of Zn is dependent on the type of thermally treated biomass.

During pyrolysis, the stable form of Pb is the solid sulphide. PbO appears in solid form and its presence in

Table 5 | The concentrations of MEs (mg/kg in dry weight) in BC produced from ligneous biomass from different sources

Type of biochar feedstock	Production temperature, °C	Pb	Cu	Zn	Cd	Cr	Ni	Refs
Wood control	400–450	0.26–6.22	2.97–85.5	20.9–252	0.04–0.27	0.62–25.8	0.83–17.8	Baltrėnaitė *et al.* (2016)
Wood control	700–750	0.55–0.58	3.71–8.86	10.7–17.9	0.01–0.90	1.02–7.47	1.49–10.8	Baltrėnaitė *et al.* (2016)
Processed wood and wood waste	400–450	8.07–39.3	22.4–382	118–1,269	0.16–6.32	24.3–227	17.9–88.1	Usevičiūtė & Baltrėnaitė (2019)
Processed wood and wood waste	700–750	9.03–33.1	32.7–452	141–1,209	0.02–0.18	17.2–97.1	18.3–52.7	Baltrėnaitė *et al.* (2016)
Wood from contaminated area	450	1.5–1.8	18.1–21.7	50.6–286.5	0.017–0.029	12.6–13.2	6.5–9.3	Chemerys & Baltrėnaitė (2018)
Basic quality biochar	–	150	100	400	1.50	90	50.0	EBC (2012)
Premium quality biochar	–	120	100	400	1.00	80	30.0	EBC (2012)

- $Cr(OH)_4CrO_4$ as an inorganic precipitates on cellulose
- Stable $CuCrO_4$ in lignin
- $CrAsO_4$ in lignin complexes
- Inorganic $CrAsO_4$ precipitates on cellulose
- $HCrO_4^-$ in lignin complexes
- Cr_2O_3
- $ZnCr_2O_4$
- Cr_2SO_4

- $ZnO \rightarrow ZnSO_4$
- $ZnCr_2O_4$
- $ZnFe_2O_4$
- $ZnAl_2O_4$
- Zn_2SiO_4
- Organic bonds with $ZnCO_3$

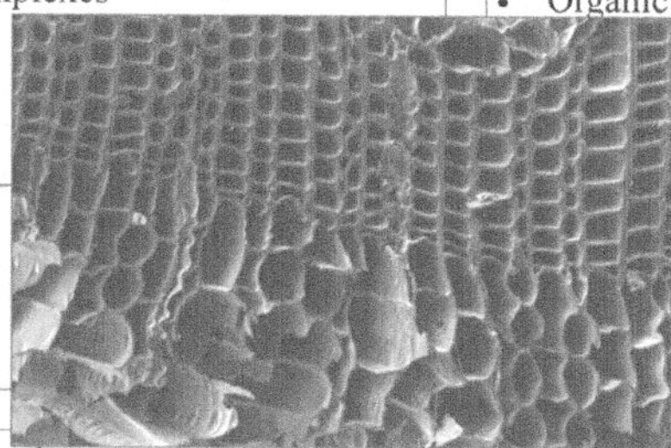

- Metallic form of Cd (highly volatile)
- CdO
- $CdSO_4$

- $Pb_2B_2O_4$
- $PbSO_4$
- PbO

- $CuCrO_4$ in lignin complexes
- Cu (II) complexes with lignin
- Physically adsorbed Cu(II) on wood constituents
- Cu(II) in cellulose complexes

- $CrAsO_4$ in lignin complexes
- Inorganic $CrAsO_4$ precipitates on cellulose

Figure 3 | List of metallic element phases which are found in biochar produced at a temperature of about 750 °C. The image in the figure presents the structural view (magnification 300×) of biochar.

liquid form is expected above the char oxidation zone, where the temperature exceeds the melting point of PbO (887 °C). These forms of $Pb_2B_2O_4$ and $PbSO_4$ were found in the solid pyrolysis product. The higher pyrolysis temperature promoted the transformation of amorphous Fe/Mn hydroxides into more crystalline forms (Janse *et al.* 2000). Lievens *et al.* (2008) concluded that Pb compounds are less receptive to volatilization at pyrolysis temperatures of 400 °C than Cd compounds.

Following Dell'Antonia (2013), Ni is a nonvolatile ME under pyrolysis conditions up to 705 °C. Its emissions can be observed only if the pyrolysis reactor used releases ME-bearing particles due to high gas velocities. With respect to release, Cu does not pose any problems. When the pyrolysis temperatures were increased to 200 °C, the organic part of biomass decomposed and Cu formed a very stable complex with organic matrixes. However, in the temperature range of 300–500 °C, thermally treated biomass' aromaticity increased and Cu could not form the Cu-organic complexes (Janse *et al.* 2000). Lievens *et al.* (2008) concluded that Cu compounds are less volatile at pyrolysis temperatures of 400 °C than Cd compounds.

Cadmium compounds were reduced to Cd^0, which volatilized at temperatures higher than 600 °C. The thermodynamically stable chemical forms of Cd (sulphide, CdS, and chloride, $CdCl_2$) were found. The most significant result refers to volatilization: Cd entirely vaporizes in the form of vapor Cd and, in a smaller quantity, as the chloride $CdCl_2$. The gaseous species can condense on the surface of particles transported in the fumes, therefore, Cd and $CdCl_2$ are likely to be found in the ashes as $CdSO_4$ (Janse *et al.* 2000).

Dell'Antonia (2013) concluded that during the pyrolysis of sewage sludge at $T > 600$ °C the transfer of Cd to the off-gas is expected. On heating at temperatures of 200–400 °C, the ligneous biomass decomposes and produces organic matter that can easily bond to Cd. When the pyrolysis temperature increased to 500 °C, the organic matter in biomass turned to more aromatic compounds and could not bond to Cd. Lievens *et al.* (2008) found that the Cd compounds are more sensitive to volatilization at pyrolysis temperatures of 400 °C than Cu, Pb and Zn compounds. The volatilization of Cd increases with increases in temperature. It has been found that a certain amount of Cr is released from the pyrolysis reactor gaseous CrO_2Cl_2, or as solid Cr_2O_3 and $ZnCr_2O_4$.

The final stable species are $ZnCr_2O_4$ and Cr_2SO_4. The Cr stays in the solid form, irrespective of the conditions of composition and temperature. No gaseous species are predicted for Cr (Mènard *et al.* 2006). As concluded by Van den Broeck *et al.* (1997) from the speciation experiments, it appears that the major part of Cr in the pyrolysis residues is bound to carbonate, oxidized wood components and organic matter. A non-negligible amount of Cr is found in the exchangeable fraction.

Temperature and residence time highly affect the amount of As released during the pyrolysis. Helsen *et al.* (2013) carried out a thermogravimetric study of the hydrated Cr arsenate and revealed that thermal degradation results in the development of solid Cr_2O_3 and gaseous O_2, H_2O and As_4O_6, also indicating that As $(^{5+})$ is reduced to As $(^{3+})$ because As_2O_3 is volatilized at a temperature below 200 °C. Char and pyrolysis vapors form a reducing environment, thereby influencing the behavior of MEs oxides.

Zhurinsh *et al.* (2005) found that, in pyrolysis of CCA-treated wood, As losses depend on the final temperature of pyrolysis; its content in the condensed pyrolysis products reaches only 2% from the total As amount, while As evaporation in the batch process begins under 200 °C.

To summarize, MEs are present in inorganic precipitates on cellulose (in the case of Cr, Cu), lignin (Cr, Cu) or in the form of oxides (for Cr, Cd, Cu, Zn, Pb), sulphates (Cd, Pb, Cr, Zn) or sulphides (for Pb). A reduction in the weight and volume of the MEs contaminated biomass is obtained, therefore, the concentration of the MEs increases. Low pyrolysis temperatures keep larger amounts of organic matter which results in higher amounts of organically bound MEs present (e.g. Zn, Cr). These conditions are favorable for less volatile MEs to be kept in the char (e.g. Pb with a melting point >887 °C, Ni, Cd which volatilize at temperatures higher than 600 °C). Most of the As and a large percentage of the Hg are released during low temperature pyrolysis because of their low temperature of volatilization, such as <200 and <500 °C respectively. The MEs stability in BC were determined during leachability studies. Low moisture content and higher oxygen content in the feedstock (about 5%) must be kept to keep BC yield higher. Keeping maximum reducing conditions in the reactor can result in 2–3 times higher amount of MEs in the BC.

Retention and leachability of metallic elements in biochar

Several studies have demonstrated that MEs present in the feedstock are stabilized during the production of BC (Lu *et al.* 2012). Based on the comparison of two leaching procedures (ASTM method vs BSI method) used to study charcoal ME content in the solution, Helsen *et al.* (1997) suggested that there was a stronger connection between Cr and the BC compared to that in the feedstock (i.e. CCA-treated wood).

According to Stals *et al.* (2010), with the increasing processing temperatures the weakest bound and chemically less stable compounds were not retained in the BC, analyses of MEs (Cd, Pb and Zn) presence in the EDTA extract from the chars produced at different pyrolysis temperatures (350, 450, and 550 °C) revealed that ME compounds are more resistant to lixiviation at 550 °C. According to Lu *et al.* (2011), the stabilization of MEs in the sludge-derived BC is associated with the pyrolysis process.

The amounts of the six examined MEs (Zn, Cr, Ni, Cu, Cd and Pb) washed out from various types of the BC found on railway sleepers were not considerable. Their amount compared to the original amount of the elements of various types of BC ranged from 0.07 to 11.28% of the amount originally present in the sleepers (Usevičiūtė & Baltrėnaitė 2019).

As found by Mancinelli *et al.* (2016), the most suitable temperature for reducing MEs leachability during pyrolysis depends on the MEs in question. The temperature of 450 °C was especially effective in stabilizing Cr and Ni in the BC as these MEs did not tend to leach. According to Lievens *et al.* (2009), it is important to note that, due to the low operating temperature, MEs are enriched in the BC/ash fraction of various pyrolysis products. This suggests that pyrolysis can be a means of reducing MEs (Zn, Pb, Cd, Cr, Ni, Cu) availability in contaminated feedstock.

Other factors (residence time, heat flux, thermal conductivity, reactor matrixes)

This research team investigated the influence of the particles' shape (spherical, cylindrical and flat) at the surface temperature of 550 °C. They found that spherical particles have the shortest conversion time, based on the higher surface/volume ratio. However, for small particle diameters (smaller than 20 mm), the rate of the reaction was dominant and various particles' shapes had a nearly equal time of conversion (Encinar *et al.* 2000).

The solid heat carrier in the pyrolysis reactor can affect the distribution of MEs in the pyrolysis products. According to Lievens *et al.* (2008), the sand matrix used in the pyrolysis reactor promoted the retention of Cd at 500 and 600 °C, and Cu at 600 °C, in the char/ash obtained from Maatheide birch contaminated by MEs. At the temperature of 400 °C, the volatilization of the compounds of MEs (Cd, Cu, Pb, and Zn) was not affected by various kinds of the solid heat matrixes, sand or fumed silica, while the solid heat matrix retains Cu and Pb.

A very large amount of moisture reduces the heating rate and, as a result, more time is required for achieving the pyrolysis temperature. In the pyrolysis of wood with 5 and 20% moisture content, it was found that the biomass with a smaller amount of moisture caused the formation of a large amount of residue (Fang *et al.* 2014). Demirbas (2007) found that the increase in the moisture content in wood reduced the BC yield. The presence of oxygen is another factor, influencing the reactivity of biomass during pyrolysis, which affects the pyrolysis product yield and its composition (Gani & Naruse 2007).

Comparing mass percentage of MEs in the bottom ash from incineration and the pyrolysis of the contaminated biomass (i.e. hyperaccumulators), Lu *et al.* (2012) found that volatile MEs (e.g. Cd, Pb and Zn) tend to accumulate in the bottom ash if the reducing conditions were used. The recoveries in the bottom ash of MEs, such as Cd, Pb and Zn, were about 2–3 times smaller under incinerations condition than those observed under pyrolysis conditions at 550 °C. Therefore, if the transfer of MEs (e.g. Cd, Zn, Pb) to the gaseous phase should be avoided, an oxidizing atmosphere should not be used.

The thermal conductivity of wood varies, depending on the direction of the heat flow with respect to the grain, as well as with the temperature, density, and moisture content. The heat transfer takes place via the complex interaction between the transfer mechanisms, such as conduction, convection, and radiation. However, when the direction of gas flow is perpendicular to the heat flux direction, the cooling effect will not occur. Grain orientation in wood needs to be taken into account due to the resulting anisotropicity (Mènard *et al.* 2006).

Effect on equilibrium, kinetics, rate limiting factors and thermodynamics during adsorption of others MEs

Information on syngenetically modified biochar and impact on adsorption of MEs is scarce while more studies are available on artificially modified BC. Artificially added metals to BC are referred to chemical modification of BC by metal salts to produce BC-based functional materials (Wu *et al.* 2017; Wei *et al.* 2018). It was determined that the chemically modified BCs have a better adsorption effect on contaminants in wastewater. These types of BCs showed high affinity for MEs. Although both Freundlich and Langmuir isotherm models (Xia *et al.* 2016) fitted the adsorption data better than other equilibrium models, there are observations which indicate that syngenetic-MEs-modified BC may precondition the double layer adsorption of MEs facilitated by varying orientation of solutes (Baltrėnaitė-Gedienė *et al.* submitted). The results of adsorption kinetics mostly followed the pseudo-second-order model, which is based on the assumption that the rate-limiting step may be chemical sorption involving valence forces through the sharing or exchange of electrons between contaminants and BC (Ho & Mckay 1999).

Based on the results of the thermodynamic studies, Van Vinh *et al.* (2015) found that Zn-modified BC exhibited the spontaneous and exothermic nature of the As(III) adsorption process (indicated by the negative free energy (ΔG°) value and positive ΔH° value (31.10 kJ/mol), respectively). The positive values of ΔS° (0.141 kJ/mol/K) showed an increase in randomness at the solid/solution interface during the adsorption of As(III) on modified BC.

INFLUENCE OF SYNGENETIC ELEMENTS ON PYROLYSIS

In the previous sections, the changes in the ME forms depending on the pyrolysis conditions were discussed. On the other hand, MEs have an impact on pyrolysis conditions as well. The typical behavior of syngenetic elements, influencing the feedstock, pyrolysis and pyrolysis products, are shown in Figure 4.

Some of the introduced MEs ions support higher temperature of cellulose decomposition: the top temperatures of

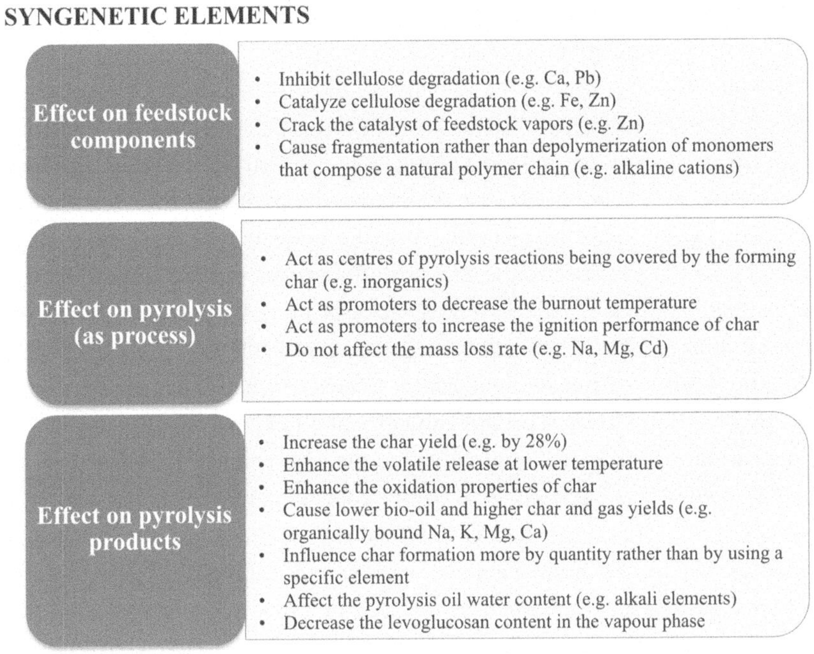

Figure 4 | The effect of syngenetic elements on the feedstock, pyrolysis and pyrolysis products.

cellulose decomposition are moved to a higher zone, when the MEs of the ligneous biomass are washed out, and to the lower temperature zones, when approximately 2% (by weight) ash is introduced to demineralized wood (Meyer *et al.* 2012): Ca^{2+} – (increase by 4 °C), Zn^{2+} – (decrease by 4 °C), Fe^{3+} – (decrease by 9 °C), Pb^{2+} – (increase by 1 °C). Fe and Zn tend to catalyze the cellulose degradation, while Ca and Pb inhibit it by moving the top temperature to higher temperature ranges (Meyer *et al.* 2012).

Inorganics not only affect BC and gas formation, but also influence the distribution of compounds forming the biooil and tars. The formation of levoglucosan (LG) (the key product of low volatility that forms tars) was found to be acid-catalyzed degradation of cellulose. The rate of cellulose-to-LG conversion was decreased when MEs were added to the feedstock, while the cellulose-to-LG conversion rate ranged between 29% (Na) and 62% (Mg).

Ash significantly increases the release of volatile BC at lower temperatures (300–350 °C) of BC pyrolysis. Both the primary and secondary reactions of volatile products are affected by the presence of inorganics (Zaror *et al.* 1985; Antal & Varhegyi 1995), and the heterogeneously catalyzed oxidation of volatiles can explain the weight loss increase in the presence of ash (Loffler *et al.* 2002). DeGroot & Shafizadeh (1983) concluded that the catalytic and inhibiting properties of inorganics affected the oxidation properties of BC, but not its structure.

Koppolu *et al.* (2004) found that BC yields are affected by the ME used to dope the feedstock, but no variations were observed depending on the type of salt (acetate or citrate) used for its preparation. The char yields can be ranked in increasing order as follows: Cu < Co < Zn < Ni < Cr < untreated feedstock. Koppolu *et al.* (2004) found a difference in the amounts of char produced from a feedstock impregnated with different MEs. The average char yield from the feedstock doped by Zn acetate (21.2%) was the highest, followed by Cu acetate (18.05%) and Ni acetate (17.15%). Hence, using different MEs changed BC yields by 4.05% at most.

Due to the presence of alkaline cations in the process of fast pyrolysis of biomass, the fragmentation rather than depolymerization of the monomers that compose the natural polymer chains can be observed (Mohan *et al.* 2006). The presence of intrinsic or ion-exchanged inorganic materials (e.g. Ca and K) in biomass catalyzes the pyrolysis reactions such as biomass decomposition and the BC forming reactions (Yaman 2004).

In the work of Stals *et al.* (2010), the lower water content of the bio-oil obtained using a hot-gas filter accounted for the decreased char particles' presence and the related alkali ME load, thus avoiding alkali ME catalysis effects on the breakdown of molecules from the condensable form into smaller ones (Fahmi *et al.* 2008).

In the process of pyrolysis, inorganics act as centers of the pyrolysis reactions and are covered by the forming char. Furthermore, a higher content of Ca, K, and Mg in the original feedstock could accelerate the pyrolytic reactions of the lignocellulosic components and the carbonization irrespective of the pyrolysis temperature, since lower oxygen and higher carbon content in the BC from *Acacia saligna* produced at 380 °C was detected compared to that obtained from sawdust at 450 °C (Agblevor & Besler 1996).

Raveendran *et al.* (1995) investigated the influence of mineral matter on the pyrolysis characteristics of wood and twelve other types of biomass. They found that demineralization increased liquid yields and decreased gas yields for all the biomass samples studied.

Kinata *et al.* (2012) concluded that treating wood with ME ions has a significant effect on its thermal degradation process. TGA and the pyrolysis experiment have shown that the presence of Cr, Cu, B in the salts of CCB results in the production of low tar and high charcoal yields.

Meyer *et al.* (2012) found that the changes in the properties of ligneous biomass and the cellulose, which is the main cell wall constituent of ligneous biomass, can bind more MEs than lignin alone. This implies for higher content of MEs to be kept in pyrolyzed biomass product which have not yet undergone thermal stages of cellulose destruction. The ME binding capacity of ligneous constituents is higher in the case of Pb and Cd than in the case of the alkali earth and alkaline MEs. The presence of Zn, Pb, Ca and Fe ions slightly changed the disintegration temperature of cellulose, while Cd, Na and Mg ions did not affect the mass loss rate during pyrolysis. The BC formation was affected mainly by the quantity rather than by the quality of the introduced ME ions. The LG yield of ligneous biomass pyrolysis was affected by the type, rather than by the total amount of the added MEs. Both ash and the added MEs acted as promoters during BC combustion by reducing the burnout temperature and increasing the ignition process of BC.

According to Beliy & Udoratina (2014), in the context of using pyrolysis as a preliminary stage of gasification, the use of Zn and Sn chlorides as catalysts seems to be the best solution because it greatly reduces the thermal degradation onset temperature and the activation energy, as well as increasing the BC yield. Sharma *et al.* (2015) described the main ways in which catalysts affect the pyrolysis processes. The degradation temperature of ligneous biomass constituents may significantly decrease when a catalyst is used. Catalysts promote the formation of coke due to dehydration reactions (mainly due to the high acidity of catalysts).

Tripathi *et al.* (2016) concluded that the acidic catalysts increase the BC yield and reduce the tar production, while the basic catalysts usually reduce the BC yield. Acidic catalysts, such as $ZnCl_2$ and ZnO, not only stimulate the BC production but also improve its characteristics. In summary, MEs presence in biomass has direct effects on pyrolysis products and indirect effects by changing pyrolysis conditions and, therefore, they affect the formation of the final pyrolysis product quality.

In relation to the BC properties for enhanced adsorption capacity, the following MEs effects are of primary interest. For example, enhancing the oxidation properties of BC can: (i) favor formation of oxygen-containing functional groups, which are the key factors for complexation reactions during adsorption; (ii) decrease the burnout temperature limiting the volatilization of MEs; (iii) enhance the volatile release at lower temperatures that can favor the development of porosity; (iv) modify the catalyzing effect on cellulose degradation (known for Fe and Zn); (v) increase of BC yield that favors retention of MEs originally precipitated in the cellulose.

The Fe and Zn catalyze the cellulose degradation by reducing degradation temperature by 4–9 °C, while Pb and Ca shift the cellulose degradation temperature to higher temperatures by 1–4 °C. Some MEs were put in the following sequence according to their impact on BC yields: Zn > Cu > Ni. It is interesting that the temperature of pyrolysis and presence of MEs in the feedstock have reciprocal effects. The lower temperature (400–450 °C), low heating rate (1–20 °C/min) and relatively long solid and vapor residence times (>30 min) are the conditions for maximizing the char yield and keeping MEs (including Zn, Cu, Mn, Cd) in the biochar (as presented above under 'Temperature effects'). On the other hand, the presence of MEs in the feedstock has impacts on the temperature of the biomass (namely, its component, cellulose) decomposition, e.g. Zn tends to catalyze (decreasing the temperature by 4 °C), while Pb suspends (increasing the temperature by 1 °C) the cellulose degradation and, therefore, increasing and decreasing, respectively, the kinetics of the pyrolysis reaction (as discussed later in this paper). High concentrations of MEs can accelerate the pyrolytic reactions and carbonization irrespective of the pyrolysis temperatures (Agblevor & Besler 1996).

IMPACT OF SYNGENETIC ELEMENTS ON THE ADSORBING CAPACITY OF BIOCHAR – FIRST FINDINGS

The changes in ligneous biomass composition undoubtedly have an influence on ligneous biomass products. Could these changes improve the physical or chemical characteristics of ligneous biomass products which are important in adsorption of water contaminants? Can these changes be associated with syngenetically accumulated MEs? Can the syngenetic elements improve the properties of ligneous biomass products or even enhance the environmental performance of the products towards removal of water pollutants?

The answers to these questions are provided with the findings presented in this paper. The authors clarified the effects of syngenetic MEs on ligneous biomass on the adsorptive properties of BCs, produced from lignin with syngenetically accumulated MEs (Cu, Cr, Pb, Zn, Ni, Cd).

Twenty test specimens were produced from lignin under two temperature regimes (400 and 700 °C) and pyrolysis durations (1, 2 and 4 h). The physico-chemical characteristics such as concentration of MEs (Cu, Cr, Zn, Pb, Ni, Cd), CEC, pH, PZC, and ash content were analyzed in the samples following methodologies described by Nguyen & Lehmann (2009) and Baltrėnaitė *et al.* (2017).

To evaluate the level of MEs concentration in the ligneous biomass, the concentration coefficients for MEs in the studied feedstock were calculated following the methods described in Markert *et al.* (2012) and Krastinytė *et al.* (2013) and using the background concentrations obtained by the authors from previous studies (Baltrėnaitė *et al.* 2012). They found that the concentration coefficients for MEs reached the value of 1.27 in the case of Pb, Cr and Zn; but they were higher than three in the case of Ni and Mn and exceeded 14 in the case of Cu. The values of concentration coefficients exceeding one indicates that the concentrations of MEs were especially pronounced in the case of Cu present in lignin and are higher than the background levels of MEs in ligneous materials.

The concentration coefficients for potentially toxic elements in the lignin indicating the elevation of elements over the background concentrations in biomass were calculated and found to be in the following order: Cu (14.28) > Ni (3.73) > Mn (3.3) > Cr (1.74) > Zn (1.65) > Pb (1.27).

The results presented in Table 6 show the concentrations of MEs determined in BC, which was produced from lignin feedstock at two temperature regimes. As expected, MEs concentration increased in most of the cases (except a slight reduction of Ni in BC at 450 °C and no difference for Cd in BC at 700 °C).

These facts separated the studied MEs into two groups:

1. The first group included MEs (Pb, Cu, Cr, Zn) where concentrations increased with the increase of the production temperature.

Table 6 | Concentration (mg/kg of dry weight, mean value ±SD, $n = 3–4$) of potentially toxic elements in BCs produced at difference temperature regimes and the concentration level of MEs in BCs as the ratio of concentration in the feedstock and the BCs produced at a particular temperature

	Pb	Cu	Ni	Cr	Zn	Cd
At 450 °C	5.54 ± 2.23	127.95 ± 23.4	7.21 ± 2.56	17.82 ± 6.01	224.0 ± 41.6	0.16 ± 0.14
Concentration level	×2.4	×1.3	×0.8	×1.5	×1.5	×3.5
At 700 °C	7.89 ± 1.32	185.53 ± 36.86	15.23 ± 9.23	28.01 ± 4.35	397.84 ± 74.19	0.05 ± 0.02
Concentration level	×3.5	×1.9	×1.6	×2.3	×2.7	×1.0

2. The second group included Cd which concentrated more in BC produced at 450 °C than in the BCs produced at 700 °C because Cd is the most volatile among the studied MEs (the boiling point of Cd is 767 °C).

Therefore, it can be concluded that most of the investigated MEs were concentrated in the BCs. The higher concentrations of MEs were typical in BC produced at 700–750 °C, but higher concentrations of MEs (including the more volatile elements) were retained in BCs at 450 °C. These findings were in line with the results obtained by Thy *et al.* (2017).

With the author's objective to study the effects of the syngenetic MEs on the adsorptive characteristics of BC, the key properties, including cation exchange capacity (CEC) (Beesley *et al.* 2015; Hilber *et al.* 2017) and point of zero charge (PZC), were analyzed because of their importance in studying the adsorption of water pollutants. The CEC indicates the level of how cations can be retained on the material of interest. The PZC is typically the pH value at which a studied material submerged in the solution exhibits zero net electrical charge on the surface. These parameters are relevant indicators of materials to quantify adsorptive capacities (Wongrod *et al.* 2019).

As shown in Table 7, in comparison to the reference BC, lignin BC was characteristic for higher MEs concentrations as well as higher amount of ash content and higher CEC value. The CEC value was higher in more than three times, the one for ash content – in about seven times. The mean concentrations of MEs in lignin BC were higher than in reference BCs by four times for Pb, 70 times for Cu, 10 times for Ni, 38 times for Cr, 1.7 times for Mn, 3.5 times for Zn and 3.8 times for Cd. The results show that lignin BC had higher-ash BCs which also had higher inorganic content that is typically formed from inorganic elements including MEs. The findings of higher MEs concentration, ash content and CEC values have correlative value (Figure 5) and confirmation of MEs influence on higher CEC level were also important. Our findings that high-ash biomass-derived BCs were characteristic for higher CEC values, were in agreement with those of Lehmann *et al.* (2011). The MEs impact on higher CEC values of BCs could be explained by the fact that much of MEs, which were present in BCs feedstock and in the BCs are in the form of oxides (Hilber *et al.* 2017) and might be replaced further (e.g. during pyrolysis) by the elements of lower charge and can increase the BCs' negative charge, therefore increasing CEC (Beesley *et al.* 2015). This phenomenon was explained for soil CEC by Birkeland (1999). The stronger correlation between Cu and CEC suggests the stronger influence of concentration level on CEC value and supports the hypothesis that MEs have impacts on the CEC as adsorptive characteristics (Figure 5). The supporting data were provided by Mohamed *et al.* (2017) who found that the increase of CEC in metal contaminated and BC amended soil is due to increased surface area and porosity.

Table 7 | Characterization of lignin and reference biochar in relation to syngenetic approach (mean values ±standard error)

Parameter	Lignin BC	Reference BC
Cation exchange capacity, $mmol_c$/kg	117.5 ± 18.5	31.8 ± 2.7
Ash content, %	23.0 ± 2.2	2.9 ± 0.2
Pb, mg/kg DW	6.7 ± 0.6	1.6 ± 0.2
Cu, mg/kg DW	243.8 ± 90.4	3.4 ± 0.3
Ni, mg/kg DW	11.2 ± 2.2	1.1 ± 0.1
Cr, mg/kg DW	22.9 ± 2.1	0.6 ± 0.1
Mn, mg/kg DW	293.4 ± 26.6	169.2 ± 15
Zn, mg/kg DW	310.9 ± 31	91.1 ± 11.9
Cd, mg/kg DW	0.105 ± 0.032	0.027 ± 0.004

The other adsorption-relevant factor was the 'Point Zero Charge' (PZC) value which is considered as the total surface charge of the materials and is determined as the point on the pH scale where the sum of anion and cation exchanges are in equilibrium (Fiol & Villaescusa 2009). As the oxidation of the material increase, the PZC appears at lower pH values (Lu *et al.* 2013). PZC was determined by the salt addition method described by Mahmood *et al.* (2011). The ΔpH (the difference between final and initial pH) values were plotted against the initial values of pH and the PZC was determined as the initial pH at which ΔpH was zero. The ΔpH values varied based upon the biochar production temperatures, residence times and biochar types as presented in Figure 6.

The larger PZC values corresponded to the case of reference wood BCs (8.11) rather than to lignin-based BCs (<8). They were also larger for the BCs of 4-hour production (8.12) than for the BCs of 1- or 2-hour production (<8) and for the temperature of 700 °C (8.13) than for 450 °C

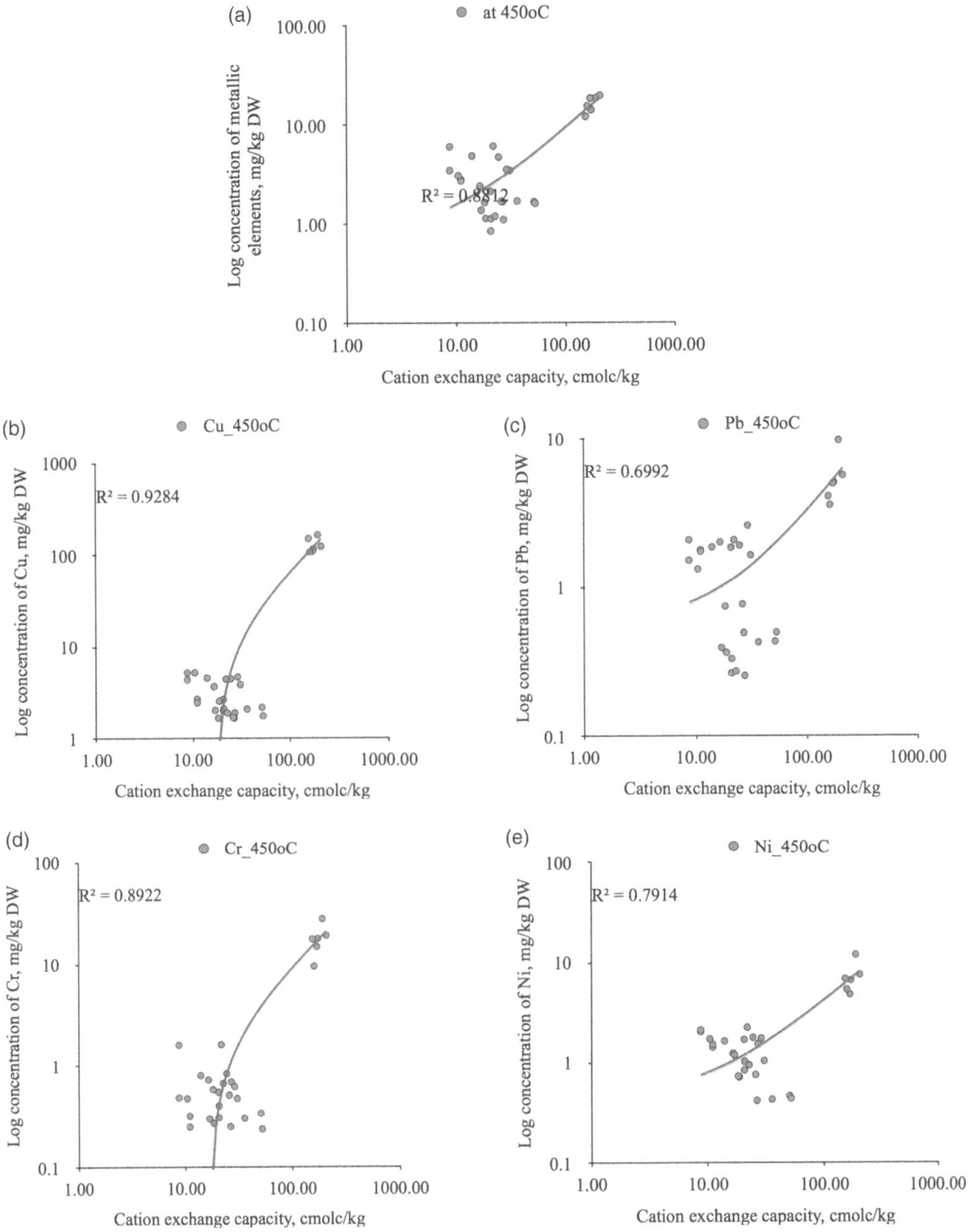

Figure 5 | Logarithmic relationship between cation exchange capacity and chemical element concentration in biochar: (a) for all chemical elements at 450 °C; (b) for Cu at 450 °C; (c) for Pb at 450 °C; (d) for Cr at 450 °C; (e) for Ni at 450 °C.

(<8). The high PZC values indicate that the adsorbent would be more positively charged and, in contrast, lower PZC values indicated that the adsorbent would be more negatively charged and therefore would attract more positively charged ions (Tang *et al.* 2019; Wongrod *et al.* 2019). In this study the BCs that have more potential to adsorb cationogenic elements (including studied MEs) were BCs obtained from lignin BCs under the pyrolysis conditions for 1 or 2 hours at 450 °C (Figure 6).

The preliminary experimental results reported in this paper documented that increased concentrations of MEs in lignocellulosic waste (lignin) are highly correlated with

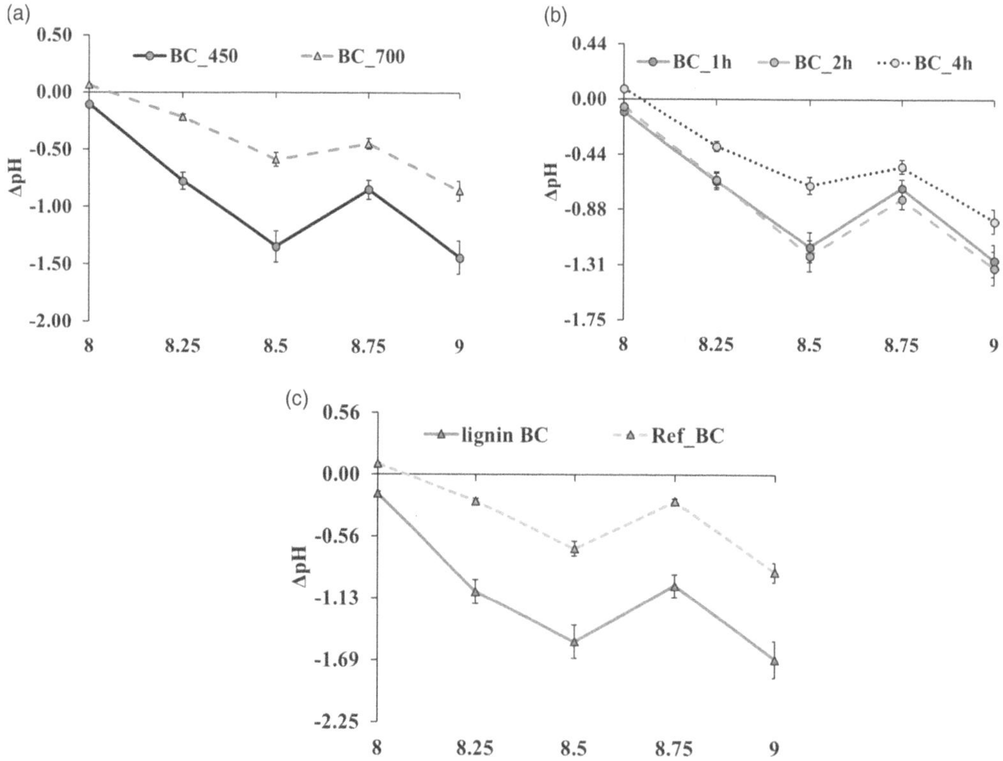

Figure 6 | The ΔpH versus initial pH of biochar produced: (a) at 450 and 700 °C; (b) at residence times of 1, 2 and 4 h; (c) for lignin BCs and reference wood BCs.

higher CEC and PZC of the produced BCs, which are important parameters to determine the adsorption capacity of BCs as well as other adsorbents.

There are at least two mechanistic dimensions involved in the dynamics of BC adsorption. First, MEs as inorganics are the principal components of the biomass ash. Increases in MEs concentration in biomass leads to higher portions of ash. Higher-ash mass leads to increases in CEC values because MEs in BCs are mostly in the form of oxides and can be easily replaced by elements of lower charge (Lehmann *et al.* 2011) and, therefore, increase the negative charge of BCs (Beesley *et al.* 2015; Hilber *et al.* 2017).

Second, the MEs have a potential to influence the pHpzc expressed in deltapH. The pHpzc allows one to hypothesize the ionization of functional groups and their interactions with MEs in solution (Bakshi *et al.* 2018). In this study BCs with higher MEs content had lower PZC values and they were able to form more negative surface charge which showed higher ability to adsorb cationogenic elements. The results of both indicators, CEC and PZC, support the potential of enhanced adsorption capacity for BCs produced from biomass with higher MEs and the syngenetic origin of MEs increases the potential of BCs with higher adsorptive features.

PRACTICAL APPLICATIONS AND FUTURE RESEARCH PERSPECTIVES

Some of the United Nations' efforts focus on working on ensuring availability and sustainable management of water, however, under the conditions of intensive technogenesis, the technophilicity of metallic elements and their entrance into water bodies increases. The need for clean drinking water and sustainable wastewater treatment facilities is high, especially in developing countries.

This research documented a new pathway of MEs entrance into biomass, therefore leading to nature-anthropogenized impacts of MEs on properties of biomass and its products.

From a practical perspective, the enhancement of BCs adsorption capacity by means of syngenetic MEs showed potential to use MEs affected BCs for adsorption purposes and increased economic benefits from the use MEs contaminated biomass. For example, instead of utilizing the MEs contaminated ligneous biomass, the products, such as BCs, with MEs impacted characteristics can be produced for removal of ME contaminants from polluted water. Further research of syngenetic MEs should be focused on other syngenetic MEs and their impacts on ligneous BCs characteristics.

The findings of this research support the observation that with increasing technogenesis, the syngenetically formed MEs will increasingly contribute to changes in the characteristics of biomass and its products, including BCs.

CONCLUSIONS

The BCs from ligneous plants are useful organic products for syngenetic elements to manifest their beneficial effects. At the pyrolysis temperature of 400–450 °C, most of the potentially toxic elements (including Zn, Cu, Mn, Cd) are retained within the BCs, and most of them are in the metal oxide forms, which in turn have an influence on BCs adsorptive properties. The increased concentrations of MEs in lignocellulosic wastes are highly correlated with higher CEC and PZC of their BCs, which results in higher adsorptive capacities of BCs. Among the studied MEs, syngenetic Cu was determined to correlate strongly with the CECs of the BCs.

ACKNOWLEDGEMENTS

The authors thank Dr Eglė Marčiulaitienė, Dr Mantas Pranskevičius and Dr Jelena Titova for providing analytical data. This research was funded by grants (No. S-MIP-17-83 and No. S-MIP-17-20) from the Research Council of Lithuania.

REFERENCES

Agblevor, F. A. & Besler, S. 1996 Inorganic compounds in biomass feedstocks. 1. Effect on the quality of fast pyrolysis oils. *Energy Fuels* **10**, 293–298.

Antal Jr., M. J. & Varhegyi, G. 1995 Cellulose pyrolysis kinetics: the current state of knowledge. *Ind. Eng. Chem. Res.* **34**, 703–708.

Azarov, V. I., Burov, A. V. & Obolenskaya, A. V. 1999 *Wood Chemistry and Synthetic Polymers*. Manual for High Schools. SPbLTA, St. Petersburg (in Russian).

Bakshi, S., Banik, C., Rathke, S. J. & Laird, D. A. 2018 Arsenic sorption on zero-valent iron-biochar complexes. *Water Res.* **137**, 153–163.

Baltrėnaitė, E. & Baltrėnas, P. 2019 Using the method of dynamic factors for assessing the transfer of chemical elements from soil to plants from various perspectives. *Environ. Sci. Pollut. Res.* (accepted) doi:10.1007/S11356-018-3866-1.

Baltrėnaitė, E., Lietuvninkas, A. & Baltrėnas, P. 2012 Use of dynamic factors to assess metal uptake and transfer in plants – example of trees. *Water Air Soil Pollut.* **223** (7), 4297–4306.

Baltrėnaitė, E., Baltrėnas, P. & Lietuvninkas, A. 2016 *The Sustainable Role of the Tree in Environmental Protection Technologies*. Monograph, Springer, Dordrecht.

Baltrėnaitė, E., Baltrėnas, P., Bhatnagar, A., Vilppo, T., Selenius, M., Koistinen, A., Dahl, M. & Penttinen, O.-P. 2017 A multicomponent approach to using waste-derived biochar in biofiltration: a case study based on dissimilar types of waste. *Int. Biodeter. Biodegrad. J.* **119**, 565–576.

Baltrėnaitė, E., Lietuvninkas, A. & Baltrėnas, P. 2018 Biogeochemical and engineered barriers for preventing spread of contaminants. *Environ. Sci. Pollut. Res.* **25** (6), 5254–5268.

Beesley, L., Moreno-Jimenez, E., Fellet, G., Melo, L. & Sizmur, T. 2015 Biochar and heavy metals. In: *Biochar for Environmental Management: Science, Technology and Implementation* (J. S. Lehmann & S. Joseph eds). Routledge, London, UK, pp. 563–594.

Begum, S. A., Golam Hyder, A. H. M. & Vahdat, N. 2018 Adsorption isotherm and kinetic studies of As(V) removal from aqueous solution suing cattle bone char. *J. Water Supply Res. Technol. Aqua* **65** (3), 244–252.

Beliy, V. A. & Udoratina, E. V. 2014 Kinetic study of wood pyrolysis in presence of metal halides. *Cent. Eur. J. Chem.* **12** (12), 1294–1303.

Birkeland, P. W. 1999 *Soils and Geomorphology*, 3rd edn. Oxford University Press, Oxford.

Brebu, M. & Vasile, C. 2010 Thermal degradation of lignin – A review. *Cellulose Cell Chem. Technol.* **44** (9), 353–363.

Brewer, C. E., Schmidt-Rohr, K., Satrio, J. A. & Brown, R. C. 2009 Characterization of biochar from fast pyrolysis and gasification systems. *Environ. Prog. Sustain. Energy* **28**, 386–396.

Chemerys, V. & Baltrėnaitė, E. 2018 Influence of intrinsic properties of lignocellulosic feedstock on adsorptive properties of biochar. *ASCE J. Environ. Eng.* **144** (9), 1–10.

Chojnacka, K. 2010 Biosorption and bioaccumulation – the prospects for practical applications. *Environ. Int.* **36**, 299–307.

DeGroot, W. F. & Shafizadeh, F. 1983 Influence of inorganic additives on oxygen chemisorption on cellulosic chars. *Carbon* **21**, 61–68.

Dell'Antonia, D. 2013 *Combustione della biomassa Panoramica delle principali tecnologie Benchmarking e analisi delle potenzialitrà (Biomass Combustion. Overview of the Main Technologies. Benchmarking and Potential Analysis).* Available from: http://smartenergyproject.eu/site/assets/files/2302/senoe_ta_biomass-combustion_disa_it_orig.pdf.

Demirbas, A. 2007 Bio-fuels from agricultural residues. *Energy Sour. A Recover Util. Environ. Eff.* **30** (2), 101–109.

Demirbas, A. & Arin, G. 2007 An overview of biomass pyrolysis. *Energy Sour.* **24**, 471–482.

Doshi, V., Vuthaluru, H. B., Korbee, R. & Kiel, J. H. A. 2009 Development of a modeling approach to predict ash formation during co-firing of coal and biomass. *Fuel Process. Technol.* **90** (9), 1148–1156.

EBC 2012 European Biochar Certificate. Guidelines for a Sustainable Production of Biochar. European Biochar Foundation (EBC), Arbaz, Switzerland. http://www.european-biochar.org/en/download. Version 8.3E of 1 September 2019, doi.10.13140/RG.2.1.4658.7043.

Encinar, J. M., Gonzalez, J. F. & Gonzalez, J. 2000 Fixed-bed pyrolysis of Cynar-acardunculus L. Product yield sand compositions. *Fuel Process. Technol.* **68**, 209–222.

Fahmi, R., Bridgewater, A. V., Darvell, L. I., Jones, J. M., Yates, N., Thain, S. & Donnison, I. S. 2008 The effect of alkali metals on combustion and pyrolysis of Lolium and Festuca grasses, switchgrass and willow. *Fuel* **86** (10–11), 1560–1569.

Fang, J., Leavey, A. & Biswas, P. 2014 Controlled studies on aerosol formation during biomass pyrolysis in a flat flame reactor. *Fuel* **116**, 350–357.

Feng, Y., Liu, Q., Yu, Y., Kong, Q., Zhou, L., Du, Y. & Wang, X. 2018 Norfloxacin removal from aqueous solution using biochar derived from luffa sponge. *J. Water Supply Res. Technol. Aqua* **67** (8), 703–714.

Fiol, N. & Villaescusa, I. 2009 Determination of sorbent point zero charge: usefulness in sorption studies. *Environ. Chem. Lett.* **7**, 79–84.

Gani, A. & Naruse, I. 2007 Effect of cellulose and lignin content on pyrolysis and combustion characteristics for several types of biomass. *Renew. Energy* **32** (4), 649–661.

Helsen, L., Van den Bulck, E., Van den Broeck, K. & Vandecasteele, C. 1997 Low-temperature pyrolysis of CCA-treated wood waste: chemical determination and statistical analysis of metal input and output; mass balances. *Waste Manage.* **17** (1), 79–86.

Helsen, L., Van den Bulck, E., Van Bael, M. K. & Mullens, J. 2013 Arsenic release during pyrolysis of CCA treated wood waste: current state of knowledge. *J. Anal. Appl. Pyrolysis* **68**, 613–633.

Hilber, I., Bastos, A. C., Loureiro, S., Soja, G., Marsz, A., Cornelissen, G. & Bucheli, T. D. 2017 The different faces of biochar: contamination risk versus remediation tool. *J. Environ. Eng. Landsc. Manage.* **25** (2), 86–104.

Ho, Y. S. & Mckay, G. 1999 Pseudo-second order model for sorption processes. *Process. Biochem.* **34**, 451–465.

Ivaska, A. & Harju, L. 1999 Analysis of inorganic constituents. In: *Analytical Methods in Wood Chemistry, Pulping, and Papermaking* (E. Sjoestroem & A. Raimo ed.). Springer Verlag, Berlin, Heidelberg, pp. 287–304.

Janse, A. M. C., Janse, A. M. C., Westerhout, R. W. J., Westerhout, R. W. J. & Prins, W. 2000 Modelling of flash pyrolysis of a single wood particle. *Chem. Eng. Process. Process Intens.* **39** (3), 239–252.

Kan, T., Strezov, V. & Evans, T. J. 2016 Lignocellulosic biomass pyrolysis: a review of product properties and effects of pyrolysis parameters. *Renew. Sustain. Energy Rev.* **57**, 1126–1140.

Keown, D. M., Ji, H. & Li, C. Z. 2008 Effects of volatile–char interactions on the volatilization of alkali and alkaline earth metallic species during the pyrolysis of biomass. *Fuel* **87** (7), 1187–1194.

Kinata, S. E., Loubar, K., Bouslamti, A., Belloncle, C. & Tazerout, M. 2012 Influence of impregnation method on metal retention of CCB-treated wood in slow pyrolysis process. *J. Hazard. Mater* **233–234**, 172–176.

Koppolu, L., Prasad, R. & Clements, L. D. 2004 Pyrolysis as a technique for separating heavy metals from hyperaccumulators. Part III: pilot-scale pyrolysis of synthetic hyperaccumulator biomass. *Biomass Bioenergy* **26**, 463–472.

Kovacs, H. & Szemmelveisz, K. 2017 Disposal options for polluted plants grown on heavy metal contaminated brownfield lands – A review. *Chemosphere* **166**, 8–20.

Krastinytė, V., Baltrėnaitė, E. & Lietuvninkas, A. 2013 Analysis of snow-cap pollution for the air quality assessment in the vicinity of oil refinery. *Environ. Technol.* **34** (6), 757–763.

Lehmann, J., Rillig, M. C., Thies, J., Masiello, C. A., Hockaday, W. C. & Crowley, D. 2011 Biochar effects on soil biota – a review. *Soil Biol. Biochem.* **43**, 1812–1836.

Leijenhorst, E. J., Wolters, W., van de Beld, W. & Prins, W. 2016 Inorganic element transfer from biomass to fast pyrolysis oil: review and experiments. *Fuel Process. Technol.* **149**, 96–111.

Lietuvninkas, A. 2012 *Aplinkos Geochemija (Environmental Geochemistry).* Technika, Vilnius, p. 310.

Lievens, C., Yperman, J., Vangronsveld, J. & Carleer, R. 2008 Study of the potential valorisation of heavy metal contaminated biomass via phytoremediation by fast pyrolysis: Part I. Influence of temperature, biomass species and solid heat carrier on the behavior of heavy metals. *Fuel* **87**, 1894–1905.

Lievens, C., Carleer, R., Cornelissen, T. & Yperman, J. 2009 Fast pyrolysis of heavy metal contaminated willow: influence of the plant part. *Fuel* **88**, 1417–1425.

Loffler, G., Wargadalam, V. J. & Winter, F. 2002 Catalytic effect of biomass ash on CO, CH_4 and HCN oxidation under fluidised bed combustor conditions. *Fuel* **81**, 711–718.

Lu, H., Zhang, W., Yang, Y., Huang, X., Wang, S. & Qiu, R. 2011 Relative distribution of Pb^{2+} sorption mechanisms by sludge-derived biochar. *Water Res.* **46** (3), 854–862.

Lu, S., Du, Y., Zhong, D., Zhao, B., Li, X., Xu, M., Li, Z., Luo, Y., Yan, J. & Wu, L. 2012 Comparison of trace element emissions from thermal treatments of heavy metal hyperaccumulators. *Environ. Sci. Technol.* **46** (9), 5025–5031.

Lu, H., Zhang, W., Wang, S., Zhuang, L., Yang, Y. & Qiu, R. 2013 Characterization of sewage sludge-derived biochars from different feedstocks and pyrolysis temperatures. *J. Anal. Appl. Pyrolysis* **102**, 137–143.

Mahmood, T., Saddique, M. T., Naeem, A., Westerhoff, P., Mustafa, S. & Alum, A. 2011 Comparison of difference methods for the point of zero charge determination of NiO. *Ind. Eng. Chem. Res.* **50**, 10017–10023.

Mancinelli, E., Baltrėnaitė, E., Baltrėnas, P., Paliulis, D. & Passerini, G. 2016 Trace metals in biochars from biodegradable byproducts of industrial processes. *Water Air Soil Pollut.* **227** (6), 1–21.

Markert, B., Wunschmann, S. & Baltrėnaitė, E. 2012 Aplinkos stebėjimo naujovės. Bioindikatoriai ir biomonitoriai: apibrėžtys, strategijos ir taikymas. [Innovative observation of the environment: bioindicators and biomonitors: definitions, strategies and applications]. *J. Environ. Eng. Landsc. Manage.* **20** (3), 221–239.

Markert, B., Fränzle, S. & Wünschmann, S. 2015 *Chemical Evolution – The Biological System of the Elements*. Springer, Heidelberg, New York, Dordrecht, London.

Mènard, Y., Asthana, A., Patisson, F., Sessiecq, P. & Ablitzer, D. 2006 Thermodynamic study of heavy metals behavior during municipal waste incineration. *Process Saf. Environ. Prot.* **84** (4), 290–296.

Meyer, Z. A., Apfelbacher, A. & Hornung, A. 2012 A comparative study of the pyrolysis of metal- and ash-enriched wood and the combustion properties of the gained char. *J. Anal. Appl. Pyrolysis* **96**, 196–202.

Mohamed, I., Zhang, G. S., Li, Z. G., Liu, Y., Chen, F. & Dai, K. 2017 Ecological restoration of an acidic Cd contaminated soil using bamboo biochar application. *Ecol. Eng.* **84**, 67–76.

Mohan, D., Pittman Jr., C. U. & Steele, P. H. 2006 Pyrolysis of wood/biomass for bio-oil: a critical review. *Energy Fuels* **20**, 848–889.

Nguyen, B. T. & Lehmann, J. 2009 Black carbon decomposition under varying water regimes. *Org. Geochem.* **40** (8), 846–853.

Oluwatuyi, O. E., Ashaka, E. C. & Ojuri, O. O. 2019 Cement stabilization treatment of lead and naphthalene contaminated lateritic soils. *J. Environ. Eng. Landsc. Manage.* **27** (1), 41–48.

Rai, M. K., Giri, B. S., Nath, Y., Bajaj, H., Soni, S., Singh, R. P., Singh, R. S. & Rai, B. N. 2018 Adsorption of hexavalent chromium from aqueous solution by activated carbon prepared from almond shell: kinetics, equilibrium and thermodynamics study. *J. Water Supply Res. Technol. Aqua* **67** (8), 724–737.

Rauch, J. N. & Pacyna, J. M. 2009 Earth's global Ag, Al, Cr, Cu, Fe, Ni, Pb and Zn cycles. *Glob. Biogeochem. Cycles* **23**, GB2001.

Raveendran, K., Ganesh, A. & Khilar, K. C. 1995 Influence of mineral matter on biomass pyrolysis characteristics. *Fuel* **74**, 1812–1822.

Rene, E. R., Shu, L. & Jegatheesn, V. 2018 Environmentally friendly (bio)technologies for the removal of emerging organic and inorganic pollutants from water. *J. Water Supply Res. Technol. Aqua* **67** (8), 697–702.

Saet, I. E., Revich, B. A., Ianin, E. P. & Smirnova, R. S. 1990 *Geokhimiia Okruzhaiushchei Sredy (Environmental Geochemistry)*. Nedra, Moksva, 335s (Monograph).

Sharma, A., Pareek, V. & Zhang, D. 2015 Biomass pyrolysis – a review of modelling, process parameters and catalytic studies. *Renew. Sustain. Energy Rev.* **50**, 1081–1096.

Shen, K. 2019 Experimental study on pressure sensitive properties of copper contaminated soil solidified by modified red mud. *J. Environ. Eng. Landsc. Manage.* **27** (2), 93–100.

Srinivasan, P. & Sarmah, A. K. 2014 Characterization of agricultural waste-derived biochars and their sorption potential for sulfamethoxazole in pasture soil: a spectroscopic investigation. *Sci. Total Environ.* **502**, 471–480.

Stals, M., Thijssen, E., Vangronsveld, J., Carleer, R., Schreurs, S. & Yperman, J. 2010 Flash pyrolysis of heavy metal contaminated biomass from phytoremediation: influence of temperature, entrained flow and wood/leaves blended pyrolysis on the behavior of heavy metals. *J. Anal. Appl. Pyrolysis* **87**, 1–7.

Stamm, A. J. 1964 *Wood and Cellulose Science*. The Ronald Press Company, New York.

Tang, Q., Shi, C., Shi, W., Huang, X., Ye, Y., Jiang, W., Kang, J., Liu, D., Ren, Y. & Li, D. 2019 Preferable phosphate removal by nano-La(III) hydroxides modified mesoporous rice husk biochars: role of the host pore structure and point of zero charge. *Sci. Total Environ.* **662**, 511–520.

Thy, P., Barfod, G. H., Cole, A. M., Brown, E. L., Jenkins, B. M. & Lesher, C. E. 2017 Trace metal release during wood pyrolysis. *Fuel* **203**, 548–556.

Tripathi, M., Sahu, J. N. & Ganesan, P. 2016 Effect of process parameters on production of biochar from biomass waste through pyrolysis: a review. *Renew. Sustain. Energy Rev.* **55**, 467–481.

Usevičiūtė, L. & Baltrėnaitė, E. 2019 Application of pyrogenic carbonaceous product for immobilization of potentially toxic elements in railway sleepers and polluted soil. *Int. J. Environ. Sci. Technol.* **16** (1), 23–36.

Van den Broeck, K., Helsen, L., Vandecasteele, C. & Van den Bulck, E. 1997 Determination and characterization of copper, chromium and arsenic in chromated copper arsenate (CCA) treated wood and its pyrolysis residues by inductively coupled plasma mass spectrometry. *Analyst* **122**, 695–700.

Van Vinh, N., Zafar, M. & Behera, S. K. 2015 Arsenic(III) removal from aqueous solution by raw and zinc-loaded pine cone biochar: equilibrium, kinetics, and thermodynamics studies. *Int. J. Environ. Sci. Technol.* **12**, 1283–1294.

Vassilev, S. V., Baxter, D., Andersen, L. K., Vassileva, C. G. & Morgan, T. J. 2012 An overview of the organic and inorganic phase composition of biomass. *Fuel* **94**, 1–33.

Vassilev, S. V., Vassileva, C. G. & Baxter, D. 2014 Trace element concentrations and associations in some biomass ashes. *Fuel* **129**, 292–313.

Wei, D., Li, B., Huang, H., Luo, L., Zhang, J., Yang, Y., Guo, J., Tang, L., Zeng, G. & Zhou, Y. 2018 Biochar-based functional materials in the purification of agricultural wastewater: fabrication, application and future research needs. *Chemosphere* **197**, 165–180.

WHO (World Health Organization) 2017 *Progress on Drinking Water, Sanitation and Hygiene: 2017 Update and SDG Baselines*. World Health Organization (WHO) and the United Nations Children's Fund (UNICEF), Geneva, p. 116. Licence: CC BY-NC-SA 3.0 IGO. Available from: https://apps.who.int/iris/bitstream/handle/10665/258617/9789241512893-eng.pdf;jsessionid=EB284213FF896D09B4FC5F23B7CC32C8?sequence=1.

Wongrod, S., Simon, S., van Hullebusch, E. D., Lens, P. N. L. & Guibaud, G. 2019 Assessing arsenic redox state evolution in solution and solid phase during As(III) sorption onto chemically-treated sewage sludge digestate biochars. *Bioresour. Technol.* **275**, 232–238.

Woolf, D., Amonette, J. E., Street-Perrott, F. A., Lehmann, J. & Joseph, S. 2010 Sustainable biochar to mitigate global climate change. *Nat. Commun.* **1**, 56.

Wu, H., Zeng, G., Liang, J., Chen, J., Xu, J., Dai, J., Li, X., Chen, M., Xu, P., Zhou, Y. & Li, F. 2016 Responses of bacterial community and functional marker genes of nitrogen cycling to biochar, compost and combined amendments in soil. *Appl. Microbiol. Biotechnol.* **100**, 8583–8591.

Wu, W., Li, J., Tian, L., Müller, K., Niazi, N. K., Xin, C., Song, X., Zheng, L., Chu, Y. & Li, J. 2017 Unraveling sorption of lead in aqueous solutions by chemically modified biochar derived from coconut fiber: a microscopic and spectroscopic investigation. *Sci. Total Environ.* **576**, 766–774.

Xia, D., Tan, F., Zhang, C., Jiang, X., Chen, Z., Li, H., Zheng, Y., Li, Q. & Wang, Y. 2016 ZnCl2-activated biochar from biogas residue facilitates aqueous As(III) removal. *Appl. Surf. Sci.* **377**, 361–369.

Yaman, S. 2004 Pyrolysis of biomass to produce fuels and chemical feedstock. *Energy Convers. Manag.* **45**, 651–671.

Yao, Y., Gao, B., Chen, J., Zhang, M., Inyang, M., Li, Y., Alva, A. & Yang, L. 2013 Engineered carbon (biochar) prepared by direct pyrolysis of Mg-accumulated tomato tissues: characterization and phosphate removal potential. *Bioresour. Technol.* **138**, 8–13.

Zaror, C. A., Hutchings, I. S., Pyle, D. L., Stiles, H. N. & Kandiyoti, R. 1985 Secondary char formation in the catalytic pyrolysis of biomass. *Fuel* **64**, 990–994.

Zeng, J., Zhou, S., Ligang, L. V., Su, Q. & Wang, J. 2018 Soil heavy metal contamination in rural land consolidated areas in the Yahgtze river delta. *China. J. Environ. Eng. Landsc. Manage.* **26**, 28–37.

Zhurinsh, A., Zandersons, J. & Dobele, G. 2005 Slow pyrolysis studies for utilization of impregnated waste timber materials. *J. Anal. Pyrolysis* **74**, 439–444.

Zobel, B. J. & van Buijtenen, J. P. 1989 *Wood Variation. Its Causes and Control*. Springer-Verlag, Heidelberg, Germany.

First received 11 June 2019; accepted in revised form 6 October 2019. Available online 2 December 2019

Bacterial community analysis of sulfate-reducing granular sludge exposed to high concentrations of uranium

Taotao Zeng, Shiqi Zhang, Wei Liao, Hualong Ma, Piet N. L. Lens and Shuibo Xie

ABSTRACT

Sulfate-reducing granular sludge was used for uranium-contaminated wastewater treatment and the responsible microbial community was analyzed. Four feedings, with 6 days for every feeding and 20 mg/L initial uranium, were provided at 27.5 (±2.5) °C. After the four feedings, a uranium removal efficiency of 94% was obtained. Environmental scanning electron microscopy (ESEM) showed that short rod bacteria were the dominant microorganisms in the granular sludge. X-ray energy dispersive spectroscopy (EDS) confirmed the presence of uranium on the granular surface. High-throughput sequencing was carried out for analyzing the bacterial diversity and community structure. The total data set comprised 8,290 high quality sequences, which could be divided into 605 operational taxonomic units (OTUs). The library coverage was 0.96 and the alpha diversity indices of ACE, Chao1, Shannon and Simpson were 2,255.40, 1,346.12, 4.03 and 0.05, respectively. There were 13 bacterial genera present with a ratio of more than 1% of the total 124 genera, among which *Desulfovibrio* (16.48%), *Clostridium* IV (9.29%), *Bacteroides* (3.46%) and *Citrobacter* (1.41%) were assumed as the functional bacteria, with a cumulative proportion of 30.64% of the total bacterial population. The results provide insights into the bacterial community of sulfate-reducing granular sludge exposed to high concentrations of uranium (20 mg/L).

Key words | bacterial population, community structure, high-throughput sequencing, sulfate-reducing granular sludge, uranium exposure

Taotao Zeng (corresponding author)
Shiqi Zhang
Wei Liao
Hualong Ma
Shuibo Xie
Hunan Province Key Laboratory of Pollution Control and Resources Reuse Technology,
University of South China,
Hengyang,
China
E-mail: *biowater@126.com*

Piet N. L. Lens
UNESCO-IHE Institute for Water Education,
Delft,
The Netherlands

Shuibo Xie
Key Discipline Laboratory for National Defence for Biotechnology in Uranium Mining and Hydrometallurgy,
University of South China,
Hengyang,
China

INTRODUCTION

Nuclear research, nuclear fuel production, mining activities, and weapon manufacturing result in pervasive environmental contamination by uranium (Suriya *et al.* 2017). One of the major sources of uranium contamination is uranium tailings and mining greatly affects the natural groundwater flow (Hu *et al.* 2016; Lipansky 2017). The most common valence states of uranium in nature are hexavalent [U(VI)] and tetravalent [U(IV)] uranium. The reduced form of U(IV) is insoluble and therefore less toxic to microbial cells, while U(VI) exerts great toxicity due to its solubility and mobility (Yan & Luo 2015). The U(VI) can also cause displacement and/or substitution of essential ions from cellular sites or blocking of enzymes and essential nutrient transport systems (Cason *et al.* 2012). Uranium accumulation in the human body is hazardous, as it is toxic to renal cells and osteoblasts (Gouget 2018) and can even be fatal (Sankhla *et al.* 2016). The radioactive compounds in water should be treated to avoid potential health impacts (Rodriguez *et al.* 2009).

Physical-chemical methods are conventionally used in uranium wastewater treatment, which include ion-exchange, coagulation, precipitation, extraction, and reverse osmosis.

doi: 10.2166/aqua.2019.027

However, physicochemical methods are environmentally intrusive and fairly expensive. Additionally, chemical methods create large quantities of toxic sludge and chemical by-products, which also pollute the environment (Mtimunye & Chirwa 2014). Given these limitations, developing safe, effective and environmentally friendly methods to treat uranium-contaminated wastewater are required. Some microorganisms are found to be uranium resistant, which is an alternative approach for uranium remediation.

Sulfate-reducing bacteria (SRB) can use sulfate as the terminal electron acceptor for the degradation of organic compounds and hydrogen, converting SO_4^{2-} to H_2S, which converts soluble U(VI) to insoluble U(IV) under anaerobic conditions (Zhang *et al.* 2017b). *Desulfococcus*, *Desulfobacterium* and *Desulfovibrio* are the common SRB genera. U(VI) immobilization by SRB occurs by two processes, enzymatically or chemically, by reacting with microbially generated H_2S (Beyenal *et al.* 2004). Uranium concentrations in the contaminated groundwater are usually below 1 mg/L (Hu *et al.* 2016). Aida *et al.* (2010) used batch experiments inoculated with methanogenic granule sludge to investigate their U(VI) removal capacity. The anaerobic removal activity could be sustained after several re-spikes of U(VI). However, there is little knowledge about the removal capacity of high uranium concentrations by sulfate-reducing granular sludge, let alone the bacterial community involved.

In this study, anaerobic granular sludge was collected from a citric acid wastewater treatment plant and cultivated under sulfate-reducing conditions. Then, the biomass was used to investigate the removal capacity of high uranium concentrations (20 mg/L) in synthetic wastewater. The microstructure was analyzed by environmental scanning electron microscopy (ESEM) and X-ray energy dispersive spectroscopy (EDS), whereas the microbial diversity and community composition were investigated by high-throughput sequencing.

MATERIALS AND METHODS

Source of biomass

Anaerobic granular sludge, collected from a full-scale upflow anaerobic sludge blanket (UASB) reactor in Yueyang (Hunan, China), was used to enrich sulfate-reducing granular sludge under anaerobic conditions. The microbial community of the inoculum anaerobic granular sludge was already characterized by Zeng *et al.* (2016a).

A batch reactor was used to enrich SRB for three months, which was fed with a culture medium containing (mg/L): $C_6H_{12}O_6 \cdot H_2O$ (3,300), Na_2SO_4 (4,437), NH_4Cl (230), KH_2PO_4 (65), |$NaHCO_3$ (1,250), yeast extract (500), $FeSO_4 \cdot 4H_2O$ (32), $CaCl_2 \cdot 2H_2O$ (38), $MgSO_4 \cdot 7H_2O$ (42), $CuCl_2 \cdot 2H_2O$ (0.5), $ZnCl_2$ (0.5), $MnCl_2 \cdot 4H_2O$ (0.5), $CoCl_2 \cdot 6H_2O$ (10), $NiCl_2 \cdot 6H_2$ O (0.5) (Xie *et al.* 2015). In this culture medium, the chemical oxygen demand (COD): N: P ratio equaled 200:5:1 and the COD:SO_4^{2-} ratio was 1:1, with an initial pH value of 7.2 (±0.2) and temperature of 35 °C. Then, the cultured sulfate-reducing anaerobic granular sludge was used as the inoculum for uranium removal in this experiment.

Batch experiments for uranium removal

Serum bottles, with 100 mL culture medium, were used for batch assays in duplicate. The standard 20 mg/L U(VI) was prepared according to the oxidization of U_3O_8 by nitrohydrochloric acid of GB6768-86 (SEPAC 1986). The chemical reaction was set as follows:

$$U_3O_8 + H_2O_2 + 2HCl + 4HNO_3UO_2Cl_2 + 2UO_2(NO_3)_2 + 4H_2O$$

After thoroughly washing using deionized water thrice, 10% w/v of the granular sludge was added to the bottles. The final volatile suspended solids/total suspended solids (VSS/TSS) of the granular sludge was 0.706. The pH of the medium was adjusted to 6.0, which was previously found to be the optimum pH for uranium removal by the sulfate-reducing granular sludge (Xie *et al.* 2015). The batch bottles were put on an orbital shaker at 150 rpm and 27.5 (±2.5) °C. The control test was set as incubating bottles with only synthetic culture medium and uranium, but without any anaerobic granular sludge. After 6 days' incubation, the new uranium (20 mg/L) containing medium was transferred to the anaerobic granular sludge. A total of four replacements were carried out as feeding I to IV (Table 1).

Table 1 | Operation of uranium removal by sulfate-reducing granular sludge in batch tests

Phase	Influent U(VI) concentration (mg/L)	Time (d)	Temperature (°C)	pH	Rotating speed in shaker (rpm)	Average U(VI) removal (%)	Average residual U(VI) concentration (mg/L)
I	20	6	27.5 ± 2.5	6	150	97.4	0.53
II	20	6	27.5 ± 2.5	6	150	92.4	1.53
III	20	6	27.5 ± 2.5	6	150	94.0	1.20
IV	20	6	27.5 ± 2.5	6	150	93.2	1.37

ESEM and EDS

The sulfate-reducing granular sludge was collected together after the final uranium exposure and ESEM-EDS was used to observe the microstructure and morphology according to Zeng *et al.* (2016b). Briefly, the sludge was washed twice in deionized water for 2 min, and then centrifuged at 8,000 *g* and 4 °C for 10 min. Then, the samples were frozen at −80 °C for 24 h (SANYO, MDF-U32 V). After drying by a vacuum freeze-drying machine (Freeze Dryer, FD5-series) for 24 h, gold sputter coating was carried out on the samples. The micromorphology and elemental composition were measured by an ESEM (FEI QUANTA 200) and X-ray EDS (EDAX Genesis 2000), respectively.

DNA extraction and polymerase chain reaction (PCR)

At the end of the incubation (Feeding IV), 5 g granular sludge was collected and homogenized for bacterial DNA extraction. Samples were rinsed twice with phosphate buffer saline (PBS). After centrifugation at 5,000 g for 5 min, the pellet was collected for DNA extraction using the E.Z.N.A.® Soil DNA Kit (OMEGA, USA) (Huang *et al.* 2016). The DNA quality was detected using 1% agarose gel electrophoresis. After purification of the DNA, the concentrations were determined using a Qubit 2.0 fluorometer (Invitrogen, USA).

The extracted DNA samples were used as template for 16S rRNA gene amplification, with 341F/805R primers targeting the V3–V4 region (Antwi *et al.* 2017). The PCR reaction system contained 0.5 μL forward primer (50 μM), 0.5 μL reverse primer (50 μM), 0.5 μL Plantium Taq (5 U μL^{-1}, Thermo, USA), 10 ng genomic DNA, 5 μL 10 × PCR buffer and 0.5 μL dNTP (10 mM each), and ultrapure water was added to a final volume of 50 μL. The PCR protocol was as follows: denatured at 94 °C for 3 min; 25 cycles of denatured at 94 °C for 30 s, 50 °C annealed for 20 s, 72 °C extended for 30 s; and a final extension at 72 °C for 5 min. PCR products were detected by 2% agarose gel electrophoresis and purified using a SanPrep DNA Fragment Purification Kit (Sangon Biotech, China).

High-throughput sequencing and bioinformatics analysis

High-throughput sequencing was carried out three times by an Illumina Miseq sequencer at Sangon Biotech Co., Ltd (Shanghai, China) as described in a previous report (Cui *et al.* 2016). For bioinformatics information analysis, Mothur was used for sequences matching the specific barcodes. Quality control was performed using Prinseq software to remove the sequence with an average quality score of less than 20, or with any unknown bases. Operational taxonomic units (OTUs) were assigned by Uclust software with more than 97% sequence similarity. Alpha indices of Coverage, ACE, Chaos, Shannon and Simpson were calculated by Mothur software. Taxonomic classification of OTUs was conducted by Ribosomal Database Project (RDP) Classifier (version 2.6) at a confidence threshold of 80% (Zhu *et al.* 2018). The sequences were submitted to the Sequence Read Archive (SRA) database in the National Coalition Building Institute (NCBI) and obtained the access number PRJNA490812.

Analytical methods

TSS and VSS of the granular sludge were characterized following APHA (2012). The U(VI) concentrations were measured daily in triplicate using the 2-(5-bromo-2-pyridylazo)-5-(diethylamino) phenol (5-Br-PADAP) spectrophotometric method at a wavelength of 578 nm (Ganesh *et al.* 2014), which had a detection limit of 50 μg/L for U(VI).

RESULTS AND DISCUSSION

Uranium removal in batch experiments

The U(VI) removal efficiency of the sulfate-reducing granular sludge is shown in Figure 1. During the first feeding, the U(VI) removal efficiency reached 95.6% on the first day, and further increased to 98% on the sixth day, which indicated that the sulfate-reducing granular sludge performed well upon uranium exposure. On the first day of feeding II, the U(VI) removal efficiency decreased to 83.1% with a residual uranium concentration of 3.38 mg/L after adding fresh culture medium containing 20 mg/L uranium. Two days later, the U(VI) removal efficiency increased to above 95.1% and the residual uranium concentration decreased to less than 0.98 mg/L in the last four days of feeding II. During feedings III and IV, a similar phenomenon was observed with higher uranium removal efficiencies achieved on the third day after fresh uranium containing medium was added. When four feedings were given, 94% U(VI) removal efficiency was retained at the end of feeding IV.

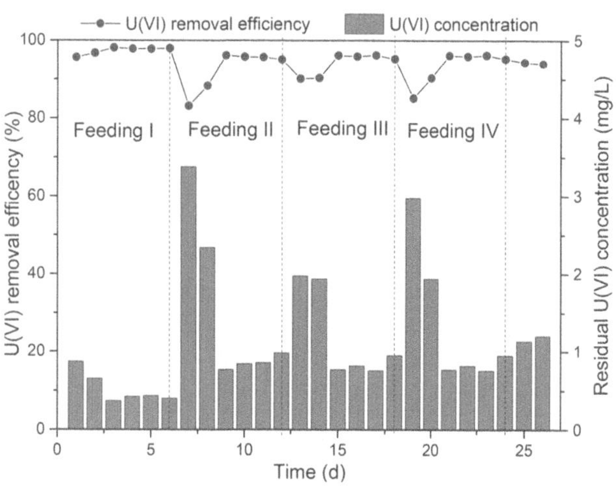

Figure 1 | Batch feeding of experiment on U(VI) removal by sulfate-reducing granular sludge exposed to U(VI) concentration of 20 mg/L.

Aida *et al.* (2010) showed that methanogenic granular sludge had a stable U(VI) removal capacity exceeding 99.8% using batch experiments. However, little information was given about the bacterial community of the methanogenic granular sludge. Beyenal *et al.* (2004) reported that U(VI) was immobilized by *Desulfovibrio desulfuricans* G20 (SRB). Hence, SRB were assumed to contribute to U(VI) removal in the present study. Our previous study (Chen *et al.* 2014) found that 93.72% of 10 mg/L U(VI) could be adsorbed by dried powder of crushed anaerobic granular sludge (2 g/L) within 60 min at pH 6.0 and 30 °C, wherein the equilibrium data fitted well with the pseudo-second-order model and Freundlich model ($R^2 > 0.99$). Hydroxy, amino and carbonyl groups were the major functional groups involved in the adsorption process. However, long-term performance studies are still lacking. In previous reports, the uranium concentrations were often less than 1 mg/L in the groundwater of contaminated sites (Hu *et al.* 2016; Wu *et al.* 2018). In this study, 92.4–97.4% of the supplied 20 mg/L U(VI) was removed during four feedings, which confirmed granular sludge can be used for the treatment of high concentrations of uranium-contaminated wastewater.

Microscopic observations of sulfate-reducing granular sludge

ESEM-EDS was used to reveal the morphology and elemental composition of the sulfate-reducing granular sludge upon uranium exposure for four successive feedings (Figure 2). Under 500-times magnification (Figure 2(a)), a honeycomb structure was observed. At 5,000-times magnification (Figure 2(b)), the major proportion of microorganisms had a short, rod shape. These bacteria retained a complete cell morphology, and the morphology was different from Zeng *et al.* (2016b), who observed cells with a spherical shape. The probable reason is the different microbial composition of the inoculum anaerobic granular sludge used in both studies. EDS showed that the major elements on the cell surface were O (19.45%), P (4.63%), and S (3.91%) (Figure S1(a) and (b) in Supplementary Materials). The U element ratio was 1.85%, which confirmed uranium was precipitated on the sulfate-reducing granular surface.

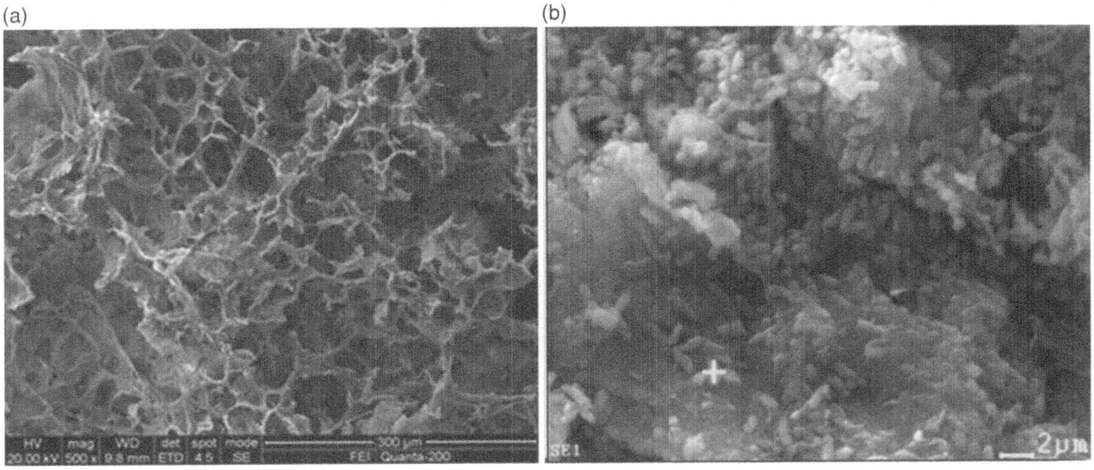

Figure 2 | SEM observation of granular sludge at (a) 500 and (b) 5,000 times magnification after exposure to U(VI) at a concentration of 20 mg/L with temperature of 27.5 (±2.5) °C.

Table 2 | Alpha index of microbial diversity in uranium-fed sulfate-reducing granular sludge

Sequencing number	Filtered number	Mean length (bp)	OTU number	Shannon index	ACE index	Chao1 index	Simpson index	Coverage
9,839	8,290	416.6	605	4.03	2,255.40	1,346.12	0.05	0.96

Microbial diversity assessment

After high through-put sequencing 9,839 raw sequences were obtained. Quality control measures were carried out for removing barcode sequences, terminal primers and low-quality sequences. Then 8,290 sequences were available for further analysis, with an average length of 416.6 bp, coverage of 0.96 and OTUs of 605 (Table 2). The abundance-rank curve showed that more than 97% of bacterial OTUs had less than 1% abundance in the total bacterial sequence (Figure 3(a)). Only 2.8% of the OTUs had an abundance >1%, with 17 of the total 605 OTUs; 97.2% of the OTUs had abundances less than 1%, showing an even bacterial distribution in the sulfate-reducing granular sludge.

Alpha diversity is an indicator for evaluating the richness and diversity within communities (Yan *et al.* 2016), which is expressed by indices of ACE, Chao1, Shannon and Simpson. The ACE and Chao1 indices were 2,255.40 and 1,346.12, respectively. The corresponding ACE and Chao1 rarefactions are listed in Figure 3(b). The above curves tendencies were quickly close to saturation, which indicated a high richness available in the sulfate-reducing granular sludge. The Shannon and Simpson indices were 4.03 and 0.05, respectively, and the corresponding curves leveled up quickly (Figure 3(c)), which indicated high community diversity. Rarefaction curves nearly reached a plateau, which also demonstrated appropriate sequencing depths (Suriya *et al.* 2017). Alpha indices and the rarefaction curves revealed high microbial abundance and diversity in granular sludge, which benefits resistance to toxicity in radioactive wastewater (Edberg *et al.* 2012).

Microbial community composition

Based on the taxonomic assignment at phylum level using the RDP database, a total of 20 phyla were identified, which contained 8,290 reads (sequences). The results revealed that seven bacterial phyla had a relative abundance of more than 0.5% (Figure 4(a)), with a cumulative contribution of 97.89%. The total proportions were 1.16% for the bacteria whose ratios were less than 0.5% separately, marked as 'Others' in Figure 4(a). The sulfate-reducing granular sludge comprised 0.93% unassigned sequences derived from bacterial phyla.

The Bacteroidetes phylum had the largest part in the composition, which accounted for 42.25% with 3,504

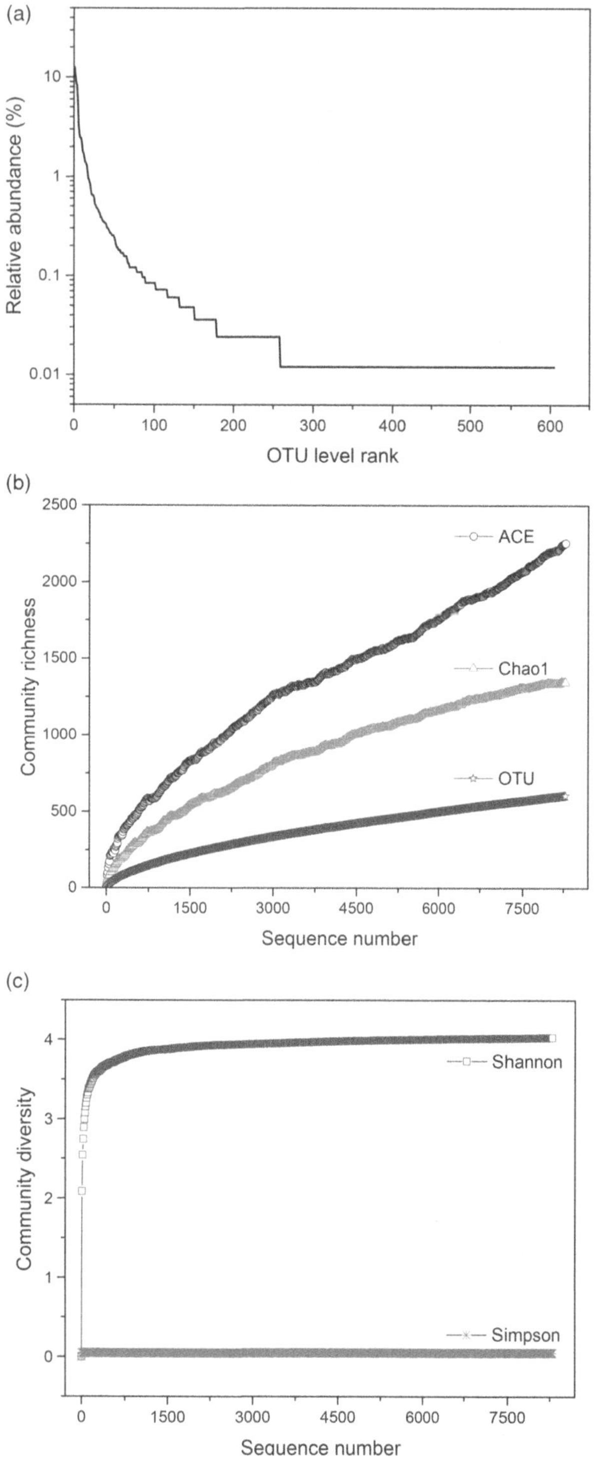

Figure 3 | (a) Abundance-rank, (b) community richness, and (c) diversity charts of bacterial community in sulfate-reducing granular sludge after exposure to 20 mg/L uranium.

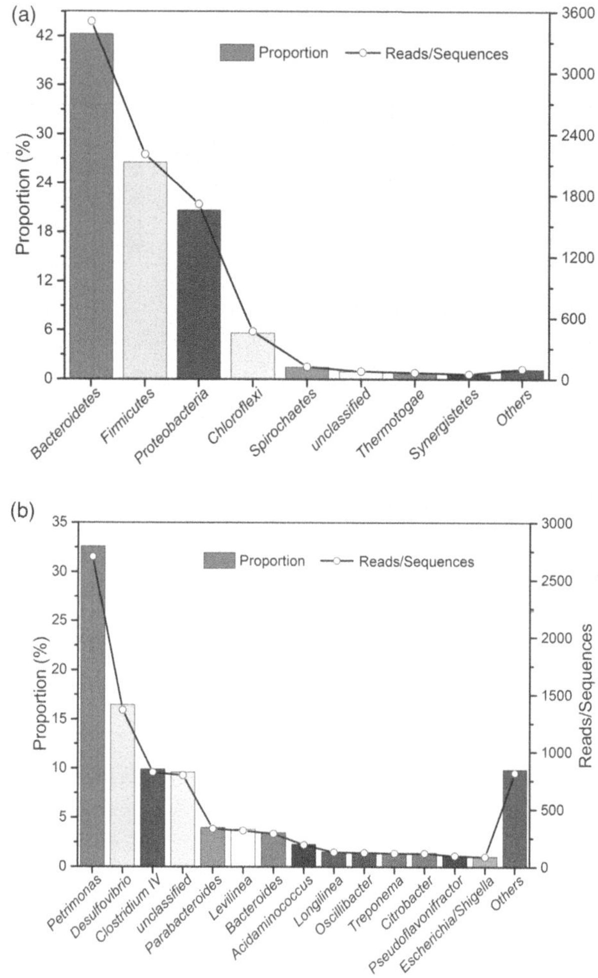

Figure 4 | Bacterial composition of anaerobic granular sludge after exposure to U(VI) at a concentration of 20 mg/L based on (a) phylum and (b) genus level.

reads. The second most dominant phylum was Firmicutes (26.55%) followed by Proteobacteria (20.66%). The top three dominant phyla were the same as those present in a uranium-contaminated sediment derived from the sacred Cauvery River, India: Proteobacteria (47.49%), Bacteroidetes (22.36%) and Firmicutes (14.58%) (Suriya *et al.* 2017). Furthermore, a relatively high proportion of Chloroflexi (5.62%) was observed with 466 reads. Phyla of Spirochaetes, Thermotogae and Synergistetes also emerged, with a ratio of 1.46%, 0.77% and 0.58%, respectively.

The eubacterial community of the uranium-fed sulfate-reducing granular sludge was further analyzed at the genus level. There were 124 bacterial genera present, among which 13 bacterial genera were present with ratios more than 1% (Figure 4(b)). The total proportions were 9.86% for the bacterium whose ratio was less than 1% separately, marked as 'Others' in Figure 4(b). The sulfate-reducing granular sludge also comprised 9.6% unclassified sequences derived from the bacterial genus level analysis.

Petrimonas was the most abundant genus in the sulfate-reducing granular sludge, which accounted for 32.59% with 2,702 reads (sequences). *Petrimonas* has been previously found to be an abundant genus in an anaerobic-oxic-sedimentation bioreactor for diazo dyes decolorization and mineralization (Zhu *et al.* 2018). *Desulfovibrio* was the second dominant genus, which occupied 16.48% with 1,366 sequences. In a previous report, Stylo *et al.* (2015) investigated the reduction and immobilization of uranium by *Desulfovibrio vulgaris* biofilms. They concluded that biotic reduction was the most important role for uranium immobilization. Compared with Zeng *et al.* (2016a) who found an abundance of *Desulfovibrio* of below 0.5% in the original anaerobic granular sludge, *Desulfovibrio* was a major population in the sulfate-reducing granular sludge after repeated uranium exposure. Therefore, the *Desulfovibrio* genus was assumed to be one of the functional bacteria in this study. *Clostridium* IV, which belongs to the Firmicutes phylum, accounted for 9.29% and included 820 sequences. *Clostridium* sp. has been reported to be effective in uranium bioreduction and precipitation in a synthesized ionic liquid aqueous solution (Zhang *et al.* 2013). Therefore, *Clostridium* IV could also contribute to the uranium removal within the sulfate-reducing granular sludge.

The *Parabacteroides* genus included 330 sequences and accounted for 3.98% of the total reads. The *Bacteroides* genus was represented by 287 sequences and accounted for 3.46% of the total reads. A small proportion of *Bacteroides* was previously found in a uranium mill tailings at Gittersee/Coschutz, Germany (Akhtar *et al.* 2017). *Bacteroides* sp. and *Parabacteroides* belong to the order Bacteroidales, which are typical fermentative bacteria or cellulose degraders. Maize straw could be degraded by these bacteria, thus providing suitable electron donors to the SRB (Zhang & Wang 2016). Therefore, it can be assumed that *Parabacteroides* and *Bacteroides* might degrade organic matter to provide electron donors for SRB (e.g. *Desulfovibrio*), promoting the uranium removal efficiency.

Levilinea and *Longilinea* accounted for 3.8% and 1.53% of the total bacteria, respectively. They belong to the Chloroflexi phylum (Yamada & Sekiguchi 2009), which were found to play important roles in degrading organic matter. *Acidaminococcus* included 191 sequences and occupied 2.3% of the total reads. *Oscillospira* and *Treponema* included 123 and 117 sequences, respectively, accounting for 1.48% and 1.41% of the total reads. The genera of *Acidaminococcus*, *Oscillospira* and *Treponema* have been previously found in rivers and marine ecosystems (Ming & Zhen 2014).

The *Citrobacter* genus was represented by 117 sequences and occupied 1.41% of the total reads. *Citrobacter* has been previously reported to be present in the Shihongtan, a sandstone-type uranium tailings in Xinjiang, China (Chen *et al.* 2012), which is also possibly a functional genus in the present study. *Pseudoflavonifractor* (1.16%) and *Escherichia/Shigella* (1.05%) were represented by 96 and 87 sequences, respectively. Both genera have not yet been found in uranium rich environments.

Thus, compared with the control anaerobic granular sludge by Zeng *et al.* (2016a), an abundance variation was evident at the genus level (Table 3). The abundance of *Desulfovibrio* and *Parabacteroides* genera increased significantly from below 1% to 16.48% and 3.98%, respectively. The abundance of the *Clostridium* genus also increased from 5.95% to 9.29%. Therefore, the above three genera were thought to contribute to uranium removal in the sulfate-reducing anaerobic granular sludge, which could be

Table 3 | Abundance variation between the uranium exposure granular sludge and the control sludge at genus level

Control	Ratio (%)	Uranium exposure granular sludge	Ratio (%)
Petrimonas	32.14	*Petrimonas*	32.59
Levilinea	11.42	*Desulfovibrio*	16.48
Erysipelotrichaceae	7.68	*Clostridium* IV	9.29
Paludibacter	7.17	*Parabacteroides*	3.98
Clostridium	5.95	*Levilinea*	3.8
Methanosaeta	3.44	*Bacteroides*	3.46
Syntrophomonas	2.75	*Acidaminococcus*	2.3
Longilinea	2.2	*Oscillospira*	1.48
Phascolarctobacterium	1.63	*Treponema*	1.41
Thermovirga	1	*Citrobacter*	1.41
		Pseudoflavonifractor	1.16
		Escherichia/Shigella	1.05

supported by previous reports (Zhang *et al.* 2013; Stylo *et al.* 2015; Zhang & Wang 2016). In the control, the abundance of *Levilinea* decreased dramatically from 11.42% to 3.8%. A dramatic change of abundance also happened for other bacterial genera upon uranium exposure (Table 3). The decrease in abundance of other genera was probably due to their sensitivity to uranium exposure.

The microbial community plays an important role in uranium recovery from contaminated sites, which is also beneficial for the control of uranium contamination. Abhilash & Pandey (2013) reported a 98.3% uranium recovery in 14 days for Turamdih uranium ore (Jharkhand, India) with an enriched culture of *Acidithiobacillus ferrooxidans*, using Fe(II) at pH 2.0, 35 °C and 20% (w/v) pulp density. Coral *et al.* (2018) evaluated microbial communities present in uranium in-situ recovery from acid mine drainage; the dominant bacteria were *Sulfobacillus* sp., *Leptospirillum* sp. and *Acidithiobacillus* sp.

Practical applications

The present study demonstrated that sulfate-reducing granular sludge has a strong uranium-resistant ability. The removal efficiency of 92.4–97.4% from a 20 mg/L U(VI) solution during the batch incubation was most likely due to the natural occurrence of U(VI)-reducing microorganisms in the inoculum sludge (Aida *et al.* 2010). Nancharaiah *et al.* (2006) reported almost complete removal of uranium by aerobic granules in the range of 6–100 mg/L in less than 1 h in an acidic pH range (1 to 6), which indicated the aerobic granular biomass was an effective biosorbent material for recovering/removing uranium from dilute nuclear waste. This study demonstrates the potential of utilizing anaerobic (sulfate-reducing) granular sludge for biological removal of U(VI) from contaminated run-off, groundwater and acid mine drainage.

Desulfovibrio, *Clostridium* IV, *Bacteroides* and *Citrobacter* were the dominant taxa in the present study. As there is little information about uranium removal by anaerobic granular sludge (Aida *et al.* 2010; Zhang *et al.* 2017a), it is necessary to investigate the functional microorganisms by metagenomics in future research, especially to evaluate the roles of Archaea in uranium removal.

CONCLUSIONS

Sulfate-reducing granular sludge exhibited high (92.4–97.4%) removal efficiencies during four feedings for a total of 24 days from synthetic 20 mg/L uranium acid mine drainage, which confirmed granular sludge can be used for the treatment of uranium-contaminated wastewater. Short rod-shaped bacteria occupied a major proportion of the microorganisms in the anaerobic granular sludge. The granular sludge had a high microbial abundance and diversity in spite of the high uranium concentration exposure. Microbial community analysis indicated seven bacterial phyla in the uranium-exposed sulfate-reducing granular sludge, with Bacteroidetes (42.25%), Firmicutes (26.55%) and Proteobacteria (20.66%) accounting for the top three phyla. Thirteen bacterial genera were dominant, wherein the taxa of *Desulfovibrio* (16.48%), *Clostridium* IV (9.29%), *Bacteroides* (3.46%) and *Citrobacter* (1.41%) were the indicators for uranium removal, with a cumulative proportion of 30.64%.

ACKNOWLEDGEMENTS

This work was supported by the National Natural Science Foundation of China (51408293 and 11475080), the Research Program for Young Scholars in Hunan Province of 2018 and the Foundation of China Scholarship Council (201608430243).

SUPPLEMENTARY MATERIAL

The Supplementary Material for this paper is available online at https://dx.doi.org/10.2166/aqua.2019.027.

REFERENCES

Abhilash & Pandey, B. D. 2013 Bioreactor leaching of uranium from a low grade Indian silicate ore. *Biochem. Eng. J.* **71** (0), 111–117.

Aida, T. R., Antonia, L. V., Jim, A. F. & Reyes, S. A. 2010 Anaerobic bioremediation of hexavalent uranium in groundwater by reductive precipitation with methanogenic granular sludge. *Water Res.* **44** (7), 2153–2162.

Akhtar, S., Yang, X. & Pirajno, F. 2017 Sandstone type uranium deposits in the Ordos Basin, Northwest China: a case study and an overview. *J. Asian Earth Sci.* **146**, 367–382.

Antwi, P., Li, J., Boadi, P. O., Meng, J., Shi, E., Chi, X., Deng, K. & Ayivi, F. 2017 Dosing effect of zero valent iron (ZVI) on biomethanation and microbial community distribution as revealed by 16S rRNA high-throughput sequencing. *Int. Biodeterior. Biodegrad.* **123**, 191–199.

APHA 2012 Standard Methods for the Examination of Water and Wastewater, American Public Health Association/American Water Works Association/Water Environment Federation, Washington DC, USA.

Beyenal, H., Sani, R. K., Peyton, B. M., Dohnalkova, A. C., Amonette, J. E. & Lewandowski, Z. 2004 Uranium immobilization by sulfate-reducing biofilms. *Environ. Sci. Technol.* **38** (7), 2067–2074.

Cason, E. D., Piater, L. A. & Heerden, E.v. 2012 Reduction of U(VI) by the deep subsurface bacterium, *Thermus scotoductus* SA-01, and the involvement of the ABC transporter protein. *Chemosphere* **86** (6), 572–577.

Chen, Z., Cheng, Y. J., Pan, D. M., Wu, Z. X., Li, B., Pan, X. H., Huang, Z. P., Lin, Z. & Guan, X. 2012 Diversity of microbial community in Shihongtan Sandstone-type uranium deposits, Xinjiang, China. *Geomicrobiol. J.* **29** (3), 255–263.

Chen, H., Xie, S., Liu, J., Xiao, S., Zeng, T., Ling, H. & Wang, J. 2014 Characteristics and mechanism of uranium(VI) absorbed by anaerobic granular sludge. *Chin. J. Nonfer. Met.* **24** (9), 2418–2425. (in Chinese).

Coral, T., Descostes, M., De Boissezon, H., Bernier-Latmani, R., de Alencastro, L. F. & Rossi, P. 2018 Microbial communities associated with uranium in-situ recovery mining process are related to acid mine drainage assemblages. *Sci. Total Environ* **628–629**, 26–35.

Cui, M., Cui, D., Lee, H., Liang, B., Wang, A. & Cheng, H. 2016 Effect of electrode position on azo dye removal in an up-flow hybrid anaerobic digestion reactor with built-in bioelectrochemical system. *Sci. Rep.* **6**, 25223.

Edberg, F., Andersson, A. F. & Holmström, S. J. 2012 Bacterial community composition in the water column of a lake formed by a former uranium open pit mine. *Microb. Ecol.* **64** (4), 870–880.

Ganesh, S., Velavendan, P., Pandey, N. K., Mudali, U. K. & Natarajan, R. 2014 On-site monitoring of uranium in low level liquid waste streams using U-Br-PADAP strip indicator paper technique. *J. Radioanal. Nucl. Chem.* **302** (3), 1513–1518.

Gouget, B. 2018 Uranium: Toxicity to Renal Cells and Osteoblasts [M]. Reference Module in Earth Systems and Environmental Sciences, Elsevier.

Hu, N., Ding, D. X., Li, S. M., Tan, X., Li, G. Y., Wang, Y. D. & Fei, X. 2016 Bioreduction of U(VI) and stability of immobilized uranium under suboxic conditions. *J. Environ. Radioact.* **154**, 60–67.

Huang, X., Mu, T., Shen, C., Lu, L. & Liu, J. 2016 Effects of bio-surfactants combined with alkaline conditions on volatile fatty acid production and microbial community in the anaerobic fermentation of waste activated sludge. *Int. Biodeterior. Biodegrad.* **114**, 24–30.

Lipansky, T. 2017 Regional impact of uranium mining on piezometric surfaces in a multi-layered water-bearing system, Bohemian Cretaceous Basin, Czech Republic. *Mine Water Environ.* **36** (1), 4–17.

Ming, L. X. & Zhen, L. P. 2014 Characterization of bacterial communities in sediments receiving various wastewater effluents with high-throughput sequencing analysis. *Microb. Ecol.* **67** (3), 612–623.

Mtimunye, P. J. & Chirwa, E. M. 2014 Characterization of the biochemical-pathway of uranium (VI) reduction in facultative anaerobic bacteria. *Chemosphere* **113**, 22–29.

Nancharaiah, Y. V., Joshi, H. M., Mohan, T. V. K., Venugopalan, V. P. & Narasimhan, S. V. 2006 Aerobic granular biomass: a novel biomaterial for efficient uranium removal. *Curr. Sci.* **91** (4), 503–509.

Rodriguez, C., Devine, B., Cook, A., Weinstein, P. & Buynder, P. V. 2009 Gross alpha and gross beta particle activity in recycled water for augmentation of drinking water supplies. *J. Water Supply Res. T. -Aqua* **58** (3), 191–202.

Sankhla, M. S., Kumari, M., Nandan, M., Kumar, R. & Agrawal, P. 2016 Heavy metals contamination in water and their hazardous effect on human health: a review. *Int. J. Curr. Microbiol. App. Sci* **5** (10), 759–766.

SEPAC (State Environmental Protection Administration of China) 1986 *Methods of Analysing Microquantity of Uranium in Water*. China Environmental Science Press, Beijing, pp. 296–301 (in Chinese).

Stylo, M., Neubert, N., Roebbert, Y., Weyer, S. & Bemier-Latmani, R. 2015 Mechanism of uranium reduction and immobilization in *Desulfovibrio vulgaris* biofilms. *Environ. Sci. Technol.* **49** (17), 10553–10561.

Suriya, J., Shekar, M. C., Nathani, N. M., Suganya, T., Bharathiraja, S. & Krishnan, M. 2017 Assessment of bacterial community composition in response to uranium levels in sediment samples of sacred Cauvery River. *Appl. Microbiol. Biotechnol.* **101** (2), 831–841.

Wu, Y., Li, J., Wang, Y. & Xie, X. 2018 Variations of uranium concentrations in a multi-aquifer system under the impact of surface water-groundwater interaction. *J. Contam. Hydrol.* **211**, 65–76.

Xie, S., Ma, H., Tan, Z., Ling, H., Zeng, T., Chen, S. & Wang, J. 2015 Study of U(VI) removal by sulfate reducing granular sludge under micro-aerobic condition. *At. Energ. Sci. Techno.* **49** (1), 26–32. (in Chinese).

Yamada, T. & Sekiguchi, Y. 2009 Cultivation of uncultured *Chloroflexi Subphyla*: significance and ecophysiology of formerly uncultured *Chloroflexi 'Subphylum i'* with natural and biotechnological relevance. *Micro. Environ.* **24** (3), 205–216.

Yan, X. & Luo, X. 2015 Radionuclides distribution, properties, and microbial diversity of soils in uranium mill tailings from southeastern China. *J. Environ. Radioact.* **139**, 85–90.

Yan, X., Luo, X. G. & Zhao, M. 2016 Metagenomic analysis of microbial community in uranium-contaminated soil. *Appl. Microbiol. Biotechnol.* **100** (1), 299–310.

Zeng, T., Liao, W., Xie, S., Rong, L., Li, S., Jiang, X. & Ma, H. 2016a Analysis of bacterial community in anaerobic granular sludge for citric acid wastewater treatment. *J. Harbin Inst. Technol.* **48** (8), 115–120. (in Chinese).

Zeng, T., Lu, H., Liu, Y., Liu, J., Rong, L., Ma, H. & Xie, S. 2016b Analysis of microbial community structure of uranium-resistant granular sludge. *The Chinese J. Nonfer. Met.* **26** (1), 233–241. (in Chinese).

Zhang, M. & Wang, H. 2016 Preparation of immobilized sulfate reducing bacteria (SRB) granules for effective bioremediation of acid mine drainage and bacterial community analysis. *Miner. Eng.* **92**, 63–71.

Zhang, C., Dodge, C. J., Malhotra, S. V. & Francis, A. J. 2013 Bioreduction and precipitation of uranium in ionic liquid aqueous solution by *Clostridium* sp. *Bioresour. Technol.* **136**, 752–756.

Zhang, H., Cheng, M., Liu, W., Huang, F., Ding, H., Li, S., Guo, W., Wang, Y. & Huang, H. 2017a Characterization of uranium in the extracellular polymeric substances of anaerobic granular sludge used to treat uranium-contaminated groundwater. *Rsc. Adv.* **7** (85), 54188–54195.

Zhang, P., He, Z. L., Van Nostrand, J. D., Qin, Y. J., Deng, Y., Wu, L. Y., Tu, Q. C., Wang, J. J., Schadt, C. W., Fields, M. W., Hazen, T. C., Arkin, A. P., Stahl, D. A. & Zhou, J. Z. 2017b Dynamic succession of groundwater sulfate-reducing communities during prolonged reduction of uranium in a contaminated aquifer. *Environ. Sci. Technol.* **51** (7), 3609–3620.

Zhu, Y., Xu, J., Cao, X., Cheng, Y. & Zhu, T. 2018 Characterization of functional microbial communities involved in diazo dyes decolorization and mineralization stages. *Int. Biodeterior. Biodegrad.* **132**, 166–177.

First received 5 March 2019; accepted in revised form 1 July 2019. Available online 30 July 2019

Removal of cyanotoxins in drinking water using ozone and ozone-hydrogen peroxide (peroxone)

Guhankumar Ponnusamy, Lijo Francis, Kavithaa Loganathan, Oluwaseun O. Ogunbiyi, Saad Jasim and Jayaprakash Saththasivam

ABSTRACT

Presence of cyanotoxins in drinking water poses a great risk to public health. Elevated levels of cyanotoxins in drinking water can lead to acute gastroenteritis, liver diseases, and neurotoxicity. In this study, drinking water samples were collected across the eastern part of Qatar and screened using a rapid assay to detect the presence of microcystins and nodularins. The results showed that the toxin concentrations in all the water samples were below the WHO prescribed limit of 1 μg/L. Considering a worst-case scenario, toxin removal efficiencies were evaluated using ozone and ozone-hydrogen peroxide by spiking drinking water samples with microcystin-LR (MC-LR) at different oxidant dosages, toxin concentrations, water temperatures, and total organic carbon. It was found that peroxone-treated water samples have better MC-LR removal efficiency than molecular ozone at lower oxidant dosages. Nevertheless, at higher oxidant dosages, both ozonation and peroxone oxidation methods showed a similar removal efficiency. The experimental results also clearly indicated that variation in water temperature between 22 °C and 35 °C has minimal effect on the removal efficiency in both the treatment methods. It was also confirmed that the presence of organic carbon has a more profound detrimental impact than water temperature for toxin removal.

Key words | advanced oxidation process, cyanotoxin, drinking water, enzyme-linked immunosorbent assay (ELISA), microcystin-LR, ozonation

Guhankumar Ponnusamy
Lijo Francis
Kavithaa Loganathan
Oluwaseun O. Ogunbiyi
Jayaprakash Saththasivam (corresponding author)
Qatar Environment and Energy Research Institute, Hamad Bin Khalifa University, Qatar Foundation, Doha, 34110, Qatar
E-mail: *jsaththasivam@hbku.edu.qa*

Saad Jasim
SJ Environmental Consultants (Windsor) Inc., 4483 Cherry Hill Road, Windsor, ON N9G 2W3, Canada

INTRODUCTION

The presence of harmful cyanotoxins such as hepatotoxins and neurotoxins in drinking water is a serious concern for human health. Microcystins (MCs) are one of the common yet highly potent hepatotoxins that are found in water bodies during algal blooms. MCs are monocyclic heptapeptides that are produced by species like *Microcystis aeruginosa*, *Microcystis* spp., *Anabaena*, *Planktothrix*, *Nostoc*, *Oscillatoria*, and *Anabaenopsis* (Van der Merwe 2014). In a recent study from Ethiopia (Tilahun *et al.* 2019), very high concentrations of MCs in the drinking water source was detected. Exposure to MCs via skin contact, inhalation, or ingestion of toxin-contaminated water sources can lead to breathing problems, nausea, diarrhea, skin irritation, and even acute liver damage at a high level of exposure (Hunter 1998). The cyclic structure of MCs renders them chemically stable in water and heat resistant (Lawton & Robertson 1999), hence boiling of toxin-contaminated water prior to drinking is ineffective in degrading MCs. In addition, MCs are difficult to physically detect in water due to their colorless, odorless, and tasteless characteristics.

Complete removal of cyanotoxin could be a challenging process for conventional water treatment plants as there is a possibility of dissolved extracellular toxin passing through the treatment process (Kull *et al.* 2006). The potential regrowth of toxin-producing cyanobacteria in distribution pipelines and storage tanks also cannot be ruled out. One

doi: 10.2166/aqua.2019.028

of the effective methods to remove cyanotoxins from drinking water is by using ozone (O_3) and ozone-hydrogen peroxide (peroxone). Ozone has an oxidation potential of 2.07 V and is widely used as an effective oxidant and disinfectant in water treatment industries (Jasim & Saththasivam 2017). The efficiency of ozone treatment process largely depends on the source water pH, contact time, and other constituents present in water such as dissolved organic carbon, iron, and manganese. The oxidation strength of a conventional ozonation system can be further enhanced when hydrogen peroxide is used in conjunction with ozone to promote the formation of free hydroxyl radicals (•OH), which is a superior oxidant when compared to molecular ozone alone (Bourgin *et al.* 2017). Several studies were conducted in the past to evaluate the efficiency of ozone and peroxone in removing cyanotoxins from various water sources. Rositano *et al.* (1998) reported that 0.2 mg/L of ozone with a mere contact time of 15 seconds was sufficient to remove 1 mg/L of microcystin-LR (MC-LR). Similar removal efficiency was observed when peroxone at a H_2O_2/O_3 ratio of 0.5 was used. An ozonation study conducted by Hall *et al.* (2000) to evaluate the removal efficiency of MC-LR in river water reported that 2 mg/L ozone dosage was sufficient to remove both the intra- and extracellular toxin. Another study that evaluated the efficiency of ozonation in removing MC-LR from lake water indicated that approximately 95% of oxidation efficiency can be achieved at ozone dosage of 0.25 mg/L (Rodríguez *et al.* 2007). Al Momani *et al.* (2008) claimed that ozone dosage of 2.4 mg/L was required to completely degrade 5 mg/L of MC-LR. At a lower MC-LR concentration of 1 mg/L, an ozone dosage of 0.6 mg/L was required to eliminate the toxin within 90 seconds of reaction time. For peroxone-related experiments at initial ozone and MC-LR concentrations of 0.1 mg/L and 1 mg/L, respectively, the study reported that the toxin removal efficiency improved from 65% to 100% when H_2O_2 concentration was increased from 0.001 to 0.01 mg/L. A study by Miao *et al.* (2010) showed that at a high initial MC-LR concentration of 50 mg/L, the toxin removal efficiency increased from 29.5% to 92.1% when ozone dosage to MC-LR ratio was increased from 1 to 6, respectively.

It has to be emphasized that most of the reported MC-LR removal studies using ozone and peroxone were conducted at a very high MC-LR concentration (over 1 mg/L). The presence of such a high toxin level in water is rare, even during the bloom period. Turner *et al.* (2018) reported that 122 of the total 137 water samples analyzed for MCs in cyanobacterial blooms from freshwater bodies was less than 50 µg/L. A summary by the Michigan Department of Environmental Quality (MDEQ) that reviewed several studies concluded that MCs' level in the 232 Michigan inland lakes from 2002 to 2012 did not exceed 4 µg/L (Birbeck *et al.* 2019). Another study reported that MC-LR in the range of 0.7–1.3309 µg/L was detected in the water storage tanks of the urban area of Qatar (Chatziefthimiou *et al.* 2016). As higher toxin concentration in water requires higher oxidant dosage for complete removal, it is important to conduct experimental studies at an appropriate toxin concentration in order to accurately determine the oxidant dosing range. All the experiments in this study were conducted at lower initial MC-LR concentrations, ranging between 3 µg/L and 50 µg/L. This range was chosen to approximate the concentration of MC-LR that could exist in treated water. Munoz *et al.* (2019) treated real surface water and observed a drop in cyanotoxin removal rate due to the presence of natural organics when compared with toxins spiked in deionized water. In addition, to the best of our knowledge, there are very limited experimental studies that investigate the combined effect of high water temperature and dissolved organic carbon in drinking water using ozone and peroxone. This specific scenario is particularly important for a hot and arid country like Qatar, where the water temperature can easily go beyond 35 °C during the summer months. The following questions were the key motives behind this study: (i) What is the MC-LR removal efficiency at varying ozone dosages and initial toxin concentrations? (ii) What is the effectiveness of MC-LR removal with peroxone process? (iii) How is MC-LR removal efficiency affected by water temperature and total organic carbon (TOC) considering the high water temperature during the summer season in Qatar? The first part of the study involved the collection of drinking water samples from various residential, public, and commercial sites in the eastern part of Qatar. The water samples were tested for basic water qualities such as pH, conductivity, TOC, and algal cells. In addition, the samples were also screened to assess and quantify the presence of specific cyanotoxins, namely, microcystin and nodularin. A qualitative toxin-screening tool, enzyme-linked immunosorbent assay (ELISA), which detects the presence of toxins based on the Adda moiety, was used (Loganathan 2016). The second part of the study evaluated the effect of initial toxin concentrations, oxidant dosages, water temperature, and quality on the removal efficiency of MC-LR using ozone and peroxone.

MATERIALS AND METHODS

Drinking water sampling

Drinking water samples were collected from residential villas, public mosques, and commercial sites (shopping malls, cafeterias, and public water dispenser units) located in the eastern part of Qatar. Figure 1 shows the sampling locations that include major areas such as Al Wakrah, Industrial Area, Abu Hamour, Al Sadd, Ar-Rayyan, and Al-Gharrafa. Water samples were collected in amber bottles and were kept at 4 °C during transportation. All the samples were collected and analyzed within 24 hours of collection. Conductivity, pH, and turbidity of the water samples were measured using portable meters (HACH CDC401, HACH PHC201, and HACH 2100Q, USA). The TOC content and chlorophyll cell counts in the water samples were measured using TOC analyzer (SHIMADZU TOC-L, Japan) and flow cytometer (BD ACCURI C6, USA).

Enzyme-linked immunosorbent assay (ELISA)

ELISA method was adopted for the rapid detection of toxins in the drinking water samples (Agrawal *et al.* 2012). The Microcystins-Adda ELISA Microtiter Plate was purchased

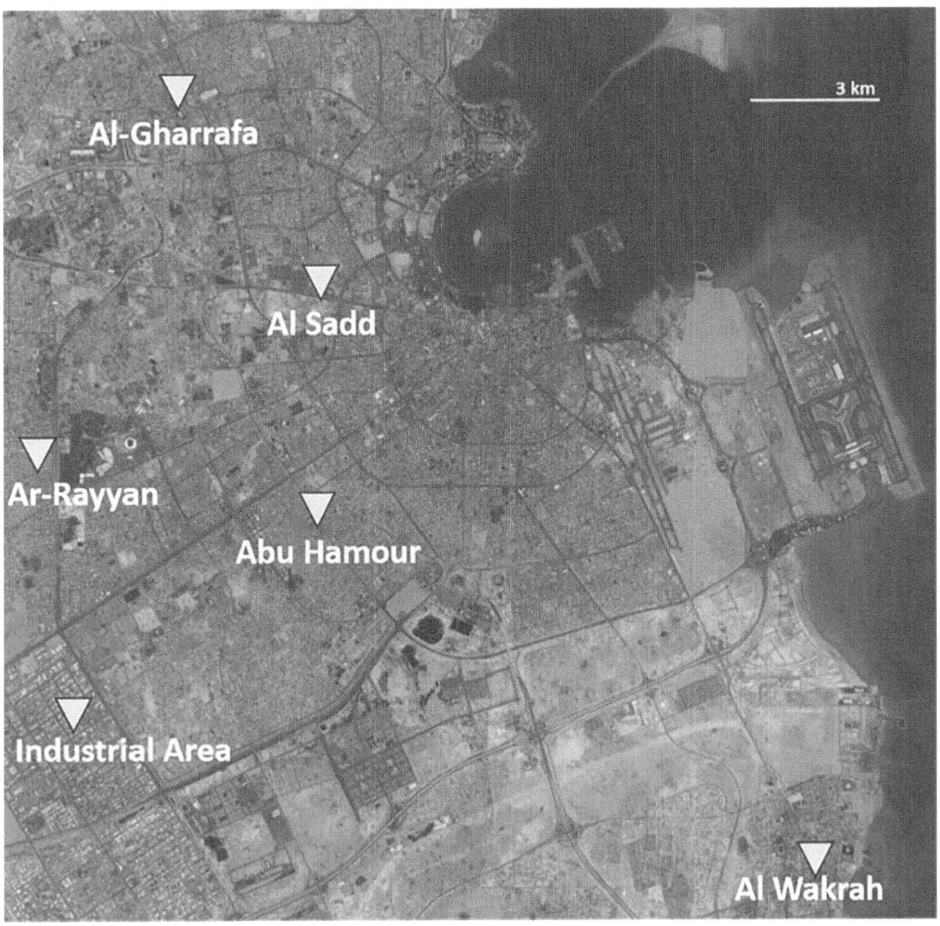

Figure 1 | Locations of the drinking water samples collected in the eastern part of Qatar.

from ABRAXIS, USA and the calibration standard solutions (0, 0.15, 0.40, 1.0, 2.0, and 5.0 μg/L) and a positive control sample (0.75 ± 0.185 μg/L) provided along with the kit were used for all measurements. In addition to extracellular toxins, cell-bound toxins were also measured followed by the cell lysis procedure using QuickLyse kit purchased from ABRAXIS, USA (Chorus 2012). All the calibration standards, positive controls used, and the samples were assayed in duplicate, and results were reported as mean values. Toxins that present in the samples will bind to the antibodies developed against Adda moiety which is similarly found in all the MC variants and nodularin. Absorbance was measured at 450 nm using an absorbance microplate reader (BioTek ELx800UV, USA).

Analytical methods

Microcystin-LR (MC-LR – $C_{49}H_{74}N_{10}O_{12}$) analytical standard purchased from ABRAXIS was used for UHPLC-PDA calibration and quantification. UHPLC (Thermo Scientific Dionex UltiMate 3000, USA) equipped with 1.9 μm; 100 × 2.1 mm C18 UHPLC Column (Sigma Aldrich, Hypersil GOLD™, USA) was used for the calibration and quantification. The UHPLC operating conditions are provided in Table 1.

Based on the calibration points used (0.01–0.5 mg/L), the calculated limit of detection (LOD) was 9.7 μg/L and the limit of quantitation (LOQ) was 29.4 μg/L. The treated water samples after the oxidation experiments were concentrated up to 500 times using a solid-phase extraction (SPE) instrument (Dionex™ AutoTrace™ 280, USA) prior to UHPLC analyses. Hydrophilic polymer phase SPE cartridge (Sigma Aldrich, Supel™ Select HLB 6 mL/200 mg, USA) was used for concentration. The cartridge was conditioned using 4 mL HPLC grade methanol followed by 8 mL deionized water prior to sample loading, where 500 mL of water sample was passed through the cartridge at a flow rate of 10 mL/min. The cartridge was then dried for 15 min using nitrogen gas and subsequently soaked and eluted with 9 mL of methanol at a flow rate of 1 mL/min. Finally, the eluent was concentrated by evaporating to dryness under a gentle stream of nitrogen gas and re-suspended using 1 mL of 20% methanol prior to UHPLC analysis.

Table 1 | UHPLC operating condition

Parameters	Conditions
Solvent A	Acetonitrile (≥99.93% purity, HPLC grade)
Solvent B	Water with 0.05% H_3PO_4
Injection volume (μL)	10
Run time (min)	17
Flow rate (mL/min)	0.3
Temperature (°C)	40
Wavelength (nm)	238
Gradient	0 to 1 min – 85% B, 1 to 10 min – 40% B, 10 to 15 min – 85% B followed by a stable 85% B for 2 min

Ozone and peroxone experimental studies

Ozone was produced using an oxygen-fed 4 g/hr corona discharge ozone generator (BMT 802 N, Germany). The high concentration ozone gas was then bubbled and dissolved in a glass reactor that was prefilled with 1 liter of deionized water. The reactor temperature was maintained at 5 °C using chilled water jackets to maximize ozone dissolution. After 1 hour of bubbling, dissolved ozone stock concentration in the reactor was measured spectrophotometrically (258 nm, $\varepsilon O_3 = 3{,}000\ M^{-1}\ cm^{-1}$). Once a stable dissolved ozone stock concentration was achieved in the glass chamber, an appropriate volume of the dissolved ozone was transferred to aliquots of drinking water samples as per the required ozone dosages. For peroxone studies, hydrogen peroxide was added and dispersed uniformly prior to ozone dosage. After the addition of oxidants, the water samples were stirred momentarily and allowed to react with the oxidants for 5 minutes. Ozone residuals in the treated water were measured using the indigo-trisulfonate method (Bader & Hoigné 1981) and quenched by purging the samples using nitrogen prior to SPE and UHPLC analyses.

Drinking water samples collected from two of the sampling locations were spiked with MC-LR and used for oxidation efficiency studies. These samples (Sample ID: 15 and 26) were chosen in order to study the effect of low and high TOC on the oxidation efficiency, respectively. The TOC values of the chosen samples were 0.58 and 5.42 mg/L, respectively. The low TOC drinking water samples were used to investigate the effect of different initial MC-LR concentration (3, 5, 20, and 50 μg/L) at a fixed ozone dosage of 0.1 mg/L. This study was then followed by varying the ozone and peroxone dosages from 0.1 mg/L O_3 to 0.75 mg/L O_3 at a fixed MC-LR concentration of 50 μg/L. The hydrogen peroxide dosages for peroxone studies were maintained at H_2O_2/O_3 ratio of 0.25 throughout all the studies. For example, 1 mg/L peroxone dose is equivalent to 1 mg/L of O_3 and 0.25 mg/L of H_2O_2. Additional studies were conducted to investigate the effect of temperature on the oxidant efficiency in the low TOC drinking samples (Sample ID: 26) by conducting the experiments at 22 and 35 °C. The temperature of the water samples was regulated by placing the reactors in a temperature regulated bath. Two different oxidant dosages (0.5 mg/L and 0.75 mg/L) at an initial MC-LR concentration of 50 μg/L were used for this particular study. Another important experimental study that evaluated the combined effect of high DOC and high temperature on the MC-LR removal efficiency was conducted by spiking the drinking water samples (Sample ID: 15 and 26) at 50 μg/L of MC-LR and oxidant dosage of 0.5 mg/L.

RESULTS AND DISCUSSION

Characteristics of the drinking water samples

A total number of 35 samples collected from public and private locations were analyzed in this study. The pH values ranged from 7.12 to 7.75 while the electrical conductivity ranged from 118.6 μS/cm to 240.20 μS/cm (equivalent to total dissolved solids (TDS) levels of 77.1 mg/L–156.1 mg/L). The turbidity of the samples varied between 0.20 NTU and 0.41 NTU. TOC values of several samples exceeded the maximum requirement of 4.0 mg/L set by the local authority, Kahramaa, for water quality in distribution systems (Al Naamaa 2014). TOC value as high as 6.10 mg/L was recorded for one of the collected water samples. Among the samples collected from the different mosques, there were six samples with higher TOC values than the guideline limit. Out of the six samples collected from the malls, cafeterias, and public water dispensers, four of the samples showed TOC values of 5.4 mg/L or higher. A summary of the measured basic water quality parameters of the samples collected against the local drinking water requirements are provided in Table 2.

Based on the ELISA analyses, it was found that the cyanotoxin concentrations (equivalent of all microcystin variants and nodularin) in the water samples collected from the mosques (Sample ID: 1 to 19) ranged from 0.01 μg/L to 0.45 μg/L. As for the samples collected from residential villas (Sample ID: 20 to 28), toxin levels of a minimum of 0.05 μg/L and a maximum of 0.33 μg/L were detected. A similar range of toxin concentrations were also measured for the samples collected from shopping malls (Sample ID: 29 to 31), cafeterias, and public water dispensing units (Sample ID: 32 to 35) where the levels vary between 0.03 μg/L and 0.28 μg/L. This study confirms that the concentration of the toxins (i.e., all the MC variants and nodularin) in all the drinking water samples were below the 1 μg/L MC-LR limit set by the World Health Organization (WHO) (Roegner *et al.* 2014). It can be confirmed that the drinking water supplies in the above-sampled areas are free from the two potent hepatotoxic cyanotoxins. The overall results for the basic water quality parameters and cyanotoxins are provided in Table 3.

Degradation of MC-LR using ozone and peroxone

The MC-LR removal efficiency at different initial toxin concentrations at a fixed ozone dosage of 0.1 mg/L O_3 is shown

Table 2 | Summary of drinking water quality results against Qatar requirements

Parameters	KAHRAMAA requirements in distribution system	Mosque		Villa		Others	
		Min	Max	Min	Max	Min	Max
pH	6.5–8.5	7.12	7.72	7.26	7.75	7.26	7.69
Conductivity (μS/cm)	150–500	118.6	240.2	167.9	210.9	178.5	235.3
Turbidty (NTU)	<4.0	0.23	0.39	0.24	0.41	0.20	0.35
TOC (mg/L)	<4.0	0.01	5.91	0.03	4.44	0.05	6.12

Table 3 | Water quality results of the collected samples

Sample ID	pH	Conductivity (μS/cm)	Turbidity (NTU)	TOC mg/L	Chlorophyll cell/μL	Toxin concentration (μg/L)	Location
1	7.12	178.8	0.3	0.92	<1	0.14	Mosque
2	7.25	180.0	0.28	0.32	<1	0.03	Mosque
3	7.62	204.7	0.3	1.23	<1	0.10	Mosque
4	7.58	187.9	0.39	0.97	<1	0.45	Mosque
5	7.49	202.7	0.29	0.83	<1	0.10	Mosque
6	7.36	240.2	0.32	0.63	<1	0.01	Mosque
7	7.19	179.5	0.29	0.76	<1	0.14	Mosque
8	7.51	171.7	0.35	1.75	<1	0.19	Mosque
9	7.61	184.7	0.29	0.01	<1	0.16	Mosque
10	7.33	201.5	0.35	5.09	<1	0.29	Mosque
11	7.28	182.0	0.28	5.91	<1	0.10	Mosque
12	7.54	230.5	0.32	0.68	<1	0.26	Mosque
13	7.71	175.0	0.25	0.17	<1	0.14	Mosque
14	7.69	135.2	0.34	4.86	<1	0.32	Mosque
15	7.72	177.1	0.32	5.42	<1	0.04	Mosque
16	7.49	122.4	0.27	4.47	<1	0.17	Mosque
17	7.52	118.6	0.23	0.61	<1	0.26	Mosque
18	7.48	174.5	0.31	0.42	<1	0.05	Mosque
19	7.52	118.8	0.29	5.37	<1	0.10	Mosque
20	7.40	181.8	0.29	1.12	<1	0.06	Residential villas
21	7.39	180.3	0.35	1.24	<1	0.23	Residential villas
22	7.28	202.0	0.24	1.15	<1	0.10	Residential villas
23	7.42	179.5	0.34	0.03	<1	0.33	Residential villas
24	7.26	178.1	0.32	1.04	<1	0.13	Residential villas
25	7.41	210.9	0.24	0.07	<1	BDL	Residential villas
26	7.74	175.6	0.41	0.58	<1	0.07	Residential villas
27	7.68	177.0	0.33	4.44	<1	0.05	Residential villas
28	7.75	167.9	0.35	0.13	<1	0.28	Residential villas
29	7.53	233.7	0.34	0.05	<1	0.19	Mall
30	7.42	231.4	0.2	6.12	<1	0.19	Mall
31	7.38	179.9	0.35	5.48	<1	0.03	Mall
32	7.69	235.3	0.28	5.58	<1	0.24	Cafeteria
33	7.41	207.6	0.31	5.8	<1	0.28	Water dispenser
34	7.26	208.2	0.31	0.08	<1	0.11	Water dispenser
35	7.37	178.5	0.31	0.11	<1	BDL	Water dispenser

BDL, below detection limit.

in Figure 2. It can be observed that MC-LR removal efficiency decreased from 74.1% to 66.6% as the toxin concentration in the water increased from 3 μg/L to 50 μg/L. Ozone degrades MC-LR by primarily cleaving the conjugated diene structure of the Adda group in MC-LR, as shown in Figure 3. The oxidation of the Adda moiety is important as it governs the toxicity of MC-LR (Sharma *et al.* 2012). Another destruction pathway of MC-LR during

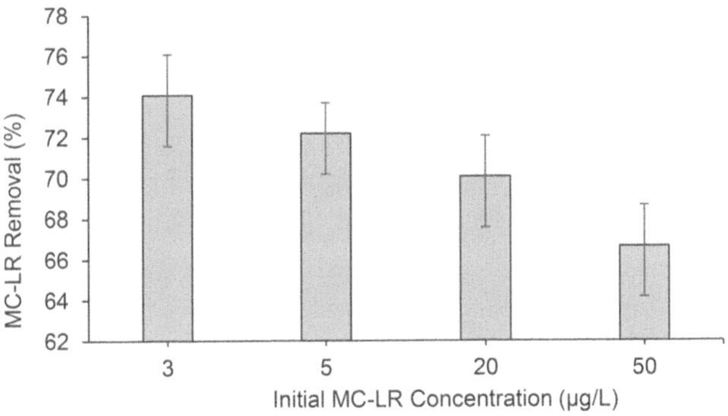

Figure 2 | Removal efficiency at varying initial MC-LR concentrations. Experimental conditions: pH, 7.3; MC-LR concentration, 3, 5, 20, and 50 µg/L; ozone dose, 0.1 mg/L; contact time, 5 min.

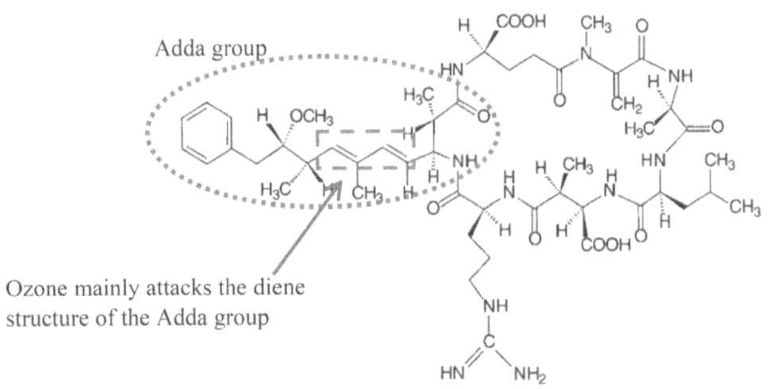

Figure 3 | Ozone oxidation of Adda group of MC-LR (figure adapted from Newcombe & Nicholson 2004).

ozonation is by fragmenting its Mdha-Ala peptide bond (Miao *et al.* 2010). Apart from the initial toxin concentration of 3 µg/L, ozone dosage of 0.1 mg/L was insufficient to oxidize the other higher MC-LR concentrations (e.g., 5, 20, and 50 µg/L) below the WHO guideline of 1 µg/L. The presence of other competing compounds such as organic carbon in the drinking water sample used in this study could contribute to the additional ozone demand and hence lead to the poor toxin removal efficiency at such a low ozone dosage of 0.1 mg/L. Similar observations were made by Rositano *et al.* (1998), who reported that 220 µg/L of pure MC-LR was degraded beyond the detection limit using ozone dosage of 0.2 mg/L while ozone dosage as high as 1.0 mg/L was required to remove a similar amount of toxin from an algal extract solution.

Additional studies were carried out to determine the required oxidant dosage for the removal of other higher initial MC-LR concentrations below the WHO limit. Figure 4 shows the MC-LR removal efficiency using ozone and peroxone at varying dosages from 0.1 mg/L to 0.75 mg/L. It can be seen

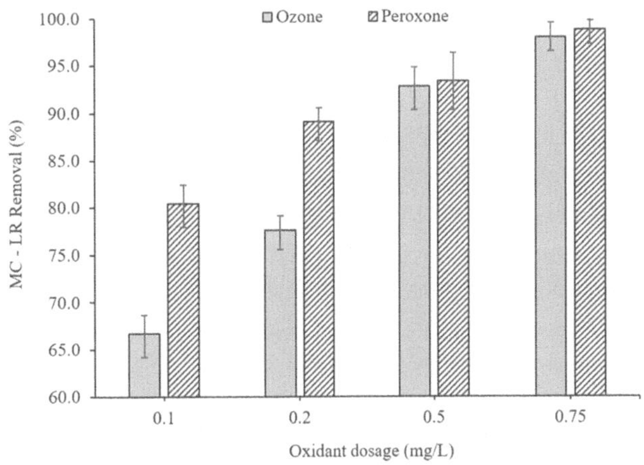

Figure 4 | Removal efficiency of MC-LR at different ozone and peroxone dosages. Experimental conditions: pH, 7.28; MC-LR concentration, 50 µg/L; oxidant dosage, 0.1, 0.2, 0.5, and 0.75 mg/L; contact time, 5 min.

that the toxin degradation efficiency improved from 66.6% to over 98% when the ozone dosage was increased from 0.1 mg/L to 0.75 mg/L. As explained earlier, the oxidation of double bonds and peptide rings by ozone and hydroxyl radical leads to degradation of MC-LR. These results show that a minimum of 0.75 mg/L O_3 is required to remove the MC-LR in the tested water sample below the safe limit of 1 µg/L. The trend of our result is in agreement with another reported study that claimed that 0.5 mg/L of ozone was sufficient to completely degrade 10 µg/L MC-LR within a short contact time (Hoger *et al.* 2002).

It can be seen from Figure 4 that peroxone has better removal efficiency when compared to conventional ozonation at lower oxidant dosage. The addition of hydrogen peroxide accelerates the oxidation of MC-LR due to the formation of •OH, which is a stronger oxidant than ozone (Rodríguez *et al.* (2007). The following reactions (Equation (1)) to (Equation (3)) explain the ozone decomposition and •OH production and the overall reaction in a peroxone system is provided in Equation (4) (Deng & Zhao 2015):

$$H_2O_2 \rightarrow HO_2^- + H^+ \tag{1}$$

$$HO_2^- + O_3 \rightarrow HO_2^\bullet + O_3^{\bullet -} \tag{2}$$

$$HO_2^- + O_3 \rightarrow \bullet OH + O_2^- + O_2 \tag{3}$$

The overall reaction of peroxone system is indicated below:

$$2O_3 + H_2O_2 \rightarrow 2 \bullet OH + 3O_2 \tag{4}$$

At a low peroxone dosage of 0.1 mg/L, the addition of hydrogen peroxide improved the MC-LR degradation efficiency to 80.4% when compared to a mere 66.6% of removal using conventional ozonation. A similar trend was also observed for 0.2 mg/L oxidant dosages where peroxone had better MC-LR removal efficiency than ozone. It is worth noting that the pH and TOC of the drinking water samples (Sample 1D: 26) used for this particular experiment was 7.74 and 0.58 mg/L, respectively. As ozone decomposes to form •OH at alkaline pH, the availability of molecular ozone for toxin oxidation is limited. In addition, the oxidation potential of ozone in alkaline water is lower than 2.07 V (under acidic conditions). These conditions could lead to lower oxidation efficiency of MC-LR. Although •OH, a superior oxidant than molecular ozone, is expected to improve the toxin degradation via the indirect pathway at alkaline pH, the presence of other competing organics and scavengers such as organic carbon and bicarbonates could further contribute to the lower MC-LR removal efficiency. On the other hand, the addition of hydrogen peroxide in the peroxone process accelerates ozone decomposition rate promoting the formation of •OH. The higher yield of •OH with a stronger oxidation potential of 2.8 V leads to better MC-LR removal efficiency even at lower oxidant dose. At higher dosage (e.g., 0.5 and 0.75 mg/L), it can be observed from Figure 4 that MC-LR removal efficiencies for both ozone and peroxone were more or less identical. A maximum removal efficiency of 98% and 98.8% was achieved for ozone and peroxone at oxidant dosage of 0.75 mg/L. These results are in agreement with the findings of Rositano *et al.* (1998), who reported that at higher oxidant dosages, both ozone and peroxone exhibited similar MC-LR removal efficiency.

The effect of water temperature on the MC-LR oxidation efficiency using ozone and peroxone is shown in Figure 5. The result shows that the toxin removal efficiency using ozone was unaffected when the water temperature was increased from 22 °C to 35 °C. As it is often speculated that ozone is inefficient at elevated water temperature due to high off-gas, poor solubility, and short half-life, it is important to understand the degradation kinetics of MC-LR is enhanced at high temperature if compared with ozone half-life. For the peroxone treatment, there is an improvement of 5% in the toxin removal efficacy at a dosage of 0.5 mg/L while the removal efficiency remained the same at 0.75 mg/L peroxone dosage. In summary, the increase in water temperature either maintains or improves the MC-LR removal efficiency in all the case studies.

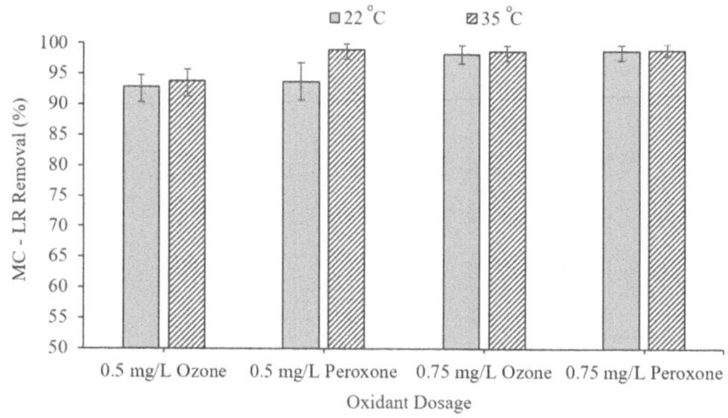

Figure 5 | MC-LR removal at different water temperatures. Experimental conditions: pH, 7.32; MC-LR concentration, 50 μg/L; oxidant dosage, 0.5 and 0.75 mg/L; contact time, 5 min. Temperature was maintained using a water bath.

Figure 6 shows the MC-LR removal efficiency at two different temperatures and TOCs at an oxidant dosage of 0.5 mg/L and initial MC-LR concentration of 50 μg/L. It is obvious from Figure 6 that TOC content in the drinking water samples used in this study plays a detrimental role in the oxidation of MC-LR. The oxidation efficiency of the ozone and peroxone dropped drastically from 92.8% and 93.8% to 67.6% and 66.4%, respectively, due to the difference in the TOC of the two different samples. A similar response was observed when the oxidation experiments were carried out at a higher water temperature for the two drinking water samples. Data shown in the graph were provided with error bars and are statistically significant. This

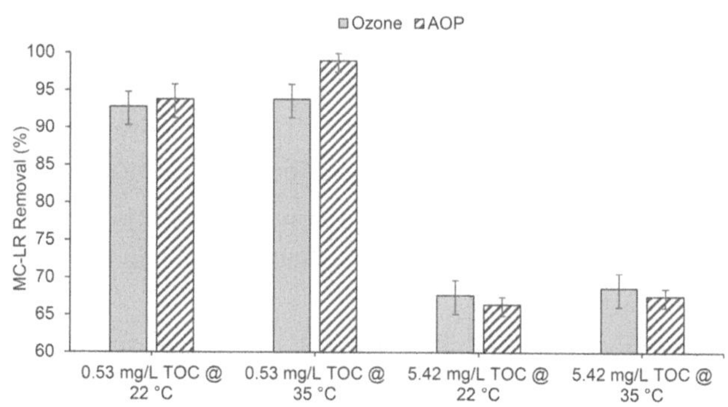

Figure 6 | MC-LR removal at different DOC and temperature. Experimental conditions: pH, 7.35; MC-LR concentration, 50 μg/L; oxidant dosage, 0.5 mg/L; contact time, 5 min. Temperature was maintained using water baths at 22 °C and 35 °C.

study clearly shows that organic carbon content in the water plays a more critical role than water temperature during the oxidation of MC-LR. The competitive reaction between the organic material and MC-LR increased the oxidant demand of the water and hence led to the poor toxin removal efficiency (Al Momani *et al.* 2008). Table 4 summarizes the results obtained in this research against published data related to the removal of MC-LR using ozone and peroxone.

Further evaluation of the oxidation by-products and understanding the toxicity of the transformation products using advanced characterization tools such as LC-QTOF MS would be necessary (Leon *et al.* 2019). This helps in ensuring clean and safe water, which is of utmost importance to public health. Especially in the developing countries, providing contamination-free drinking water despite the global risks such as water scarcity, rising temperature, and harmful cyanobacteria bloom is of serious concern. The presence of cyanotoxin in drinking water is a major threat to humans as it could lead to various health issues such as gastroenteritis, liver damage, and neurotoxicity. Hence, monitoring of cyanotoxins in drinking water at appropriate intervals using rapid screening tests such as ELISA is important to ensure a safe supply of drinking water. In addition, one of the feasible ways of degrading cyanotoxins is by using ozone and other advanced oxidation

Table 4 | Comparison study of the MC-LR removal using ozone and peroxone

Oxidant	Oxidant dose (mg/L)	Initial MC-LR Conc. (mg/L)	Treatment conditions	Removal efficiency (%)	Reference
Ozone	0.2	0.166	Medium: Pure microcystin in water; Reaction time: 4 min	100	Rositano *et al.* (1998)
Ozone	0.2	0.22	Medium: Microcystin in algal extract water; Reaction time: 4 min	100	Rositano *et al.* (1998)
Ozone	2.0	0.0035–0.0086	Medium: Raw lowland river water	100	Hall *et al.* (2000)
Ozone	0.25	1.0	Medium: Lake water; pH = 8; DOC = 3.6 mg/L	95	Rodríguez *et al.* (2007)
Ozone	0.6	1.0	Medium: Water spiked with 2 mg/LNOM (humic acid); pH = 7; Temperature = 20 °C; Reaction time = 90 s	75	Al Momani *et al.* (2008)
Ozone	0.1	0.003	Medium: Purified toxin spiked in drinking water; pH = 7.74; Reaction time = 5 min; Temperature = 25 °C; DOC = 0.58 mg/L	74.1	Current study
Ozone	0.1	0.05	Medium: Purified toxin spiked in drinking water; pH = 7.74; Reaction time = 5 min; Temperature = 25 °C; DOC = 0.58 mg/L	66.6	Current study
Peroxone	0.1 O_3 + 0.025 H_2O_2	0.05	Medium: Purified toxin spiked in drinking water; pH = 7.74; Reaction time = 5 min; Temperature = 25 °C; DOC = 0.58 mg/L	80.4	Current study
Ozone	0.75	0.05	Medium: Purified toxin spiked in drinking water; pH = 7.74; Reaction time = 5 min; Temperature = 25 °C; DOC = 0.58 mg/L	98	Current study
Peroxone	0.75 O_3 + 0.188 H_2O_2	0.05	Medium: Purified toxin spiked in drinking water; pH = 7.74; Reaction time = 5 min; Temperature = 25 °C; DOC = 0.58 mg/L	98.8	Current study
Ozone	0.5	0.05	Medium: Purified toxin spiked in drinking water; pH = 7.72; Reaction time = 5 min; Temperature = 25 °C; DOC = 5.42 mg/L	67.6	Current study
Peroxone	0.50 O_3 + 0.125 H_2O_2	0.05	Medium: Purified toxin spiked in drinking water; pH = 7.72; Reaction time = 5 min; Temperature = 25 °C; DOC = 5.42 mg/L	66.4	Current study

processes such as peroxone. Considering the negative impacts of algal bloom and based on the source of water, finished drinking water from conventional treatment plants can be treated further using peroxone, an advanced oxidation process to ensure it is cyanotoxin-free before supply.

CONCLUSION

The cyanotoxin analyses of our study using ELISA test kits on the 35 drinking water samples that were collected from various public locations such as malls, cafeterias, and private residential villas showed that the toxin levels in all the samples were below the WHO limit. Our experimental results using ozone and peroxone to remove MC-LR in drinking water showed that oxidant dosages are governed by the initial toxin concentrations and organic carbon content in water samples. Higher TOC values led to poor removal of MC-LR due to competition kinetics between the toxin and organic carbon during the oxidation process. As toxin concentration in real cases exists at extremely low concentration, an ozone dose of 0.75 mg/L would be sufficient to degrade MC-LR. Although ozone and peroxone are capable of removing MC-LR below the regulation limit, the formation of by-products during the oxidation process is a concern and requires a human risk assessment. In addition, the presence of bromide in drinking water is also a major concern as it could lead to the formation of bromate, a suspected human carcinogen.

ACKNOWLEDGEMENTS

This publication was made possible by NPRP grant #NPRP9-159-2-087 from the Qatar National Research Fund (a member of Qatar Foundation). The contents of this article are solely the responsibility of the author(s).

REFERENCES

Agrawal, M., Yadav, S., Patel, C., Raipuria, N. & Agrawal, M. K. 2012 Bioassay methods to identify the presence of cyanotoxins in drinking water supplies and their removal strategies. *European Journal of Experimental Biology* **2** (2), 321–336.

Al Momani, F., Smith, D. W. & Gamal El-Din, M. 2008 Degradation of cyanobacteria toxin by advanced oxidation processes. *Journal of Hazardous Materials* **150** (2), 238–249. https://doi.org/10.1016/j.jhazmat.2007.04.087.

Al Naamaa, A. R. 2014 *Overview on Kahraamaa Drinking Water Quality Requirements*. KAHRAMAA Publications, Qatar.

Bader, H. & Hoigné, J. 1981 Determination of ozone in water by the indigo method. *Water Research* **15** (4), 449–456. https://doi.org/10.1016/0043-1354(81)90054-3.

Birbeck, J. A., Westrick, J. A., O'Neill, G. M., Spies, B. & Szlag, D. C. 2019 Comparative analysis of microcystin prevalence in Michigan lakes by online concentration LC/MS/MS and ELISA. *Toxins* **11** (1), 13. https://doi.org/10.3390/toxins11010013.

Bourgin, M., Borowska, E., Helbing, J., Hollender, J., Kaiser, H.-P., Kienle, C., McArdell, C. S., Simon, E. & von Gunten, U. 2017 Effect of operational and water quality parameters on conventional ozonation and the advanced oxidation process O_3/H_2O_2: kinetics of micropollutant abatement, transformation product and bromate formation in a surface water. *Water Research* **122**, 234–245. https://doi.org/10.1016/j.watres.2017.05.018.

Chatziefthimiou, A. D., Metcalf, J. S., Glover, W. B., Banack, S. A., Dargham, S. R. & Richer, R. A. 2016 Cyanobacteria and cyanotoxins are present in drinking water impoundments and groundwater wells in desert environments. *Toxicon* **114**, 75–84. https://doi.org/http://dx.doi.org/10.1016/j.toxicon.2016.02.016.

Chorus, I. 2012 *Current Approaches to Cyanotoxin Risk Assessment, Risk Management and Regulations in Different Countries*. Text 63/2012. Federal Environment Agency, Germany.

Deng, Y. & Zhao, R. 2015 Advanced oxidation processes (AOPs) in wastewater treatment. *Current Pollution Reports* **1** (3), 167–176. https://doi.org/10.1007/s40726-015-0015-z.

Hall, T., Hart, J., Croll, B. & Gregory, R. 2000 Laboratory-scale investigations of algal toxin removal by water treatment. *Water and Environment Journal* **14** (2), 143–149. https://doi.org/10.1111/j.1747-6593.2000.tb00241.x.

Hoger, S., Dietrich, D. & Hitzfeld, B. 2002 Effect of ozonation in drinking water treatment on the removal of cyanobacterial toxins. *Environmental Health Perspectives* **110**, 1127–1132.

Hunter, P. R. 1998 Cyanobacterial toxins and human health. *Journal of Applied Microbiology* **84** (S1), 35S–40S. https://doi.org/10.1046/j.1365-2672.1998.0840s135S.x.

Jasim, S. Y. & Saththasivam, J. 2017 Advanced oxidation processes to remove cyanotoxins in water. *Desalination* **406**, 83–87. https://doi.org/10.1016/j.desal.2016.06.031.

Kull, T. P. J., Sjövall, O. T., Tammenkoski, M. K., Backlund, P. H. & Meriluoto, J. A. O. 2006 Oxidation of the cyanobacterial hepatotoxin microcystin-LR by chlorine dioxide: influence of natural organic matter. *Environmental Science & Technology* **40** (5), 1504–1510. https://doi.org/10.1021/es051729 g.

Lawton, L. A. & Robertson, P. K. J. 1999 Physico-chemical treatment methods for the removal of microcystins (cyanobacterial hepatotoxins) from potable waters. *Chemical Society Reviews* **28** (4), 217–224. https://doi.org/10.1039/a805416i.

Leon, C., Boix, C., Beltran, E., Peneula, G., Lopez, F., Sancho, J. & Hernandez, F. 2019 Study of cyanotoxin degradation and evaluation of their transformation products in surface waters by LC-QTOF MS. *Chemosphere* **229**, 538–548. https://doi.org/10.1016/j.chemosphere.2019.04.219.

Loganathan, K. 2016 Ozone-based advanced oxidation processes for the removal of harmful algal bloom (HAB) toxins: a review. *Desalination and Water Treatment* **59**, 65–71. https://doi.org/10.5004/dwt.2016.0346.

Miao, H. F., Qin, F., Tao, G. J., Tao, W. Y. & Ruan, W. Q. 2010 Detoxification and degradation of microcystin-LR and -RR by ozonation. *Chemosphere* **79** (4), 355–361. https://doi.org/10.1016/j.chemosphere.2010.02.024.

Munoz, M., Sandoval, J., Cires, S., Pedro, Z., Quesada, A. & Casas, J. 2019 Degradation of widespread cyanotoxins with high impact in drinking water (microcystins, cylindrospermopsin, anatoxin-a and saxitoxin) by CWPO. *Water Research* **163**, 114853. https://doi.org/10.1016/j.watres.2019.114853.

Newcombe, G. & Nicholson, B. 2004 Water treatment options for dissolved cyanotoxins. *Journal of Water Supply: Research and Technology-Aqua* **53** (4), 227–239. https://doi.org/10.2166/aqua.2004.0019.

Rodríguez, E., Onstad, G. D., Kull, T. P. J., Metcalf, J. S., Acero, J. L. & Von Gunten, U. 2007 Oxidative elimination of cyanotoxins: comparison of ozone, chlorine, chlorine dioxide and permanganate. *Water Research* **41** (15), 3381–3393. https://doi.org/10.1016/j.watres.2007.03.033.

Roegner, A. F., Brena, B., González-Sapienza, G. & Puschner, B. 2014 Microcystins in potable surface waters: toxic effects and removal strategies. *Journal of Applied Toxicology* **34** (5), 441–457. https://doi.org/10.1002/jat.2920.

Rositano, J., Nicholson, B. C. & Pieronne, P. 1998 Destruction of cyanobacterial toxins by ozone. *Ozone: Science and Engineering* **20** (3), 223–238. https://doi.org/10.1080/01919519808547273.

Sharma, V. K., Triantis, T. M., Antoniou, M. G., He, X., Pelaez, M., Han, C., Song, W., O'Shea, K. E., de la Cruz, A. A., Kaloudis, T., Hiskia, A. & Dionysiou, D. D. 2012 Destruction of microcystins by conventional and advanced oxidation processes: a review. *Separation and Purification Technology* **91**, 3–17. https://doi.org/10.1016/j.seppur.2012.02.018.

Tilahun, S., Kifle, D., Zewde, T. W., Johansen, J. A., Demissie, T. B. & Hansen, J. H. 2019 Temporal dynamics of intra-and extra-cellular microcystins concentrations in Koka reservoir (Ethiopia): implications for public health risk. *Toxicon* **168**, 83–92. https://doi.org/10.1016/j.toxicon.2019.06.217.

Turner, A. D., Dhanji-Rapkova, M., O'Neill, A., Coates, L., Lewis, A. & Lewis, K. 2018 Analysis of microcystins in cyanobacterial blooms from freshwater bodies in England. *Toxins* **10**, 1. https://doi.org/10.3390/toxins10010039.

Van der Merwe, D. 2014 Freshwater cyanotoxins. In: *Biomarkers in Toxicology* (R. Gupta ed.). Elsevier, Amsterdam, The Netherlands, pp. 539–548.

First received 5 March 2019; accepted in revised form 6 August 2019. Available online 19 September 2019

Photocatalytic degradation of Irgalite violet dye using nickel ferrite nanoparticles

Shiljashree Vijay, Raj Mohan Balakrishnan, Eldon R. Rene and Uddandarao Priyanka

ABSTRACT

Nanotechnologies have prominent applications in the field of science and technology owing to their size-tunable properties providing a promising approach for degradation of various pollutants. In this scenario, the present work aims to study the effect of nickel ferrite nanoparticles on the degradation of Irgalite violet dye by Fenton's reaction using oxalic acid as an oxidizing agent in the presence of sunlight. The effect of pH and adsorbent dosage on the rate of dye degradation was monitored. Based on these studies it was observed that 99% dye degradation was achieved for catalyst dosage of 0.2 g, 400 ppm dye concentration and 2.0 mM oxalic acid at pH 3.0 within 60 min. The studies reveal that the degradation follows pseudo-first-order kinetics and the catalyst reusability remained constant almost for five cycles. Further, nickel ferrite nanoparticles are proven to be an efficient alternative for the removal of dyes from coloured solutions.

Key words | fenton reaction, Irgalite violet dye, nickel ferrite, oxalic acid, photocatalytic degradation

Shiljashree Vijay
Raj Mohan Balakrishnan (corresponding author)
Uddandarao Priyanka
Department of Chemical Engineering,
National Institute of Technology Karnataka,
Surathkal
575025,
India
E-mail: *rajmohanbala@gmail.com*

Eldon R. Rene
UNESCO-IHE Institute for water Education,
Westvest 7, 2601 DA Delft,
The Netherlands

INTRODUCTION

Water is a fundamental and valuable source for all forms of life on Earth. This vital source is being contaminated due to urbanization and industrialization where effluents are let into water bodies without proper treatment. Large quantities of wastewaters with high colour intensity, high chemical oxygen demand and fluctuating pH are being discharged into water bodies causing serious environmental issues affecting flora and fauna. Dyes have a complex aromatic structure and are resistant towards chemicals, heat and light, while their biodegradation is a significantly slow process. From a toxicological as well as from an ecological point of view it is essential to treat effluents containing these dyes. These diverse pollutants when allowed into the soil by migration and transformation modes cause huge global environmental issues.

Most of the conventional methods are ineffective due to the large degree of aromatics/phenols present in dye molecules. Further, they call for high operational and maintenance costs, and therefore concerns lie with the treatment of effluents by economically viable methodologies (Ye *et al.* 2017a, 2017b, 2019a, 2019b). Various physico-chemical methods like coagulation, flocculation, ion exchange, membrane separation, oxidation, nanofiltration, adsorption, catalytic ozonation (El *et al.* 2019) etc. are available for the treatment of dyes. However, some of the major limitations, such as the removal of adsorbents or the concentrated sludge produced post-remediation, are the major drawbacks of these methods. One of the alternatives to the traditional methods are the advanced oxidation processes (AOPs) which have been used to remove contaminants from drinking water and effluents. AOPs include photolysis, heterogeneous photocatalysis, electro-Fenton and photo-Fenton. Fenton is considered as an AOP that combines ferrous ions and hydrogen peroxide and other hydroxyl generating acids to promote the generation of the hydroxyl radicals in aqueous solution (Khosravi &

doi: 10.2166/aqua.2019.039

Eftekhar 2013; Lassoued *et al.* 2018). The photo-Fenton process is a typically enhanced Fenton reaction with a higher rate and faster mineralization of recalcitrant organics than the dark reaction process and can take advantage of utilizing UV irradiation from solar light. According to the light sensitivity of the photo-Fenton reaction with respect to the sunlight spectrum, the utilization of a solar light is reported to be very economic and a simple alternative compared to UV lamps, especially for treatment of wastewater at full scale. In the reaction of the photo-Fenton process, Fe^{2+} ions are oxidized by oxalic acid to Fe^{3+} to generate hydroxyl radicals. The application of Fenton's reagent as an oxidant for wastewater treatment is attractive. Among the advantages of Fenton's process relative to other oxidation techniques are the simplicity of equipment and the mild operation conditions (atmospheric pressure and room temperature); mainly for these reasons, the Fenton process has been regarded as the most economical alternative (Alalm *et al.* 2015; Deng & Zhao 2015; Santos-Juanes *et al.* 2017). In order to enhance the degradation of organic pollutants, external sources of energy have been added to the Fenton process, such as light source, and this process is called photo-Fenton process (Suna *et al.* 2007; Maaz *et al.* 2009; Shihong *et al.* 2009; Youssef *et al.* 2016). The reaction of ferrous ion (Fe^{2+}) and oxalic acid ($C_2H_2O_4$) generate hydroxyl radicals where ferric ion (Fe^{3+}) regenerates Fe^{2+} by reacting with an excess of $C_2H_2O_4$ and hydroperoxides (•OOH). The photo-Fenton process enhances the production of hydroxyl radicals by photo-reduction of Fe^{3+} to Fe^{2+} and photolysis.

Recently, various nanomaterials have been used as nanocatalysts for various applications such as sensing, detection, degradation and electrical applications (Qin *et al.* 2017; Yi *et al.* 2018a, 2018b, 2018c, 2019; Ahmadi *et al.* 2019; Baghban *et al.* 2019; Ramezanizadeh *et al.* 2019; Wang *et al.* 2019; Yang *et al.* 2019). ZnO, gold and TiO_2 are the other nanoparticles which are often reported, however, the use of magnetic $NiFe_2O_4$ nanoparticles for the photo-Fenton degradation of organic model pollutants in the presence of oxalic acid is studied. It is well known that the magnetic and electrical properties of MFe_2O_4 nanoparticles can be varied by changing the identity of the divalent M_{2+} cation or by partial substitution, while maintaining the basic crystal structure. For magnetic nanodevices and biomedical applications, MFe_2O_4 ferrites of size-tuned ferrite particles of superparamagnetic threshold were used at room temperature (Li *et al.* 2015; Weng & Huang 2015; Yahya *et al.* 2016; Behin *et al.* 2017; Shi *et al.* 2017; Sohrabi *et al.* 2017).

In this scenario, the ferrite materials based on nickel and cobalt are the subject of extensive research due to important magnetic properties like good remnant magnetization, high coercivity, magnetic anisotropy and moderate saturation magnetization. They also exhibit high electrical resistivity, and good thermal and chemical stability (Sivakumar *et al.* 2011; Sagadevan *et al.* 2018). Nickel ferrite ($NiFe_2O_4$) has an inverse spinel structure with crystal structure as AB_2O_4, in which tetrahedral sites (A) are occupied by Fe^{3+}, octahedral sites (B) by Fe^{3+} and Ni^{2+} ions formed by cubic close packing of oxygen. However, due to random distribution of divalent ions in tetrahedral and octahedral sites reverse cationic distribution in nanocrystalline spinel takes place. For instance, utilizing mesoporous $ZnFe_2O_4$-H_2O_2 under visible light for photocatalytic degradation of orange II, enhanced degradation was observed within 2 h. $NiFe_2O_4$ nanoparticles were used for the degradation of rhodamine B under visible light/oxalic system (Liu *et al.* 2012) and it was observed that they were highly stable even after seven cycles. Similarly, Fenton-like performance was also observed by which the nZVI/H_2O_2 system yielded 91.95% removal efficiency of magenta flexographic dye (Kecić *et al.* 2018). Sodium hypochlorite, hydrogen peroxide (H_2O_2) and ozone (O_3) are the other oxidizing agents used. Fenton treatment with the addition of oxalic acid has high reaction rates with ·OH and can also act as a scavenging agent in the reaction process. Due to the above-mentioned advantages, this study aims to determine the effect of nickel ferrite nanoparticles and oxalic acid using the photo-Fenton process in the degradation of Irgalite violet dye.

Irgalite violet belongs to the category of azo pigments with strong colour intensity, good transparency and rheology for aqueous, solvent-based and energy-curable inks. These characteristics of the dye make it difficult for the dye industry to decolourise or degrade the dye wash water and its effluent. Considering the unsustainable water use and water pollution due to dye effluents, in this work, the combined photo-Fenton and ferrite process was used to treat persistent dyes. Nickel ferrites due to their relatively dry and more compact physical structure than ferric hydroxide considerably reduce sludge volume in ferrite formation. Hence, nickel ferrite is considered as a heterogeneous photo-Fenton catalyst.

MATERIALS AND METHODS

Photocatalytic experimental studies

$NiFe_2O_4$ nanoparticle synthesis was carried out by co-precipitation method as mentioned in previous work (Kollarahithlu & Balakrishnan 2018). Photocatalytic degradation of Irgalite violet dye was carried out utilizing $NiFe_2O_4$ nanoparticles in the presence of sunlight between 11a.m. and 3p.m. (IST) with an average light intensity of 6.2 kWh/m^2/day, at a temperature of 28–32 °C. Prior to irradiation, the solution was magnetically stirred in the dark to attain adsorption-desorption equilibrium for 30 min between the dye and photocatalyst. The photo-Fenton degradation of the dye was performed by varying the amount of $NiFe_2O_4$ (0.05–0.20 g), oxalic acid (0.5–2 mmol), dye concentrations (200–1,000 ppm) and pH (2.0–4.0) in a 250 mL conical flask.

Degradation efficiency estimation

Aliquots of samples were withdrawn at regular intervals of time and the intensity was measured at 547 nm using Ultraviolet-visible (UV-Vis) spectroscopy (UV-2450, Shimadzu, Japan). The photodegradation efficiency was defined as follows:

$$\text{Degradation efficiency, } \eta = \frac{C_0 - C_t}{C_0} \times 100\%$$

The pseudo-first-order kinetics is given using the following equation:

$$ln\left(\frac{C}{C_o}\right) = kt$$

where, C_O is the initial concentration of dyes and C_t is the concentration of dye (mg L^{-1}) at certain reaction time t (min) and k is the rate constant.

RESULTS AND DISCUSSION

Effect of pH

The degradation of Irgalite violet dye by $NiFe_2O_4$ nanoparticles was studied between pH 2.0 and 4.0. At initial pH the degradation rate was minimal. The degradation efficiency was found to increase with increase in pH and attained a maximum of 99.9% degradation at pH 3.0 (Figure 1). This could be due to the increase in ferric/oxalate species and similar findings were reported by Liu *et al.* (2012), where the degradation of rhodamine B in the presence of oxalic acid was more effective at pH 3.0 confirming that the conducted Fenton oxidation process is optimum at pH 3.0 (Suna *et al.* 2007). Above pH 4.0 the degradation rate slows down to 78%; the probable reason might be due to the lower concentration of H_2O_2 production by the oxalic acid (Monteagudo *et al.* 2008). The acid environment is beneficial for the Fenton-like process, the working pH for the reaction is confined to 1.0–5.0 and the highest efficiency of the Fenton reaction occurs at pH 3.0. At pH higher than 4.0, the Fenton reaction appears to terminate due to the precipitation of Fe^{3+}. The amount of iron ions decreases because Fe^{3+} precipitates after use due to its extremely low solubility product constant even at the pH 4.0. The degradation rate increases as the pH decreases and reaches an optimum at initial pH 3.0. At the initial pH

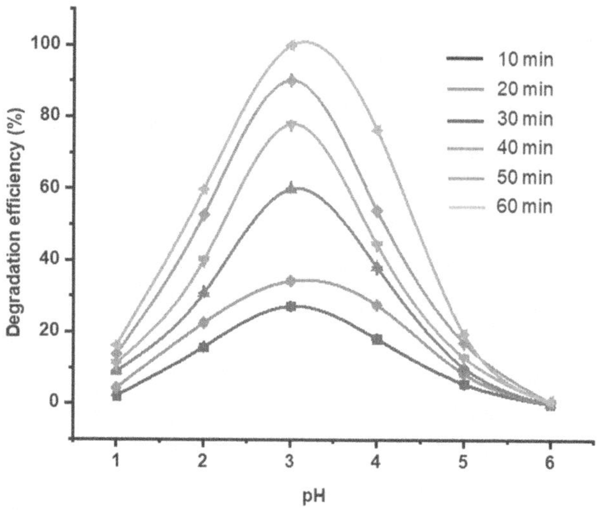

Figure 1 | Effect of pH for catalyst concentration of 0.15 g and initial dye concentration of 400 ppm.

of 2.0, a colourless solution is observed at 40 min; at pH 3.0, 99.9% of the dye is converted at 60 min reaction time and at pH 4.0 it is 78%. The oxidation potential of ˙OH is lower at pH 4.0 than that at pH 3.0. At pH > 3, the formation of hydroxyl radicals slows down because of hydrolysis of Fe^{3+} and Fe^{3+} precipitation as Fe $(OH)_3$ from the solution. Thus, the efficiency of Fenton oxidation decreases at pH higher than 3. Also, ferric hydroxide formation could decompose oxalic acid into O_2 or H_2O and, consequently, the oxidation rate is decreased as a result of the low concentration of hydroxyl radicals. Similar performance in the oxidation of dye solutions using iron-alumina catalyst in the heterogeneous Fenton process was reported where a detrimental effect on decolouration at pH > 3 was found (Barrera-Salgado *et al.* 2016; Zhou *et al.* 2016; Yi *et al.* 2018a, 2018b, 2018c).

Effect of oxalic acid concentration

The effect of initial concentration of the oxidizing agent oxalic acid was studied to determine the rate of degradation from 1.0 to 3.0 mM for the experimental condition of 0.15 g of catalyst loading at dye concentration 400 ppm and pH 3.0 for 60 min. It can be observed that the degradation rate increased with the increase in oxalic acid concentration up to 2 mM (99.5%), reached a plateau, and decreased at 2 mM by 1% for various dye concentrations (Figure 2). As

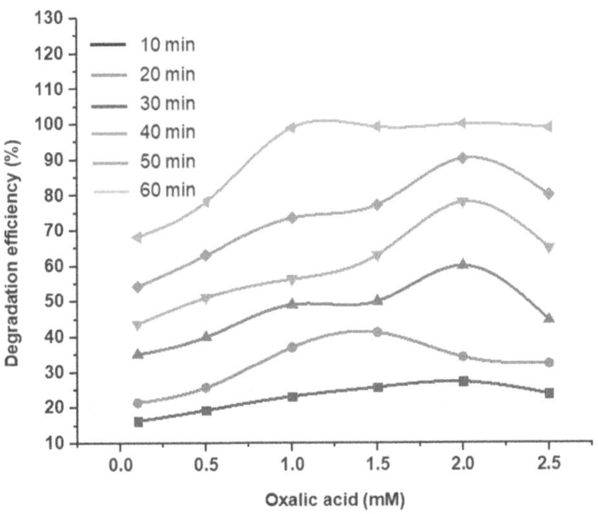

Figure 2 | Effect of oxalic concentration (1 mM, 2 mM and 3 mM) for the catalyst loading of 0.15 g for 400 ppm dye concentration at pH 3.0.

the dye concentration increased, the degradation efficiency decreased; this is because at higher concentrations of dye the number of active sites decreases, which is due to absorption of more dye molecules on the catalyst surface. This could be attributed to the hindering effect of the dye molecules on the production of •OH radicals and H_2O_2 leading to reduction in the rate of degradation. A concentration of 1.0 mM of magnetic $NiFe_2O_4$ and oxalic acid increased to 2 mM concentration leads to a high degradation rate under irradiation, indicating that $NiFe_2O_4$ is a heterogeneous Fenton catalyst. The degradation rate increases with increased oxalic concentration from 0.1 mM to 2.5 mM and then reaches a plateau at concentrations greater than 1.0 mM. This phenomenon can be attributed to forming of the predominant species $[FeIII(C_2O_4)n]^{3-2n}$ on the solid catalyst surface; its concentration reaches a maximum when the concentration of oxalic acid is 2.0 mM in solution. The reason for the optimum degradation at 2 mM of oxalic acid could be due to the generation of more OH radicals and Fe^{2+} ions (Liu *et al.* 2012) With the addition of oxalic acid, the degradation ratio of the dye is only 3.9% without irradiation. A 5% increase in decolourisation ratio is observed in the solution of oxalic acid and Irgalite violet without $NiFe_2O_4$ under irradiation. However, 99.99% degradation was achieved in the presence of the magnetic catalyst $NiFe_2O_4$ with the addition of 2 mM of oxalic acid crystals to the solution. The major role of oxalic acid is to enhance the oxidation process, and the reason for decreases at 2 mM may be because the excess production of OH radicals might compete among themselves and inhibit the efficiency of the degradation process.

Effect of catalyst loading

Figure 3 shows the dependence of Irgalite violet dye 400 ppm degradation on the loading of the $NiFe_2O_4$ catalyst in the presence of oxalic acid with an initial concentration of 2.0 mM. The degradation ratio at 20 min is 68.4%, 75.0%, 78.2%, 79.3% and 81.5% when the dosage of the $NiFe_2O_4$ catalyst is 0.05 g, 0.10 g, 0.15 g, 0.20 g and 0.25 g, respectively. When the concentration of the catalyst is further increased a decrease in degradation efficiency is observed. At the initial stage, the increase of the dosage of the catalyst leads to an obvious rise of the degradation

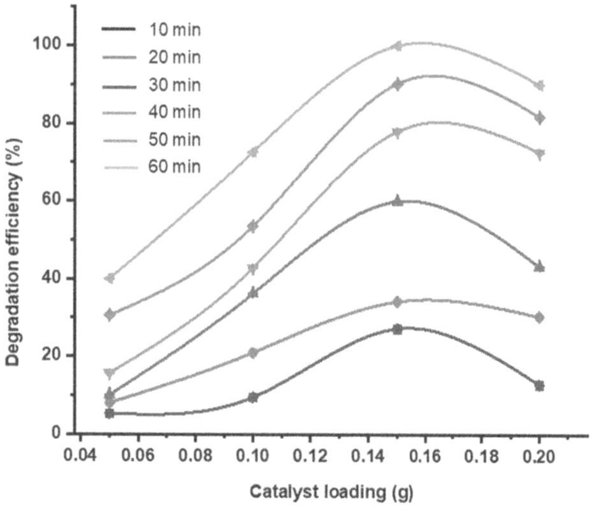

Figure 3 | Effect of time for catalyst dosage (0.05 g, 0.1 g, 0.15 g and 2.0 g) for 400 ppm of the dye concentration and 2 mM of oxalic acid maintained at pH 3.0.

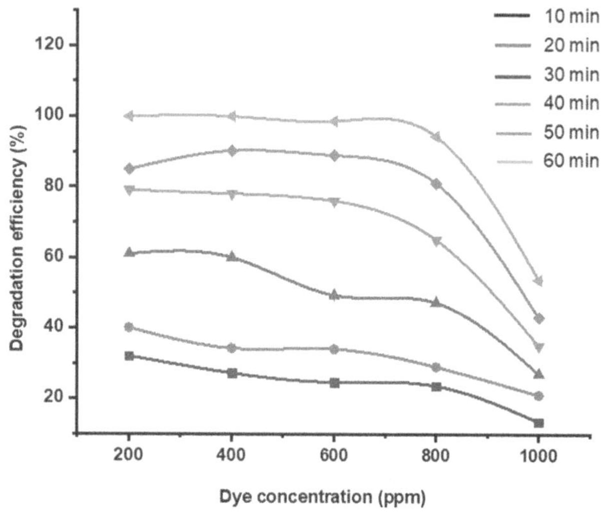

Figure 4 | Effect of dye concentration at 0.15 g catalyst loading and 2 mM of oxalic acid maintained at pH 3.0.

ratio. However, the dosage of nickel ferrite from 0.20 g to 0.25 g only results in a small change of 2% and this difference almost disappears at 30 min. The observed trend is explained on the basis that on increasing the Fe^{2+} ions in the reaction mixture, there is an enhanced generation of •OH radicals, consequently increasing the rate of photodegradation (Kakodia *et al.* 2013). After the addition of optimum Fe^{2+} ions, the higher dose of Fe^{2+} results in a brown turbidity that causes the recombination of •OH radicals and Fe^{2+} reacts with •OH as a scavenger. Therefore, on further increase, the rate decreases. Hence, 0.15 g $NiFe_2O_4$ was used in our experiments.

Effect of dye concentration

The concentrations ranging from 200 ppm to 1,000 ppm were taken individually and degradation studies carried out. The maximum degradation percentage observed for Irgalite violet is 99.995% for 400 ppm within 60 min at pH 3.0, as shown in Figure 4. There is a gradual decrease in the percentage of degradation as concentration of the dye concentration increases due to the occupancy of active sites on the catalyst by the dye molecules (Ameta *et al.* 2012). Since the amount of hydroxyl radicals generated by the given concentration of photocatalyst has been scavenged by the increase of dye molecules, a decrease in degradation efficiency is observed.

Effect of contact time

It was observed that the degradation efficiency of nickel ferrite magnetic nanoparticles over Irgalite violet dye effluent was greater in the presence of solar radiation compared to the absence of solar light, as shown in Figure 5. In the presence of light source, the photons excite the surface electrons of the catalyst, whereby these electrons move from the valence band to the conduction band leaving positive holes in the valence band, which further reacts with water

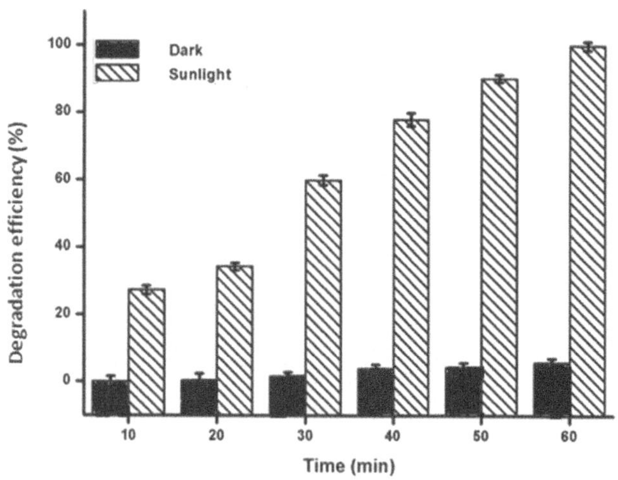

Figure 5 | Dye degradation in the presence and absence of solar light dye at 400 ppm dye concentration, 0.15 g catalyst loading and 2 mM of oxalic acid maintained at pH 3.0.

to release hydroxyl ions that degrade the dyes. Oxalic acid also contributes to the production of hydroxyl radicals which enhances the rate of degradation. As the irradiation time proceeds the degradation efficiency was found to increase. It is evident that in the presence of solar irradiation the photo catalyst (nickel ferrite) had degraded about 99.995% of Irgalite violet whereas in the absence of solar irradiation, the degradation efficiency had dropped to 29% within 60 min. In addition, the degradation rate was observed to remain almost constant after 60 min due to the competitiveness of the parent dye molecules with intermediate/short chain aliphatics during the photocatalytic degradation process (Muruganandham & Swaminathan 2006). Since the batch studies were carried out in aqueous solutions, these radicals contribute to the degradation of dye under solar irradiation in the absence of nickel ferrite nanoparticles and showed 17% degradation efficiency. This may be due to the presence of minute concentrations of hydroxyl radicals generated in water. The photons of solar light generate hydroxyl (OH) radicals due to the water molecules present in the system.

The major challenge encountered during this work is optimizing the pH, since the optimum pH is around 3.0 in the Fenton treatment which is acidic. The modified Fenton oxidation process is improved by using oxidizing agents which, in turn, increase the •OH radicals' formation to improve the efficiency of the system, and the use of $NiFe_2O_4$ nanoparticles, due to the presence of Fe^{2+}, showed strong reactivity to catalyse dyes at various conditions. The band gap of the synthesized $NiFe_2O_4$ nanoparticles for this study is 2.14 eV. Moreover, its catalytic stability can be maintained for multiple cycles and it can be a promising treatment. Even though economic feasibility was not the focus of this investigation, since this method uses solar light for the Fenton process costs, the only costs involved in the study are $NiFe_2O_4$ nanoparticles and the oxidizing agent oxalic acid. Furthermore, the use of solar energy has brought advantages since it widened the range of reactant concentrations that could be used and still achieved very satisfactory treatment performances, making it reasonable to state that the lower the number of reactants, the more economically feasible the treatment becomes (Pignatello *et al.* 2006; Hansson *et al.* 2012; Usman *et al.* 2016).

Kinetic study of photodegradation

The reaction rate constant was calculated from the slopes C/C_o against time as shown in Figure 6. The correlation constant for the fitted line was calculated to be $R^2 = 0.99$ and rate constant (k) was calculated to be 0.0223 min^{-1}. The half-life, $t_{1/2}$ of the pseudo-first-order reaction given by $t_{1/2} = 0.693/k$ was deduced to be 31.07 min.

The possible mechanism for photodegradation of Irgalite violet dye over MFe_2O_4 (M = Ni) and the active species for photodegradation of dye could be produced by three pathways. When ferrite was irradiated in the presence of solar light, the electron/hole (e^-/h^+) pair was generated on the surface:

$$MFe_2O_4 + h\upsilon \rightarrow MFe_2O_4(e^-_{CB} + h^+_{VB})$$

The photogenerated h^+_{VB} could react with water or hydroxyl ion (H_2O or OH^-) to generate hydroxyl radical:

$$h^+_{VB} + H_2O \rightarrow H^+ + OH$$

$$h^+_{VB} + OH^- \rightarrow OH^{\cdot}$$

The photogenerated electron e^-_{CB} could be captured by $C_2H_2O_4$ to yield OH; limiting the recombination of holes and electrons thus enhances the photocatalytic activity:

$$C_2H_2O_4 + e^-_{CB} \rightarrow OH^- + OH^{\bullet} + CO_2$$

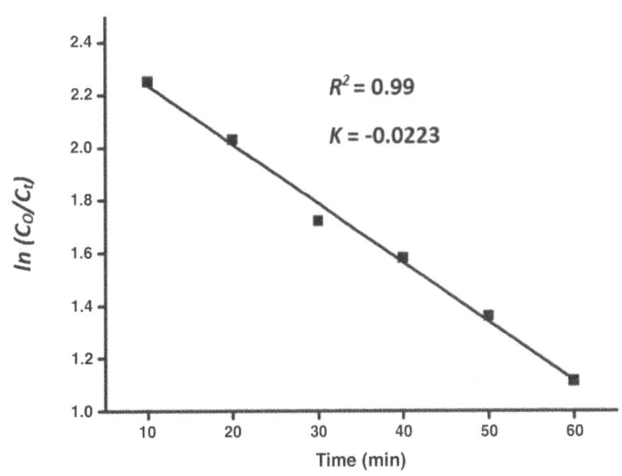

Figure 6 | Plot of pseudo-first-order kinetics of photodegradation of Irgalite violet dye mechanism for degradation of Irgalite violet dye by photo-Fenton process.

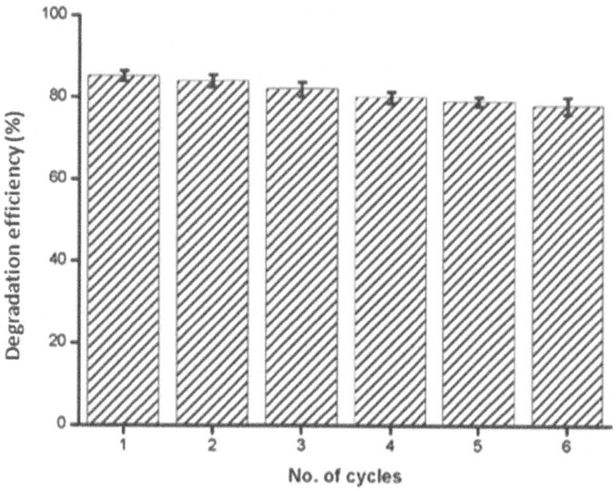

Figure 7 | Effect of catalyst reuse on Irgalite dye degradation.

The reaction of $C_2H_2O_4$ with Fe(III) on the surface of ferrite and Fe(II) generated Fenton reagent which produced hydroxyl radical:

$$Fe^{3+} + C_2H_2O_4 \rightarrow Fe^{2+} + HOO^{\bullet} + H^+ + CO_2$$

$$Fe^{2+} + C_2H_2O_4 \rightarrow Fe^{3+} + OH^{\bullet} + OH^- + CO_2$$

Since OH· was produced by three pathways, enhancement in the rate of degradation was observed. The hydroxyl radical produced by the above three pathways was the main active species for degradation of Irgalite violet dye.

Recyclability of the nanoparticles

The same catalyst was washed with ethanol and distilled water and dried at 60 °C for seven times. The catalyst exhibited activity with almost the same efficiency up to six runs, as shown in Figure 7. With the reuse of catalyst the active sites of the catalyst become occupied, therefore a decrease in the efficiency is observed after six runs. The decrease in activity could be due to the loss in the activity of the catalyst during washing and drying. Figure 8(a) and 8(b) depict the morphology of nanoparticles before and after six runs.

CONCLUSION

The nickel ferrite nanoparticles synthesized by co-precipitation method were found to be stable under acidic conditions for the degradation of Irgalite dye effluent. Further, the dye was degraded completely within 60 min in the presence of 0.15 g of the nickel ferrite catalyst ($k = 0.0223\ min^{-1}$) and adding oxalic acid (2 mM) enhanced the degradation efficiency of the nanoparticles (99.95%). The UV absorbance peaks of the dye solution after the photodegradation process showed no peaks of the dye which confirms the degradation of the dye. Therefore, it can be concluded that this study can effectively be used for the degradation of dye effluents from industries. Further, this approach to dye degradation offers a simple and

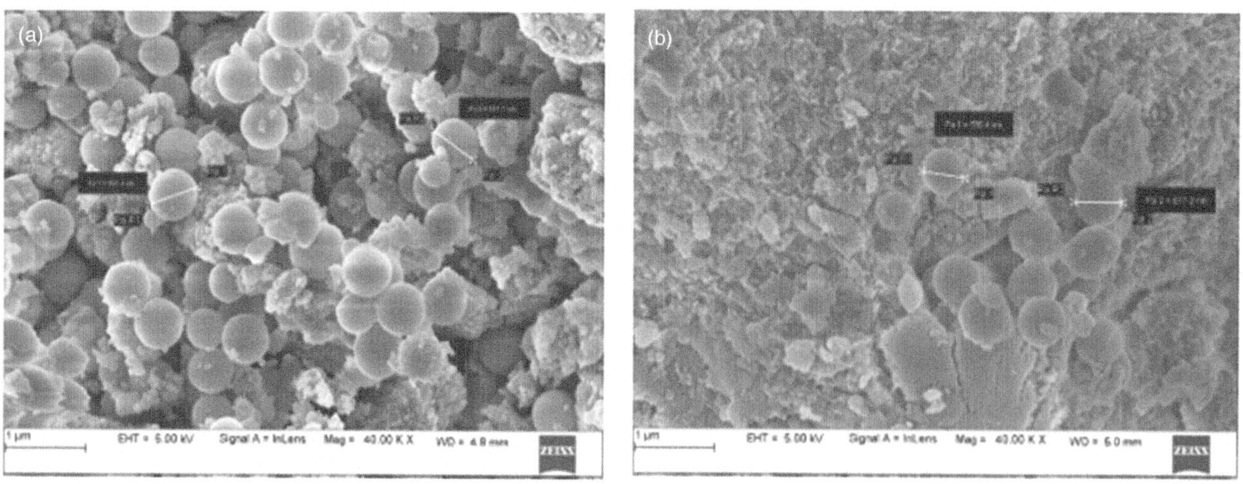

Figure 8 | SEM images of $NiFe_2O_4$ nanoparticles: (a) pure and (b) after six runs.

efficient catalyst and does not require any special conditions for the degradation process. The major limitation of this work is the pH as it is operated at acidic pH 3.0. The work can be improved by changing various heterogenous catalysts and oxidizing agents to widen the spectrum of the working pH, thereby improving the efficiency. However, the present work is a cost-effective and feasible method. Further, due to nickel ferrite's relatively dry and more compact physical structure than ferric hydroxide, ferrite formation considerably reduces sludge volume.

REFERENCES

Ahmadi, M. H., Mohseni-Gharyehsafa, B., Farzaneh-Gord, M., Jilte, R. D., Kumar, R. & Chau, K. W. 2019 Applicability of connectionist methods to predict dynamic viscosity of silver/water nanofluid by using ANN-MLP, MARS and MPR algorithms. *Engineering Applications of Computational Fluid Mechanics* **13** (1), 220–228.

Alalm, M. G., Tawfik, A. & Ookawara, S. 2015 Degradation of four pharmaceuticals by solar photo-Fenton process: kinetics and costs estimation. *Journal of Environmental Chemical Engineering* **3** (1), 46–51.

Ameta, N., Sharma, J. & Chanderia, K. 2012 Degradation of crystal violet using copper modified iron oxide as heterogeneous photo-fenton reagent. *Journal of Iranian Chemical Research* **5** (4), 241–253.

Baghban, A., Jalali, A., Shafiee, M., Ahmadi, M. H. & Chau, K. W. 2019 Developing an ANFIS-based swarm concept model for estimating the relative viscosity of nanofluids. *Engineering Applications of Computational Fluid Mechanics* **13** (1), 26–39.

Barrera-Salgado, K. E., Ramírez-Robledo, G., Álvarez-Gallegos, A., Pineda-Arellano, C. A., Sierra-Espinosa, F. Z., Hernández-Pérez, J. A. & Silva-Martínez, S. 2016 Fenton process coupled to ultrasound and UV light irradiation for the oxidation of a model pollutant. *Journal of Chemistry* **2016**, Article ID 4262530, 1–7.

Behin, J., Akbari, A., Mahmoudi, M. & Khajeh, M. 2017 Sodium hypochlorite as an alternative to hydrogen peroxide in Fenton process for industrial scale. *Water Research* **121**, 120–128.

Deng, Y. & Zhao, R. 2015 Advanced oxidation processes (AOPs) in wastewater treatment. *Current Pollution Reports* **1** (3), 167–176.

El, H. K., Kalnina, D. & Turks, M. 2019 Enhanced degradation of an azo dye by catalytic ozonation over Ni-containing layered double hydroxide nanocatalyst. *Separation and Purification Technology* **210**, 764–774.

Hansson, H., Kaczala, F., Marques, M. & Hogland, W. 2012 Photo-Fenton and Fenton oxidation of recalcitrant industrial wastewater using nanoscale zero-valent iron. *International Journal of Photoenergy* **2012**, Article ID 531076, 1–11.

Kakodia, A. K., Gurjar, L., Ameta, S. C. & Sharma, B. K. 2013 Photodegradation of acid violet 90 in aqueous solution by photo-Fenton reagent. *International Journal of Chemical Sciences* **11**, 855–864.

Kecić, V., Đurđa, K., Miljana, P., Ognjan, L., Milena, B. T., Dragana, T. P. & Božo, D. 2018 Optimization of azo printing dye removal with oak leaves-nZVI/H_2O_2 system using statistically designed experiment. *Journal of Cleaner Production* **202**, 65–80.

Khosravi, I. & Eftekhar, M. 2013 Characterization and evaluation catalytic efficiency of $NiFe_2O_4$ nano spinel in removal of reactive dye from aqueous solution. *Powder Technology* **250**, 147–153.

Kollarahithlu, S. C. & Balakrishnan, R. M. 2018 Adsorption of ibuprofen using cysteine-modified silane-coated magnetic nanomaterial. *Environmental Science and Pollution Research International* 1–10.

Lassoued, A., Mohamed, S. L., Santiago, G. G., Brahim, D., Salah, A. & Abdellatif, G. 2018 Synthesis and characterization of Ni-doped α-Fe_2O_3 nanoparticles through co-precipitation method with enhanced photocatalytic activities. *Journal of Materials Science: Materials in Electronics* **29** (7), 5726–5737.

Li, Y., Li, F., Li, F., Yuan, F. & Wei, P. 2015 Effect of the ultrasound–Fenton oxidation process with the addition of a chelating agent on the removal of petroleum-based contaminants from soil. *Environmental Science and Pollution Research* **22** (23), 18446–18455.

Liu, S. Q., Feng, L. R., Nan, X., Chen, Z. G. & Wang, X. M. 2012 Magnetic nickel ferrite as a heterogeneous photo-Fenton catalyst for the degradation of rhodamine B in the presence of oxalic acid. *Chemical Engineering Journal* **203**, 432–439.

Maaz, K., Karim, S., Mumtaz, A., Hasanain, S. K., Liu, J. & Duan, J. L. 2009 Synthesis and magnetic characterization of nickel ferrite nanoparticles prepared by co-precipitation route. *Journal of Magnetism and Magnetic Materials* **321** (12), 1838–1842.

Monteagudo, J. M., Durán, A. & López, A. C. 2008 Homogeneus ferrioxalate-assisted solar photo-Fenton degradation of Orange II aqueous solutions. *Applied Catalysis B Environmental* **83** (1–2), 46–55.

Muruganandham, M. & Swaminathan, M. 2006 Photocatalytic decolourisation and degradation of Reactive Orange 4 by TiO_2-UV process. *Dyes and Pigments* **68** (2–3), 133–142.

Pignatello, J. J., Oliveros, E. & MacKay, A. 2006 Advanced oxidation processes for organic contaminant destruction based on the Fenton reaction and related chemistry. *Critical Reviews in Environmental Science and Technology* **36** (1), 1–84.

Qin, L., Zeng, G., Lai, C., Huang, D., Zhang, C., Xu, P., Hu, T., Liu, X., Cheng, M., Liu, Y. & Hu, L. 2017 A visual application of gold nanoparticles: simple, reliable and sensitive detection of kanamycin based on hydrogen-bonding recognition. *Sensors and Actuators B: Chemical* **243**, 946–954.

Ramezanizadeh, M., Alhuyi Nazari, M., Ahmadi, M. H. & Chau, K. W. 2019 Experimental and numerical analysis of a nanofluidic thermosyphon heat exchanger. *Engineering Applications of Computational Fluid Mechanics* **13** (1), 40–47.

Sagadevan, S., Chowdhury, Z. Z. & Rafique, R. F. 2018 Preparation and characterization of nickel ferrite nanoparticles via co-precipitation method. *Materials Research* **21**, 1–5.

Santos-Juanes, L., Einschlag, F. G., Amat, A. M. & Arques, A. 2017 Combining ZVI reduction with photo-Fenton process for the removal of persistent pollutants. *Chemical Engineering Journal* **310**, 484–490.

Shi, J., Long, T., Ying, R., Wang, L., Zhu, X. & Lin, Y. 2017 Chemical oxidation of bis (2-chloroethyl) ether in the Fenton process: kinetics, pathways and toxicity assessment. *Chemosphere* **180**, 117–124.

Shihong, X., Daolun, F. & Wenfeng, S. 2009 Preparations and photocatalytic properties of visible-light-active zinc ferrite-doped TiO_2 photocatalyst. *Journal of Physical Chemistry C* **113** (6), 2463–2467.

Sivakumar, P., Ramesh, R., Ramanand, A., Ponnusamy, S. & Muthamizhchelvan, C. 2011 Synthesis and characterization of nickel ferrite magnetic nanoparticles. *Materials Research Bulletin* **46** (12), 2208–2211.

Sohrabi, M. R., Khavaran, A., Shariati, S. & Shariati, S. 2017 Removal of Carmoisine edible dye by Fenton and photo Fenton processes using Taguchi orthogonal array design. *Arabian Journal of Chemistry* **10**, S3523–S3531.

Suna, J. H., Sun, S. P., Fan, M. H., Guo, H. Q., Qiao, L. P. & Sun, R. X. 2007 A kinetic study on the degradation of p-nitroaniline by Fenton oxidation process. *Journal of Hazardous Materials* **148** (1–2), 172–177.

Usman, M., Hanna, K. & Haderlein, S. 2016 Fenton oxidation to remediate PAHs in contaminated soils: a critical review of major limitations and counter-strategies. *Science of the Total Environment* **569**, 179–190.

Wang, H., Zeng, Z., Xu, P., Li, L., Zeng, G., Xiao, R., Tang, Z., Huang, D., Tang, L., Lai, C. & Jiang, D. 2019 Recent progress in covalent organic framework thin films: fabrications, applications and perspectives. *Chemical Society Reviews* **48** (2), 488–516.

Weng, C. H. & Huang, V. 2015 Application of Fe^0 aggregate in ultrasound enhanced advanced Fenton process for decolorization of methylene blue. *Journal of Industrial and Engineering Chemistry* **28**, 153–160.

Yahya, M. S., El Karbane, M., Oturan, N., El Kacemi, K. & Oturan, M. A. 2016 Mineralization of the antibiotic levofloxacin in aqueous medium by electro-Fenton process: kinetics and intermediate products analysis. *Environmental Technology* **37** (10), 1276–1287.

Yang, Y., Zhang, C., Huang, D., Zeng, G., Huang, J., Lai, C., Zhou, C., Wang, W., Guo, H., Xue, W. & Deng, R. 2019 Boron nitride quantum dots decorated ultrathin porous g-C3N4: intensified exciton dissociation and charge transfer for promoting visible-light-driven molecular oxygen activation. *Applied Catalysis B: Environmental* **245**, 87–99.

Ye, S., Zeng, G., Wu, H., Zhang, C., Liang, J., Dai, J., Liu, Z., Xiong, W., Wan, J., Xu, P. & Cheng, M. 2017a Co-occurrence and interactions of pollutants, and their impacts on soil remediation – A review. *Critical Reviews in Environmental Science and Technology* **47** (16), 1528–1553.

Ye, S., Zeng, G., Wu, H., Zhang, C., Dai, J., Liang, J., Yu, J., Ren, X., Yi, H., Cheng, M. & Zhang, C. 2017b Biological technologies for the remediation of co-contaminated soil. *Critical Reviews in Biotechnology* **37** (8), 1062–1076.

Ye, S., Yan, M., Tan, X., Liang, J., Zeng, G., Wu, H., Song, B., Zhou, C., Yang, Y. & Wang, H. 2019a Facile assembled biochar-based nanocomposite with improved graphitization for efficient photocatalytic activity driven by visible light. *Applied Catalysis B: Environmental* **250**, 78–88.

Ye, S., Zeng, G., Wu, H., Liang, J., Zhang, C., Dai, J., Xiong, W., Song, B., Wu, S. & Yu, J. 2019b The effects of activated biochar addition on remediation efficiency of co-composting with contaminated wetland soil. *Resources, Conservation and Recycling* **140**, 278–285.

Yi, H., Jiang, M., Huang, D., Zeng, G., Lai, C., Qin, L., Zhou, C., Li, B., Liu, X., Cheng, M. & Xue, W. 2018a Advanced photocatalytic Fenton-like process over biomimetic hemin-Bi_2WO_6 with enhanced pH. *Journal of the Taiwan Institute of Chemical Engineers* **93**, 184–192.

Yi, H., Huang, D., Zeng, G., Lai, C., Qin, L., Cheng, M., Ye, S., Song, B., Ren, X. & Guo, X. 2018b Selective prepared carbon nanomaterials for advanced photocatalytic application in environmental pollutant treatment and hydrogen production. *Applied Catalysis B: Environmental* **239**, 408–424.

Yi, H., Qin, L., Huang, D., Zeng, G., Lai, C., Liu, X., Li, B., Wang, H., Zhou, C., Huang, F. & Liu, S. 2018c Nano-structured bismuth tungstate with controlled morphology: fabrication, modification, environmental application and mechanism insight. *Chemical Engineering Journal* **358**, 480–486.

Yi, H., Yan, M., Huang, D., Zeng, G., Lai, C., Li, M., Huo, X., Qin, L., Liu, S., Liu, X. & Li, B. 2019 Synergistic effect of artificial enzyme and 2D nano-structured Bi_2WO_6 for eco-friendly and efficient biomimetic photocatalysis. *Applied Catalysis B: Environmental* **250**, 52–62.

Youssef, N. A., Shaban, S. A., Ibrahim, F. A. & Mahmoud, A. S. 2016 Degradation of methyl orange using Fenton catalytic reaction. *Egyptian Journal of Petroleum* **25** (3), 317–321.

Zhou, L., Wang, L., Zhang, J., Lei, J. & Liu, Y. 2016 Well-dispersed Fe_2O_3 nanoparticles on g-C_3N_4 for efficient and stable photo-Fenton photocatalysis under visible-light irradiation. *European Journal of Inorganic Chemistry* **2016** (34), 5387–5392.

First received 25 March 2019; accepted in revised form 28 May 2019. Available online 15 July 2019

Decolorization of Reactive Blue 220 aqueous solution using fungal synthesized Co_3O_4 nanoparticles

Riya Sidhikku Kandath Valappil, Ajuy Sundar Vijayanandan and Raj Mohan Balakrishnan

ABSTRACT

In this work, the photocatalytic activity of the biosynthesized cobalt oxide (Co_3O_4) nanoparticle (NP) is investigated using a textile dye Reactive Blue 220 (RB220) and decolorization % was monitored using UV–Vis spectrophotometer. The photocatalytic activity has been observed maximum at alkaline pH of 9, NP dosage of 250 mg/L, and reaction time of 270 min. In the presence of UV light irradiation, a maximum dye concentration of 10 mg/L was treated effectively using 150 mg/L NP, and 67% decolorization was achieved. Reaction kinetics has been analyzed, and the reaction followed the pseudo kinetics model with an activation energy of -484 kJ mol^{-1}.

Key words | Co_3O_4 nanoparticles, decolorization, kinetics, photocatalysis, RB220

Riya Sidhikku Kandath Valappil
Ajuy Sundar Vijayanandan
Raj Mohan Balakrishnan (corresponding author)
Department of Chemical Engineering,
National Institute of Technology Karnataka,
Surathkal, Mangaluru 575025, Karnataka,
India
E-mail: *rajmohanbala@gmail.com*

INTRODUCTION

Energy demand, which had been outstripping supplies in the last few years, has led to an enormous increase in price. While it may be necessary to learn how to manage with much less energy, the highest priority should be given to developing energy solutions that can provide plentiful energy in the decades to come. Thus, the need of the hour is to find alternative sources of energy that are clean, cheap, and abundant, with minimal environmental impacts in energy extraction, conversion, and energy consumption. Energy harvesting and storage have nowadays attracted tremendous research efforts due to the rapid consumption of fossil fuels. While solar energy is abundant, it represents a tiny fraction of the world's current energy mix. But this is changing rapidly and is being driven by global action to improve energy access, and supply security, and to mitigate climate change (Faizal *et al.* 2015). Semiconductor nanoparticles (NPs) offer an attractive choice for increasing the efficiency of energy harvesting. Conversion of light energy into chemical energy through photocatalysts is a way of utilizing light and the same has been used for the treatment of industrial wastewaters like textile, paper, etc., contain non-biodegradable dyes in trace quantities (Sharma *et al.* 2012). Synthetic dyes are used in textile and paper industries due to the broad color spectrum (Kushwaha *et al.* 2018). Textile industries generate deep-colored effluents containing carcinogenic dye materials, causing allergic dermatitis, skin irritation, and mutation in humans, plants, and animals (Moon *et al.* 2018; Karthikeyan *et al.* 2019). These compounds also avert the sunlight to enter plants and animals in the water, thereby annihilating photosynthetically active radiation in the ecosystem (Saleh 2018). To provide a clean aquatic environment, photocatalytic degradation has been used to remove harmful recalcitrant compounds by mineralization (Manimozhi *et al.* 2018). Though methods to remove dye include coagulation, membrane separation, flocculation, chemical oxidation, photochemical degradation, anaerobic biological degradation, and electrochemical oxidation have been employed, they exhibit low removal efficiency, require stern operating conditions, and high energy demand (Moon *et al.* 2018; Xu *et al.* 2018). Due to the chemical stability of most of the dyes, advanced oxidation processes have been employed for the degradation of dyes into environmentally benign products,

doi: 10.2166/aqua.2019.086

which is based on the formation of hydroxyl radicals in the solution for mineralization of dyes through the usage of NPs (Kushwaha *et al.* 2018). For azo dyes, biological decolorization is not a quite useful technique (Saleh 2018). Peculiar optical properties and small size render semiconductor NPs as interesting candidates for applications in optoelectronics, catalysts, and fluorescence spectroscopy (Sharma *et al.* 2012). Metal oxides are impressive to detoxify organic pollutants (Manimozhi *et al.* 2018), and metal oxide photocatalysis has the potential to transform organic dyes into CO_2, H_2O, and other non-toxic compounds (Saleh 2018). Also, photocatalysis in the presence of NPs offers a great solution for the elimination of pollutants from the environment. The advantage of the process is that only NPs and dye are the two components added to the reaction mixture without the involvement of harsh chemicals and conditions (Saleh 2018). Thus, it is one of the novel ways to treat water pollution in a fast, economical, and eco-friendly manner.

Reactive Blue 220 (RB220) is vinyl sulfone azo dye mainly used in textile industries and more than 50% of cotton is dyed with reactive dyes only (Niebisch *et al.* 2010). RB220 has been degraded in only two cases when nanomaterials are used (Mahmoodi *et al.* 2006; Khanna & Shetty 2014). Tricobalt tetroxide or cobalt oxide (Co_3O_4) NPs are one of the emerging and promising materials for efficient degradation of various effluents. Most of the studies revolving around cobalt oxide NPs for dye degradation were observed for the treatment of methylene blue (MB) dye (Farhadi *et al.* 2016; Bazgir & Farhadi 2017; Mathew & Shetty 2017; Nassar *et al.* 2017; Vennela *et al.* 2019). Very few studies have been carried out on the potential of cobalt oxide NPs for other dyes namely, rhodamine B (32% decolorization), and direct red 80 (78% decolorization) (Dhas *et al.* 2015), methyl orange, and rhodamine B (Abbasi *et al.* 2018), eosine blue (Koli *et al.* 2018), and methyl orange (Saeed *et al.* 2019). In this work, the efficiency of cobalt oxide NPs synthesized through endophytic fungus was utilized for the photocatalytic decolorization of RB220 dye through the electron transfer process.

Though biological methods for synthesizing NPs have already been explored in earlier studies, the involvement of endophytic fungus for the synthesis of Co_3O_4 NPs has not been reported (Kumar *et al.* 2008; Shim *et al.* 2011; Diallo *et al.* 2015; Bibi *et al.* 2017; Khalil *et al.* 2017; Saeed *et al.* 2019). Co_3O_4 NPs have been mainly synthesized by the bacterial method and plant extracts in biological methods of synthesis. Co_3O_4 NPs prepared from sunflower extract have been used for the degradation of methyl orange (Saeed *et al.* 2019). Co_3O_4 NPs synthesized from *Punica granatum* have been used to degrade Remazol Brilliant Orange dye (Bibi *et al.* 2017). In the current study, the endophytic fungus was isolated from a native medicinal plant, *Nothapodytes foetida* and used for the first time to synthesize Co_3O_4 NPs. Further, for the first time, the biologically synthesized Co_3O_4 NPs are being utilized for the decolorization of RB220 dye. Hence, the current work can be considered as a novel work for engaging biologically synthesized Co_3O_4 NPs in the decolorization of RB220.

The photocatalytic potential of Co_3O_4 NP is being tested using the dye RB220 at different time, pH, dye concentration (mg/L), and NP concentration (mg/L). Meanwhile, the decolorization efficiency of Co_3O_4 NP has been determined, and a mechanism and kinetics of the RB220 decolorization have been elucidated.

EXPERIMENTAL

Co_3O_4 NP synthesis

The synthesis of cobalt oxide NPs using endophytic fungus, *Aspergillus nidulans* isolated from the plant *Nothapodytes foetida* as the reducing agent, and cobalt (II) acetylacetonate as precursor compound was reported by Vijayanandan & Balakrishnan (2018). The scheme has been depicted in Figure 1(a).

Materials

To assess the decolorization by cobalt oxide NPs, and since textile effluent contains RB220 dye, RB220 (Indian Fine Chemicals Pvt Ltd, Bengaluru, India) solution has been used in the present study. The structure of RB220 has been given in Figure 1(b). Stock solutions of the dye were formulated by dissolving dyestuff powder in distilled water to a concentration of 0.005% (w/v). Different concentration of the dye was prepared by taking appropriate volumes of the stock solution, and the pH of the dye solution was adjusted to the desired value using 0.1 M NaOH or 0.1 M HCl.

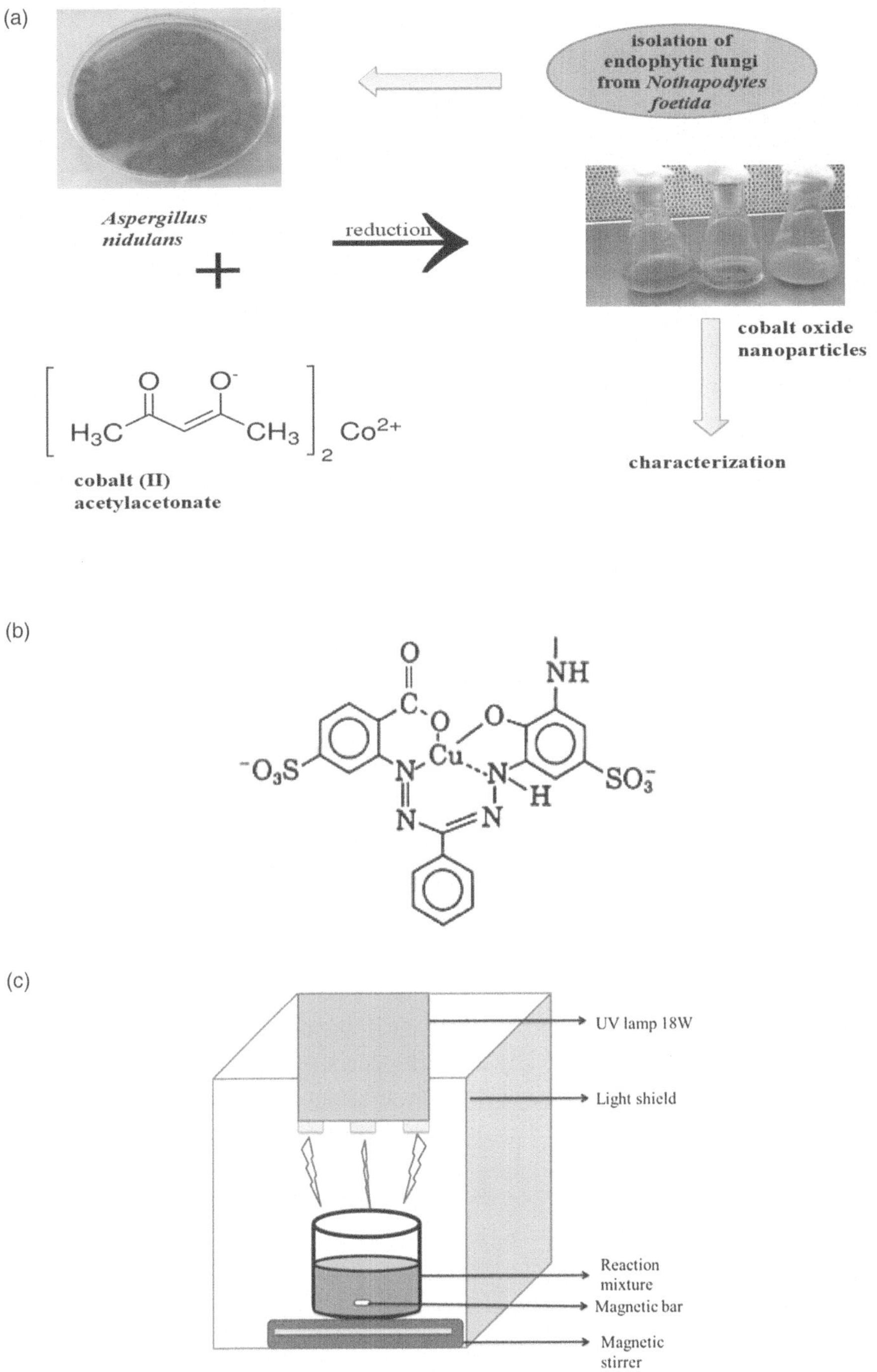

Figure 1 | (a) Synthesis of cobalt oxide nanoparticles (Vijayanandan & Balakrishnan 2018), (b) structure of RB220 dye (Boduroğlu *et al.* 2014), and (c) experimental set-up for RB220 degradation.

Method for photocatalytic degradation of RB220

Photocatalytic activity of Co_3O_4 NP was examined by the degradation of RB220 under UV light in the range of 315–400 nm as regards to time. 20 mL of RB220 dye was taken from the stock solution, and 3 mg of Co_3O_4 NP was added to the dye solution. Since the natural pH of the dye solution was 6.5–7 and to attain an adsorption–desorption equilibrium, the reaction mixture was stirred in the dark for 30 min on a magnetic stirrer (KMS 450, Kadavil Electro Mechanical Industries, Kerala, India) before exposing to the UV light. Figure 1(c) shows the schematic of the experimental set-up which consists of an aluminum chamber equipped with three UV lamps as an irradiation source (Philips TL-D 18 W, India). A 250 mL borosilicate glass beaker was used as the reactor, which was placed inside the closed aluminum chamber, and the reaction mixture was magnetically stirred constantly at room temperature. The aliquots of the sample were periodically withdrawn, centrifuged at 5,000 rpm for 10 min before measuring the absorbance for removing nanocatalysts, and the progress of the reaction was tracked by measuring absorbance at 609 nm of clear supernatants using UV–Vis spectrophotometer (Shimadzu UV-1800, Labomed, USA).

Dye decolorization efficiency of Co_3O_4 NP was estimated using the following Equation (1):

$$\text{Decolorization efficiency} = \left[\frac{A_i - A_t}{A_i}\right] \times 100\% \tag{1}$$

where A_i is the initial absorbance of the dye, and A_t is the absorbance of the dye at any time interval (Forootanfar *et al.* 2012).

Effect of parameters and measurement of chemical oxygen demand

The optimum time for dye decolorization was determined by irradiating the samples with UV light and measuring their absorbances at 609 nm for every 15 min up to 5 h. The pH of the dye solution was adjusted to 7 by adding 0.1 M NaOH and 0.1 M HCl, keeping both the concentration of the dye and NPs constant at 20 mg/L and 150 mg/L, respectively. The influence of dye concentration on photocatalytic decolorization was investigated for a range between 10 mg/L and 50 mg/L for the optimum pH, contact time, and concentration of the NP obtained from the previous experiments. The impact of NP concentration was explored by adding 150 mg/L, 250 mg/L, 350 mg/L, 450 mg/L, and 550 mg/L concentrations of NPs to dye solution at optimized pH, the concentration of dye, and contact time. The chemical oxygen demand (COD) of the samples was determined at regular intervals using the dichromatic oxidation method, which involves standard titration procedure taking distilled water as blank.

Kinetics

The kinetics model of decolorization of dye is a pseudo-first-order rate equation represented by Equation (2)

$$\frac{dC_t}{dt} = -k_1 C \tag{2}$$

where k_1 is the first-order rate constant, C_t is the concentration of the dye at t min, and t is the reactive time.

When $t = 0$, C_t is equal to C_0 and the Equation (2) becomes Equation (3) as follows:

$$C_t = C_0 e^{-k/t} \tag{3}$$

Statistical analysis

Absorbance data obtained at two different pHs (7, 9), and NP concentrations (150 mg/L and 250 mg/L) were analyzed by one-way analysis of variance (ANOVA) test using a free online calculator to determine the effect of pH and NP concentration (Tables 1 and 2). Significance level (α) was set at 0.05, and the statistical significance values (p-value) for different pH and NP concentrations were determined and compared. To find out whether the means between two

Table 1 | Absorbance data at different pH for statistical analysis – one-way ANOVA

S.No.	pH 7	pH 9
1	0.079	0.052
2	0.077	0.053

Table 2 | Absorbance data at nanoparticle concentration for statistical analysis – one-way ANOVA

S.No.	150 mg/L	250 mg/L
1	0.075	0.066
2	0.076	0.07

populations are statistically different, the ratio of variation between the sample means and variation within samples was also estimated (*F*-value).

RESULTS AND DISCUSSIONS

Effect of time

To establish an efficient photocatalyst, insight about the equilibrium time is needed as it is one of the most essential parameters. For this intent, the removal of RB220 was examined as a function of time (El-Gamal *et al.* 2015). From Figure 2(a), it is clear that the rate of decolorization increases with time. As a greater number of NPs bind to dye with a rise in contact time, there is a critical decrease in absorbance after 170 min of UV light irradiation, leading to constancy in removal efficiency. Increase in treatment span hints at more time access for the reaction of dye decolorization (Kale & Kane 2016). Further, it increases electron–hole formation, preventing recombination, which leads to higher production of hydroxyl radicals (Sari *et al.* 2017).

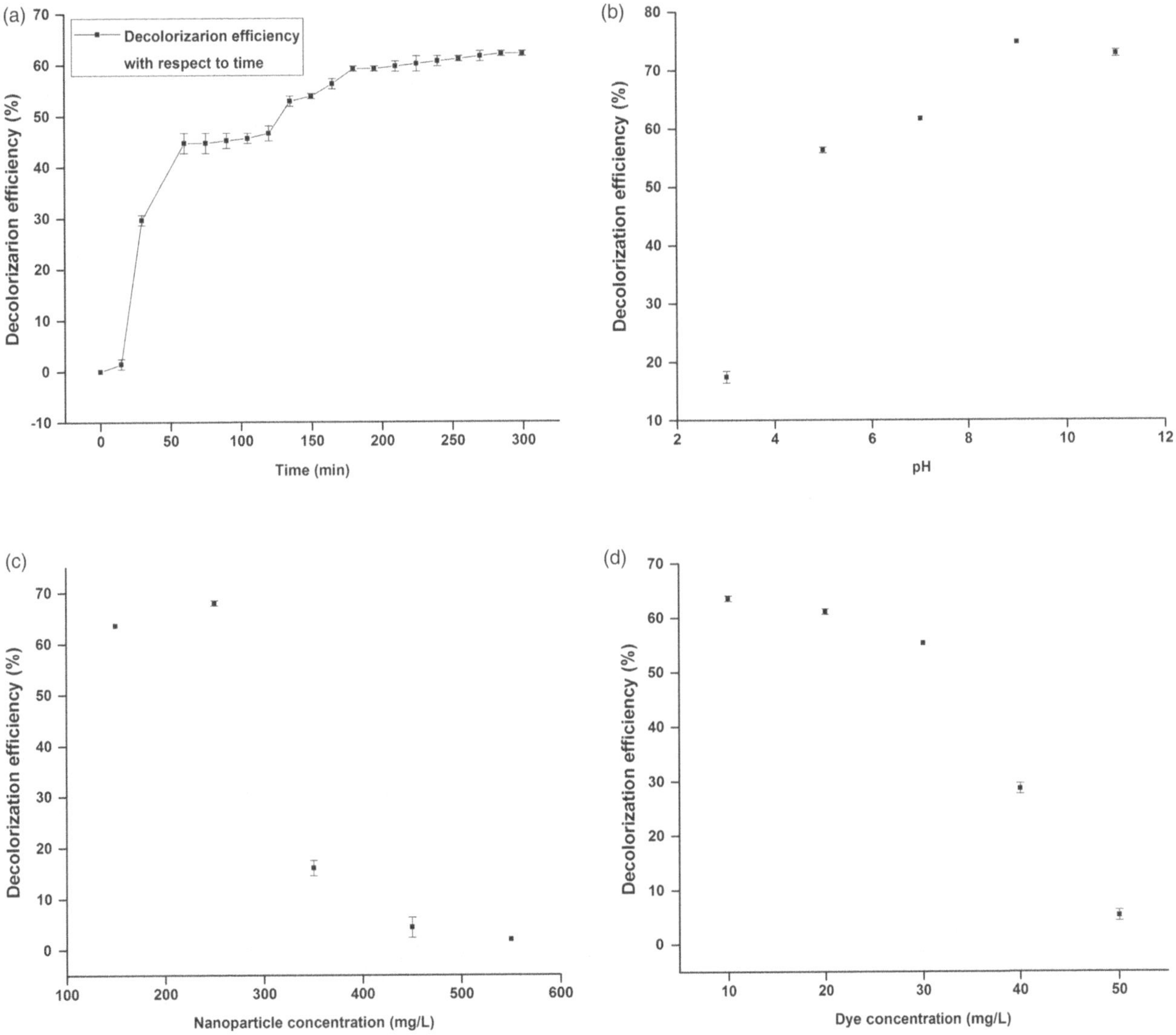

Figure 2 | Effect of parameters: (a) Decolorization efficiency of RB220 dye with respect to time. (b) Decolorization efficiency of Co_3O_4 nanoparticles with respect to pH (dye concentration = 20 mg/L, nanoparticle dosage = 150 mg/L). (c) Decolorization efficiency of nanoparticles with respect to Co_3O_4 nanoparticles' concentration (dye concentration = 10 mg/L, pH = 9). (d) Decolorization efficiency of nanoparticles with respect to dye concentration (Co_3O_4 concentration = 150 mg/L, pH = 9).

Maximum decolorization was observed to be 62% at 270 min (Figure 2(a)). This might be due to the unavailability of active sites on NPs for adsorption of the dye or due to inadequate NPs for the dye to get adsorbed (El-Gamal *et al.* 2015). Since the increase in decolorization efficiency is negligible after 270 min, the aforementioned is taken as the optimum time to carry out further experiments. The dye decolorization by Co_3O_4 NP was authenticated by a decrease in absorbance and visually detected by a gradual decrease in the intensity of the color of the dye solution (Deshmukh *et al.* 2018).

Effect of pH

The pH of the solution can have a great influence on the adsorption of dyes on the NP surface and thus is a crucial factor in photocatalysis (Thu *et al.* 2016). The effect of the pH of the dye on the decolorization efficiency has been presented in Figure 2(b). PH regulates surface characteristics, the charge of organic molecules, and the size of NPs, thereby altering the adsorption of dye molecules on the surface of NPs (Azeez *et al.* 2018). From Figure 2(b), it is evident that the decolorization efficiency progressed with the rise in pH and attained 74% at pH 9. The removal efficiency was greater in the alkaline pH when compared to the acidic pH (Figure 2(b)). The electrostatic interaction is higher between dye molecules, and the surface of NPs at pH 9, paving the way to the highest degree of oxidation, causing enhancement in photocatalytic decolorization of dye molecules (Alshabanat & Al-Anazy 2018). The isoelectric point (IEP) of cobalt oxide semiconductors is relatively high at around 8 (Kittaka & Morimoto 1980; Elhag *et al.* 2015). Thus, the surface of NPs is positively charged in an acidic environment and negatively charged in alkaline solutions. At lower pH below the IEP, H^+ ions compete effectively with dye cations, causing a reduction in decolorization efficiency. Furthermore, a low pH associated with a positively charged surface cannot provide the hydroxyl group, which is needed for hydroxyl radical formation. At higher pH above IEP, the surface of NP gets negatively charged, which enhances the adsorption of positively charged dye cations. However, the decolorization of dye molecules is hindered when the pH value is so high ($pH > 9$), because the hydroxyl ions compete with dye molecules in adsorption on the surface of NPs (Shamsipur & Rajabi 2014). Further increase in pH may augment the electron–hole recombination rate, abating the photocatalytic activity (Mohamed *et al.* 2012).

Effect of dye concentration

It is observed from Figure 2(c) that decolorization efficiency diminishes with an escalation in dye concentration. The maximum removal of dye is 10 mg/L concentration with 150 mg/L concentration of NPs. The observed decrease in decolorization % with increasing dye concentration can be due to the following reasons: (i) more dye molecules interacted over the surface of the NPs as the initial concentration of dye increased, which reduced the generation of hydroxyl ions at the surface of the NP as the active sites have been occupied by dye molecules; (ii) An increment in the light absorbed by the dye molecules may lead to a decrease in the number of photons that reach the NP surface (Shamsipur & Rajabi 2014). The sites on the surface of NPs have saturated with time, leading to competitive adsorption of oxygen on the sites, thereby slowing down the generation of radicals. Another factor is contention for photogenerated holes between adsorbed molecules of water and RB220 molecules (Alshabanat & Al-Anazy 2018).

Effect of Co_3O_4 NP dosage

As summarized in Figure 2(d), the decolorization of RB220 was found to rise with an increase in the NP concentration up to 250 mg/L for a fixed concentration of dye. The total active surface area and availability for binding increased with the enhancement in the NPs dosage. Strengthening the NPs concentration amplified to a higher number and density of NPs, which in turn caused photons to be absorbed, and hence, gain in the adsorption of dye molecules would happen. The dosage of NPs plays a key role in the efficiency of photocatalytic degradation of dyes. An increase in the NP concentration increases the number of active sites, which are charge carriers that engage in photoactivity leading to decolorization and degradation of the dye (Saleh 2018). Further, there is more possibility of a catalyst for attacking chromophores in dye (Kale & Kane 2016). All of these cases have led to enrichment in the efficiency of decolorization. A further increase in the NP count beyond the optimum dosage decreases the photo-decolorization by some margin thanks to the overlapping of adsorption sites as a result of overcrowding and owing to a collision with ground state catalyst. The opacity of the suspensions also increases with the hike in the concentration of NPs which in turn accelerates the light scattering and resulted in a reduction in the penetration depth of the photons, and hence, only a lesser number of NPs were activated (El-Gamal *et al.* 2015). Besides, higher catalyst concentration increases the turbidity of the solution, preventing the path of irradiation to reach the sample (Mohamed *et al.* 2012). Maximum decolorization of 67% has been achieved at NP concentration of 250 mg/L. Thus, with an increase in NP concentration, decolorization efficiency also enhanced. However, this relation does not extend linearly as it reached a saturation point leading to hindrance for the entire decolorization of dye. The results prove that even a smaller quantity of NPs is adequate for decolorization with fair results. In two cases only, NPs have been used to decolorize RB220 dye (Mahmoodi *et al.* 2006; Khanna & Shetty 2014). Mahmoodi *et al.* 2006 used an immobilized TiO_2 photocatalytic reactor with the addition of an H_2O_2 concentration of 450 mg/L with RB220 concentration 50 mg/L at pH 6 having irradiation time of 90 min, yet decolorization efficiency datum is unavailable. Ag@TiO_2 core–shell NPs were used by Khanna & Shetty (2014), as noble metal Ag is a core, and semiconductor dioxide TiO_2 is used as a shell. Decolorization efficiency of RB220 dye was achieved to be 98.9%, when RB220 of 50 mg/L and NP dosage of 1 g/L at pH 3 were irradiated for 240 min in UV light. Thus, a single type of NP in beakers has not been used for RB220 degradation, indicating this current study is a prime one.

COD analysis

COD values provide information about the toxicity of photocatalyzed solutions (Ertugay & Acar 2017), and since the absorbance values in spectrophotometry cannot be related to the reduction of COD, it is indispensable to know the imprint of NPs on the COD value of dye solution (Kale & Kane 2016). In Figure 3, as time proceeds, COD reduces in the supernatant as more and more dye gets accumulated over the NP surface (Nezamzadeh-Ejhieh & Shams-Ghahfarokhi 2013). This has led to the inference of NP resulting in the decolorization of dye, ultimately decrease in the value of COD, attesting the degradation of RB220 (Kale & Kane 2016).

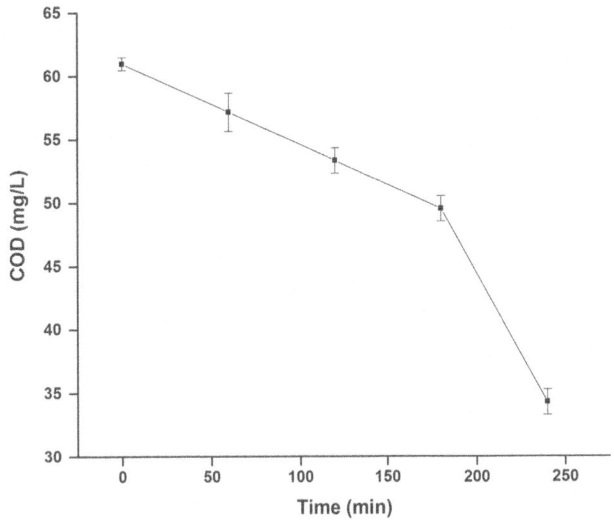

Figure 3 | COD analysis of the reaction mixture with respect to time.

Mechanism of dye removal

Photocatalytic dye degradation banks on the transfer of electrons between donor and acceptor causing an electron relay system (Nakkala *et al.* 2018). The decrease in the absorbance is related to the electron relay effect (Joshi *et al.* 2018). When cobalt oxide NPs are irradiated and bombarded against photons from the light source with energy ($h\upsilon$) energy greater than or equal to its bandgap (3.4 eV) (Vijayanandan & Balakrishnan 2018), the electrons get excited and move to the conduction band from valence band, leaving out a hole in the valence band. These electrons react with oxygen atoms to give superoxide radical anion O_2^- and holes which undergo oxidation to react with water to give hydroxyl ions and eventually hydroxyl radicals. These reactions together subsequently yield highly oxidant species like hydroperoxyl radicals, thereby yielding peroxide. Thus, the peroxide radicals attack and break down the bonds of the azo group in the dye and produce lower molecular weight compounds like formate, acetate, glyoxylate, and oxalate (Mahmoodi *et al.* 2006). Lastly, the intermediate compounds have been further mineralized into CO_2 and H_2O, leading to the decolorization of the dye (Norzaee *et al.* 2017). The blue color of the dye dimmed and turned into colorless during the process (Fairuzi *et al.* 2018). This phenomenon is attributed to the Surface Plasmon Resonance effect (oscillation of charge density propagating at the interface), in which excited surface electrons react with dissolved molecules, producing hydroxyl molecules (Selvam & Sivakumar 2015; Kumari *et al.* 2016). The above mechanism of degradation of the dye RB220 has been depicted in Figure 4.

Kinetics

Kinetics curve for the pseudo-first-order reaction representing the reaction between NPs and dye was drawn. The integrated rate law of the Equation (4) is

$$[A] = [A]_0 e^{-kt} \tag{4}$$

The rate constant for the reaction obtained from the slope (Figure 5(a)) is 0.0029 s^{-1} and the coefficient of determination (R^2) is 0.823. Activation energy obtained from the Arrhenius plot (Figure 5(b)) is equal to −484 kJ mol^{-1}, and the pre-exponential factor is found to be −190.6.

The reaction kinetics has been noticeable from the color change of the reaction as evident from Figure 5(a) and 5(b).

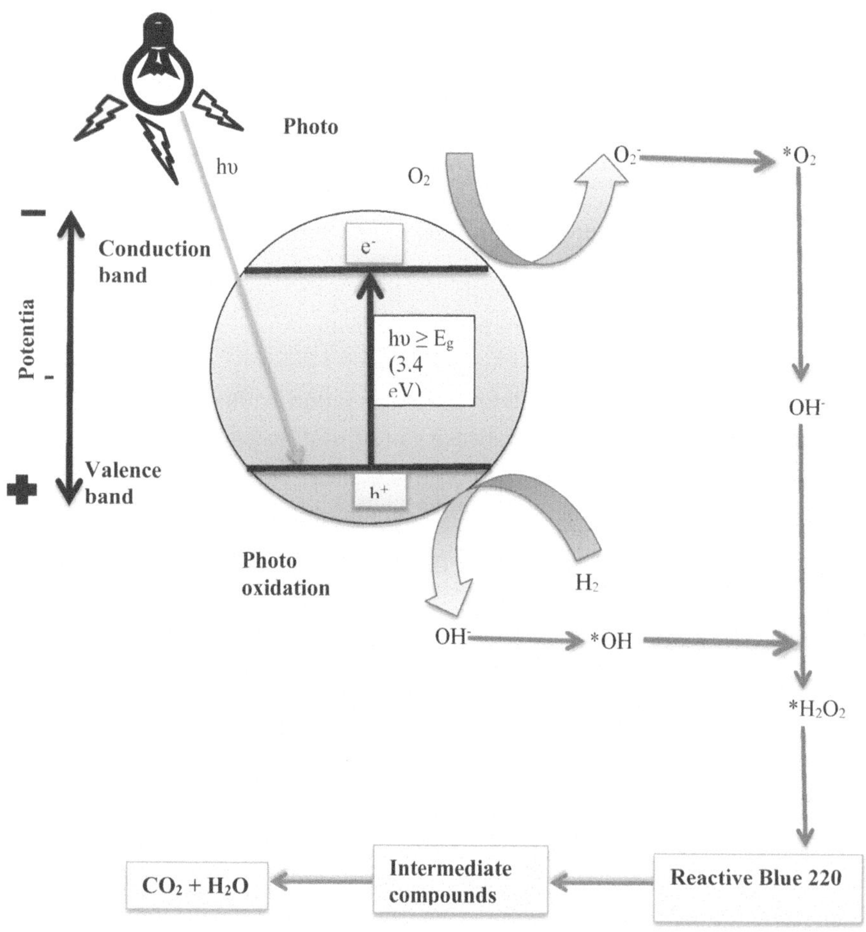

Figure 4 | Mechanism of RB220 degradation by hydroxyl radicals.

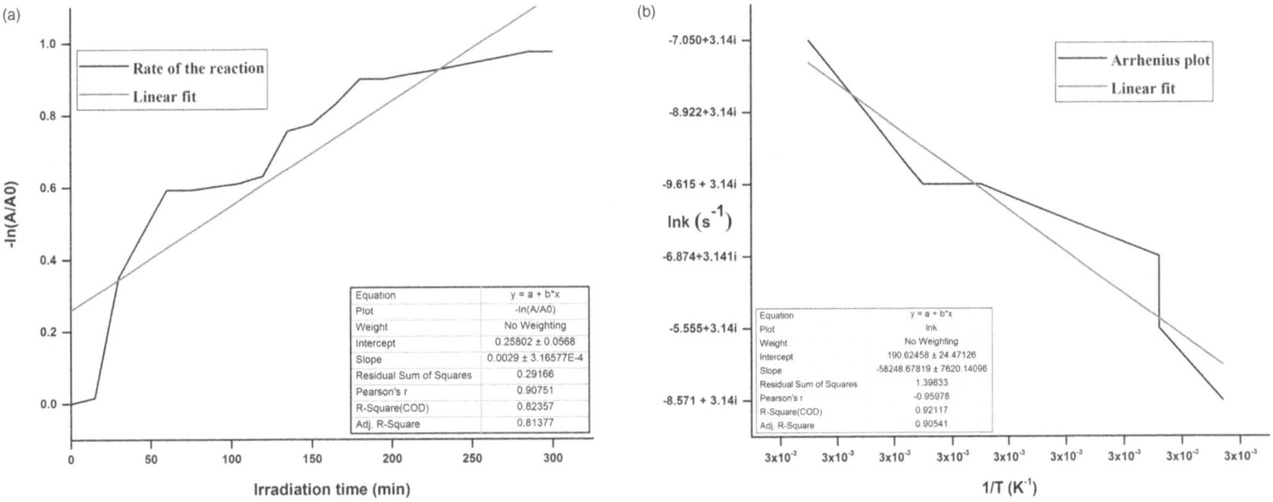

Figure 5 | Kinetics of the study: (a) pseudo-first-order decay curve, (b) Arrhenius plot of reaction.

Before the reaction, the solution was deep blue. But after the reaction with NPs, the solution became pale blue due to the decolorization phenomenon.

Statistical analysis

Tables 3 and 4 show ANOVA results, from which *p*-value and *F*-value were deduced. The *p*-value (0.0019) for the solution at different pH (7, 9) was lower than the significance level indicating that the mean absorbance values at different pH were not equal, and hence, the null hypothesis was rejected. *F*-value (530.81) was found to be greater than 1, which confirms the rejection of the null hypothesis. However, for different NP concentrations (mg/L) (150, 250), the *p*-value (0.0667) is significant, accepting the null hypothesis, though *F*-value (13.50) cannot confirm the same.

Table 3 | ANOVA for absorbance at pH (7, 9)

S.No.	pH	Mean	Std. dev.	Std. error
1	7	0.078	0.0014	0.001
2	9	0.0525	0.007	0.0005

Table 4 | ANOVA for absorbance at nanoparticle concentration (mg/L) (150, 250)

S.No.	Nanoparticle concentration (mg/L)	Mean	Std. dev.	Std. error
1	150	0.0755	0.0007	0.0005
2	250	0.068	0.0028	0.002

PRACTICAL APPLICATION AND FUTURE RESEARCH PERSPECTIVES

Increase in the textile industries has led to an increase in the utilization of the dyes. Synthetic dyes cause pollution if discharged directly into the environment without treating the effluent. Most of the dyes are resistant to biological and physical treatment methods. Co_3O_4 NPs can be used as an efficient cure for a dye solution as they behave as a photocatalytic agent. The NPs can be used to treat wastewater containing commercial RB220 discharged from cotton industries. They can detoxify pollutants and reintroduce them into the nutrient cycle, thereby reducing the amount of sludge formed. They can also be employed in leather and furniture industries where a large number of synthetic dyes are used. RB220 is a textile dye, and photocatalysis by NPs is a simple electron transfer process, using light energy which can be converted into photocatalysts, which can be used for degradation of dyes in wastewater. By scaling up this process, the content of RB220 can be photo detoxified, and effluent can be decolorized. Reactor studies were performed in 5 L immobilized TiO_2 photocatalytic reactor for the degradation of RB220 in a batch mode (Mahmoodi *et al.* 2006). Further, the feasibility of scaling-up depends on mass transfer, hydromechanics, operator-variability, mixing, and avoiding gradients. Thus, the process brings out the way for ecological health and environmental bioremediation. It is a widely accepted choice for environmental purification. So, the scheme has the potential to be a feasible solution for wastewater treatment. This work offers a sustainable solution using a promising strategy to mollify water pollution problems principally for developing countries, which experience dye effluents, especially when the financial budget is a constraint.

However, there are some limitations, which need to be rectified to establish a superior photocatalysis system. Biosynthesized Co_3O_4 NPs should be checked for photocatalytic ability in other azo dyes, as there will not be an effluent that contains only RB220 dye. Practically, there will be a mixture of dyes in the effluent generated by textile industries, and the impact of mixture on the efficiency of NPs is unknown. So, the photocatalytic potential of Co_3O_4 NPs should be checked for all other dyes in the effluent, before it is implemented for dye degradation in textile industries. Also, there is a doubt about Co_3O_4 NPs being non-specific to dyes present in the wastewater. There is a need to find a solution for the complete elimination of dyes in the effluent, so that water can be recycled and reused. Bazgir & Farhadi (2017) checked the reusability of Co_3O_4 NPs for 4 runs in the degradation of MB and a slight decline in activity was observed. Koli *et al.* (2018) checked the reusability for 4 runs for the degradation of eosine blue dye and found a decrease in activity. In addition, whether Co_3O_4 NPs will provide the same efficiency in the subsequent cycles when the same particles are washed and reused for treatment. So, the reusability of nanocatalysts has to be checked. Another problem is the durability of the NP system, and after scaling up, whether the dyebath can withstand a system continuously having Co_3O_4 NPs. The NPs can be attempted to degrade environmental pollutants like recalcitrant molecules and plastics. They can be utilized to treat oil spills. Additionally, they can also be tried to decontaminate polluted groundwater. Since factorial ANOVA involves more than two factors, it can be recommended as a future study in statistical analysis with the inclusion of different pH and NP concentrations.

CONCLUSION

Decolorization of textile dye RB220 using biosynthesized Co_3O_4 NPs has been evaluated in this current study. A maximum dye concentration of 10 mg/L of RB220 was decolorized effectively using 150 mg/L concentration of Co_3O_4 NPs. The photocatalytic activity of the NPs has also led to the degradation of the dye as indicated by the reduction in COD of the dye solution. The decolorization rate follows the pseudo-first-order kinetics model. The decolorization efficiency for RB220 is found to be 67% which is significant in the case of the biologically synthesized Co_3O_4 NPs without any further modification and can be considered as a simple solution without the involvement of sophisticated instruments, toxic chemicals, and conditions.

REFERENCES

Abbasi, A., Golsefidi, M. A., Beigi, M. M., Sadri, N. & Abroudi, M. 2018 Facile fabrication of Co_3O_4 nanostructures as an effective photocatalyst for degradation and removal of organic contaminants. *J. Nanostruct.* **8** (1), 89–96.

Alshabanat, M. N. & Al-Anazy, M. M. 2018 An experimental study of photocatalytic degradation of polymer using polymer nanocomposite films. *J. Chem.* **2018**, 8. Article 9651850.

Azeez, F., Al-Hetlani, E., Arafa, M., Abdelmonem, Y., Nazeer, A. A., Amin, M. O. & Madkour, M. 2018 The effect of surface charge on photocatalytic degradation of methylene blue dye using chargeable titania nanoparticles. *Sci. Rep.* **8** (1). Article 7104.

Bazgir, S. & Farhadi, S. 2017 Microwave-assisted rapid synthesis of Co_3O_4 nanorods from $CoC_2O_4 \cdot 2H_2O$ nanorods and its application in photocatalytic degradation of methylene blue under visible light irradiation. *Int. J. Nano Dimens.* **8** (4), 284–297.

Bibi, I., Nazar, N., Iqbal, M., Kamal, S., Nawaz, H., Nouren, S., Safa, Y., Jilani, K., Sultan, M., Ata, S., Rehman, F. & Abbas, M. 2017 Green synthesis of cobalt-oxide nanoparticle: characterization and photocatalytic activity. *Adv. Powder Technol.* **28** (9), 2035–2043.

Boduroğlu, G., Kılıç, N. K. & Dönmez, G. 2014 Bioremoval of reactive blue 220 by *Gonium* sp *Biomass. Environ. Technol.* **35** (19), 2410–2415.

Deshmukh, K. K., Hase, G. J., Gaje, T. R., Phatangare, N. D., Shilpa, G. & Ashwini, V. 2018 Titanium oxide nanoparticles and degradation of dye by nanoparticles. *Int. J. Mater. Sci.* **13** (1), 23–30.

Dhas, C. R., Venkatesh, R., Jothivenkatachalam, K., Nithya, A., Benjamin, B. S., Raj, A. M. E., Jeyadheepan, K. &

Sanjeeviraja, C. 2015 Visible light driven photocatalytic degradation of rhodamine B and direct red using cobalt oxide nanoparticles. *Ceram. Int.* **41** (8), 9301–9313.

Diallo, A., Beye, A. C., Doyle, T. B., Park, E. & Maaza, M. 2015 Green synthesis of Co_3O_4 nanoparticles via *Aspalathus linearis*: physical properties. *Green Chem. Lett. Rev.* **8** (3–4), 30–36.

El-Gamal, S. M. A., Amin, M. S. & Ahmed, M. A. 2015 Removal of methyl orange and bromophenol blue dyes from aqueous solution using Sorel's cement nanoparticles. *J. Environ. Chem. Eng.* **3** (3), 1702–1712.

Elhag, S., Ibupoto, Z. H., Nour, O. & Willander, M. 2015 Synthesis of Co_3O_4 cotton-like nanostructures for cholesterol biosensor. *Materials (Basel)* **8** (1), 149–161.

Ertugay, N. & Acar, F. N. 2017 Removal of COD and color from Direct Blue 71 azo dye wastewater by Fenton's oxidation: kinetic study. *Arabian. J. Chem.* **10** (1), S1158–S1163.

Fairuzi, A. A., Bonnia, N. N., Akhir, R. M., Abrani, M. A. & Akil, H. M. 2018 Degradation of methylene blue using silver nanoparticles synthesized from *Imperata cylindrica* aqueous extract. *IOP Conf. Ser. Earth Environ. Sci.* **105**. Article 012018.

Faizal, M., Saidur, R., Mekhilef, S., Hepbasli, A. & Mahbubul, I. M. 2015 Energy, economic and environmental analysis of a flat-plate solar collector operated with SiO_2 nanofluid. *Clean Technol. Environ. Policy* **17**, 1457–1473.

Farhadi, S., Javanmard, M. & Nadri, G. 2016 Characterization of cobalt oxide nanoparticles prepared by the thermal decomposition. *Acta. Chim. Slov.* **63** (2), 335–343.

Forootanfar, H., Moezzi, A., Aghaie-Khozani, M., Mahmoudjanlou, Y., Ameri, A., Niknejad, F. & Faramarzi, M. A. 2012 Synthetic dye decolorization by three sources of fungal laccase. *J. Environ. Health Sci. Eng.* **9** (1), 27.

Joshi, S. J., Geetha, S. J., Al-Mamari, S. & Al-Azkawi, A. 2018 Green synthesis of silver nanoparticles using pomegranate peel extracts and its application in photocatalytic degradation of methylene blue. *Nat. Pharm. Prod.* **13** (3), e67846.

Kale, R. D. & Kane, P. B. 2016 Color removal using nanoparticles. *Text. Cloth. Sust.* **2** (4). https://doi.org/10.1186/s40689-016-0015-4.

Karthikeyan, V., Ragunathan, R., Johney, J. & Kabesh, K. 2019 Green synthesis of silver nanoparticles and application in dye decolorization by *Pleurotus ostreatus* (MH591763). *J. Biosci. Biotechnol.* **8** (1), 80–86.

Khalil, N. M., Ovais, M., Ullah, M., Ali, M., Shinwari, Z. K. & Maaza, M. 2017 Physical properties, biological applications and biocompatibility studies on biosynthesized single-phase cobalt oxide nanoparticles (Co_3O_4) via *Sageretia thea* (Osbeck). *Arabian. J. Chem.* http://dx.doi.org/10.1016/j.arabjc.2017.07.004.

Khanna, A. & Shetty, V. 2014 Solar light induced photocatalytic degradation of reactive blue 220 (RB-220) dye with highly efficient Ag@TiO_2 core-shell nanoparticles: a comparison with UV photocatalysis. *Sol. Energy.* **99**, 67–76.

Kittaka, S. & Morimoto, T. 1980 Isoelectric point of metal oxides and binary metal oxides having spinel structure. *J. Colloid Interface Sci.* **75** (2), 398–403.

Koli, P. B., Kapadnis, K. H., Deshpande, U. G. & Patil, M. R. 2018 Fabrication and characterization of pure and modified Co_3O_4 nanocatalyst and their application for photocatalytic degradation of eosine blue dye: a comparative study. *J. Nanost. Chem.* **8**, 453–463.

Kumar, U., Shete, A., Harle, A. S., Kasyutich, O., Schwarzacher, W., Pundle, A. & Poddar, P. 2008 Extracellular bacterial synthesis of protein-functionalized ferromagnetic Co_3O_4 nanocrystals and imaging of self-organization of bacterial cells under stress after exposure to metal ions. *Chem. Mater.* **20** (4), 1484–1491.

Kumari, R. M., Thapa, N., Gupta, N., Kumar, A. & Nimesh, S. 2016 Antibacterial and photocatalytic degradation efficacy of silver nanoparticles biosynthesized using *Cordia dichotoma* leaf extract. *Adv. Nat. Sci.: Nanosci. Nanotechnol.* **7**. Article 045009.

Kushwaha, R., Garg, S., Bajpai, S. & Giri, A. S. 2018 Degradation of Nile blue sulphate dye onto iron oxide nanoparticles: kinetic study, identification of reaction intermediates and proposed mechanism pathways. *Asia. Pac. J. Chem. Eng.* **13** (3), e2200.

Mahmoodi, N. M., Arami, M., Limaee, N. Y. & Tabrizi, N. S. 2006 Kinetics of heterogeneous photocatalytic degradation of reactive dyes in an immobilized TiO_2 photocatalytic reactor. *J. Colloid Interface Sci.* **295** (1), 159–164.

Manimozhi, R., Kumar, D. R. & Prakash, A. P. G. 2018 Enhanced solar light driven photocatalytic degradation of organic dye using solution combustion synthesized CeO_2-ZnO nanocomposites. *J. Electron. Mater.* **47** (11), 6716–6721.

Mathew, J. & Shetty, N. 2017 Treatment of wastewater using synthesised photocatalyst cobalt oxide Co_3O_4. *Int. J. Civil Eng. Technol.* **8** (4), 1840–1844.

Mohamed, R. M., Mkhalid, I. A., Baeissa, E. S. & Al-Rayyani, M. A. 2012 Photocatalytic degradation of methylene blue by Fe/ZnO/SiO_2 nanoparticles under visible light. *J. Nanotechnol.* **2012**, 5. Article ID 329082.

Moon, S. A., Salunke, B. K., Saha, P., Deshmukh, A. R. & Kim, B. S. 2018 Comparison of dye degradation potential of biosynthesized copper oxide, manganese oxide and silver nanoparticles using *Kalopanax pictus* plant extract. *Korean J. Chem. Eng.* **35** (3), 702–708.

Nakkala, J. R., Mata, R., Raja, K., Chandra, V. K. & Sadras, S. R. 2018 Green synthesized silver nanoparticles: catalytic dye degradation, *in vitro* anticancer activity and *in vivo* toxicity in rats. *Mater. Sci. Eng. C* **91**, 372–381.

Nassar, M. Y., Aly, H. M., Abdelrahman, E. A. & Moustafa, M. E. 2017 Synthesis, characterization and biological activity of some novel Schiff bases and their Co(II) and Ni(II) complexes: a new route for Co_3O_4 and NiO nanoparticles for photocatalytic degradation of methylene blue dye. *J. Mol. Struct.* **1143**, 462–471.

Nezamzadeh-Ejhieh, A. & Shams-Ghahfarokhi, Z. 2013 Photodegradation of methyl green by nickel-dimethylglyoxime/ZSM-5 zeolite as a heterogeneous catalyst. *J. Chem.* **2013**, 1–11. Article 104093.

Niebisch, C. H., Malinowski, A. K., Schadeck, R., Mitchell, D. A., Kava-Cordeiro, V. & Paba, J. 2010 Decolorization and biodegradation of reactive blue 220 textile dye by *Lentinus crititus* extracellular extract. *J. Hazard Mater.* **180** (1–3), 316–322.

Norzaee, S., Djahed, B., Khaksefidi, R. & Mostafapour, F. K. 2017 Photocatalytic degradation of aniline in water using CuO nanoparticles. *J. Water Supply Res. Technol.* **66** (3), 178–185.

Saeed, M., Akram, N., Haq, A. U., Naqvi, S. A. R., Usman, M., Abbas, M. A., Abdeel, M. & Nisar, A. 2019 Green and eco-friendly synthesis of Co_3O_4 and Ag-Co_3O_4: characterization and photo-catalytic activity. *Green Process Synth.* **8**, 382–390.

Saleh, S. M. 2018 Metal oxide nanomaterials as photo-catalyst for dye degradation. *Res. Dev. Mater. Sci.* **9** (2), 1–8.

Sari, M. I., Agustina, T. E., Melwita, E. & Aprianti, T. 2017 Color and COD degradation in photocatalytic process of procion red by using TiO2 catalyst under solar irradiation. AIP Conf. Proc. 1903. Article 040017.

Selvam, G. K. & Sivakumar, K. 2015 Phycosynthesis of silver nanoparticles and photocatalytic degradation of methyl orange dye using silver (Ag) nanoparticles synthesized from *Hypnea musciformis* (Wulfen) J. V. Lamouroux. *Appl. Nanosci.* **5** (5), 617–622.

Shamsipur, M. & Rajabi, H. R. 2014 Study of photocatalytic activity of ZnS quantum dots as efficient nanoparticles for removal of methyl violet: effect of ferric ion doping. *Spectrochim. Acta A Mol. Biomol. Spectrosc.* **122**, 260–267.

Sharma, M., Jain, T., Singh, S. & Pandey, O. P. 2012 Photocatalytic degradation of organic dyes under UV-visible light using capped ZnS nanoparticles. *Sol. Energy* **86** (1), 626–633.

Shim, H., Jin, Y., Seo, S., Lee, S. & Kim, D. 2011 Highly reversible lithium storage in *Bacillus subtilis*-directed porous Co_3O_4 nanostructures. *ACS Nano* **5** (1), 443–449.

Thu, T. N. T., Thi, N. N., Quang, V. T., Hong, K. N., Minh, T. N. & Hoai, N. L. T. 2016 Synthesis, characterisation and effect of pH on degradation of dyes of copper-doped TiO_2. *J. Exp. Nanosci.* **11** (3), 226–238.

Vennela, A. B., Mangalaraj, D., Muthukumarasamy, N., Agilan, S. & Hemalatha, K. V. 2019 Structural and optical properties of Co_3O_4 nanoparticles prepared by sol-gel technique for photocatalytic application. *Int. J. Electrochem. Sci.* **14**, 3535–3552.

Vijayanandan, A. S. & Balakrishnan, R. M. 2018 Biosynthesis of cobalt oxide nanoparticles using endophytic fungus *Aspergillus nidulans*. *J. Environ. Manage.* **218**, 442–450.

Xu, J., Xu, D., Zhu, B., Cheng, B. & Jiang, C. 2018 Adsorptive removal of an anionic dye Congo red by flower-like hierarchial magnesium oxide (MgO)-graphene oxide composite microspheres. *Appl. Surf. Sci.* **435**, 1136–1142.

First received 14 June 2019; accepted in revised form 20 September 2019. Available online 13 November 2019

Anthropogenic markers for source tracking of fecal contamination in Bayou Lafourche: a major drinking water source in Southeast Louisiana, USA

Stacy Martinez, Marilyn Kilgen, Angie Corbin, Rajkumar Nathaniel, Balaji Ramachandran and Raj Boopathy

ABSTRACT

Bayou Lafourche, which is the sole drinking water source for 300,000 people in Louisiana, has failed to consistently meet its designated use criteria set by the Louisiana Department of Environmental Quality (LDEQ). This resulted in a total maximum daily load to be imposed on the Bayou by the Environmental Protection Agency (EPA). These designated use water quality criteria include fecal coliform (FC) levels for drinking water source, primary contact recreation, and secondary contact recreation. The goal of this study was to identify and enumerate anthropogenic nonpoint source FC contamination from malfunctioning home sewage systems in the Bayou's watershed. Thirty-four sites along the Bayou were selected for the study. Samples were analyzed for optical brightener ratios, FC CFU/100 mL (mFC), *Escherichia coli*, and three human markers, including human polyomavirus BK, the Archaeon *Methanobrevibacter smithii*, and the human-associated *Brevibacterioides* HF 183 eubacteria. Frequencies of sites with all three anthropogenic molecular markers are considered positive for human fecal contamination. This study provided data to address the problem of malfunctioning on-site sewage systems in the Bayou Lafourche watershed.

Key words | anthropogenic marker, fecal coliform, human polyomavirus, optical brightener, total maximum daily load

Stacy Martinez
Marilyn Kilgen
Angie Corbin
Rajkumar Nathaniel
Raj Boopathy (corresponding author)
Department of Biological Sciences,
Nicholls State University,
Thibodaux, LA 70310,
USA
E-mail: *ramaraj.boopathy@nicholls.edu*

Balaji Ramachandran
Department of Applied Sciences,
Nicholls State University,
Thibodaux, LA 70310,
USA

Stacy Martinez
Present Address:
Louisiana Department of Environmental Quality,
P.O. Box 4301, Baton Rouge, LA 70821,
USA

INTRODUCTION

Nonpoint source (NPS) pollution is the leading contributor to water quality problems in the United States (Gentry-Shields *et al.* 2012). Nonpoint sources of contamination may be attributed to urban and agricultural runoff, leakage from septic or sewer systems, and stormwater runoff (Bergeron *et al.* 2015, 2016, 2017). Methods have been developed in which certain microorganisms are used to indirectly identify probable sources of anthropogenic fecal pollution in Bayou Lafourche. This is known as microbial source tracking (MST) (McQuaig *et al.* 2009; Bird *et al.* 2019).

Bayou Lafourche is located in the Barataria basin in southern Louisiana and is a distributary of the Mississippi River, which starts at Donaldsonville and flows generally southeast for approximately 108 miles to the Gulf of Mexico (Bird *et al.* 2019). This Bayou serves as the main source of drinking water for a population of about 300,000 and also serves as a habitat for recreational activities, such as fishing and boating. It also provides access to the Gulf of Mexico and many other waterways. Because of this heavy usage, the quality of water in Bayou Lafourche is of vital importance. Since Bayou Lafourche waterways have failed to meet the LDEQ's designated use criteria for acceptable water quality, a total maximum daily load (TMDL) had to be developed in order to reduce fecal indicator bacteria (FIB) and potential enteric pathogens (McQuaig *et al.* 2009). TMDL plans for pathogen reduction are often

doi: 10.2166/aqua.2019.063

difficult because the plan relies on indicator bacteria, which do not give much insight into the source(s) of pollution (McQuaig *et al.* 2009). In addition, precipitation can facilitate the transport of enteric bacteria and viruses into surface waters by overwhelming wastewater treatment plants, saturating soils (thereby decreasing the efficiency of septic system drain fields), and resulting in direct runoff or groundwater base flow from urban and rural areas (Gentry-Shields *et al.* 2012). Resuspension associated with rain events may also reintroduce sediment-associated microbes into surface waters and has been shown to increase the concentration of fecal indication bacteria (FIB), coliphages, and enteroviruses in marine and estuarine waters (Gentry-Shields *et al.* 2012).

In 1972, the Clean Water Act was introduced and section 303(d) of the Act required all states to identify those bodies of water not meeting their designated water use standards and to develop a TMDL for those water bodies (McQuaig *et al.* 2009). Louisiana Department of Environmental Quality (LDEQ) sets designated water use criteria standards for its bodies of water in the state of Louisiana. These include microbial fecal coliform (FC) standards for drinking water supply, primary contact recreation, secondary contact recreation, and oyster propagation and harvesting. The current goal of the TMDL for Bayou Lafourche is to decrease anthropogenic FC levels instream by 45% during summer months (LDEQ 2012; Belding & Boopathy 2018; Bird *et al.* 2019).

Fluorometry is a chemical source tracking (CST) method, which identifies human fecal matter by detecting optical brighteners (OBs) (Hartel *et al.* 2008). OBs are fluorescent whitening agents (FWA) that are found in most household detergents for the purpose of brightening fabrics to compensate for undesired yellowing in clothing (Hagedorn *et al.* 2005; Cao *et al.* 2009). OBs have aromatic structures that are activated by light in the near UV range of 360–365 nm and emit light in the blue range of 415–445 nm (Hartel *et al.* 2008). In the United States, 97% of laundry detergents contain DSBP (4,4′-bis92-sulfostyry bisphenol) and/or DAS1 (4,4′-diamino-2,2′-stilbene-disulfonic acid), both of which are highly soluble (Hartel *et al.* 2008; Cao *et al.* 2009). Although these water-soluble compounds have a high affinity for cellulosic material, a small but substantial percentage is lost to the wastewater (Hartel *et al.* 2007, 2008). The mixing of wastewater and graywater in home plumbing systems allows for the detection of OBs in both septic and sewage systems (Cao *et al.* 2009). So, it can be implied that high OBs that decay when exposed to UV light are good indicators of human fecal contamination in the bayou. The concern with OB measurements using fluorometry is that some organic matter or aromatic compounds absorb and emit light at wavelengths similar to that of OBs which may give false-positive readings (Cao *et al.* 2009).

Fecal coliforms and *Escherichia coli* have been used as indicator organisms to assess the presence of fecal contamination of water samples (Anderson *et al.* 2005; Girones *et al.* 2010). The presence of high levels of these indicators suggest fecal pollution and potential health risks, but it is often difficult to link FIB like *E. coli* to a particular pollution source because *E. coli* is abundant in warm-blooded animals (Roslev & Bukh 2011). The recovery of FIB does not distinguish the specific source of pollution, so these indicators can sometimes underestimate the health risks associated with recreational water usage (McQuaig *et al.* 2006). The FIB found in NPS runoff like stormwater is commonly assumed to be derived from animal sources like domestic pets and local wildlife, but there is growing evidence that stormwater systems can be contaminated with sewage due to a failing infrastructure or cross-connections between the stormwater and sewage systems (Sauer *et al.* 2011). Monitoring for FIB in stormwater does not provide definitive information on the possible sources of contamination, which is a major shortcoming of only stormwater evaluations (Sidhu *et al.* 2013). Because identifying sources of pollution is challenging, MST and CST methods have been used to distinguish between human and nonhuman sources of fecal pollution in environmental waters (McQuaig *et al.* 2006; Sidhu *et al.* 2013).

The detection of sewage has been demonstrated with anthropogenic molecular markers from microorganisms and viruses, which would be a more definitive indicator of fecal contamination (Leskinen *et al.* 2010). The human polyomavirus BK (HPyV-BK), human-associated anaerobic bacteria *Bacteroidales* and archaebacteria *Methanobrevibacter smithii* have been used to track anthropogenic waste in many water systems (Bower *et al.* 2005; Fong *et al.* 2005; Gordon *et al.* 2013). The HPyV-BK is nonpathogenic, maintains lifelong productive infections in the kidneys, and is excreted in urine from human populations; thus high titers have been documented in municipal sewage (McQuaig *et al.* 2006; Harwood *et al.* 2009). The HPyV-BK sensitivity, specificity, and correlation with other human-associated markers imply that using HPyV-BK could be a useful predictor of human fecal pollution in environmental waters and a useful component of MST (McQuaig *et al.* 2006).

Methanobrevibacter smithii is an anaerobic archaeon and the most prominent methanogen in the human gastrointestinal tract, which has been found at concentrations of 10^7 to 10^{10} organisms per gram in feces (Ufnar *et al.* 2006; McQuaig *et al.* 2012). Fecal anaerobes make up the majority of fecal bacteria in the gastrointestinal tract of humans and may be present at 1,000-fold higher densities than the FC group, making these organisms highly sensitive indicators of fecal pollution (Bower *et al.* 2005). Harwood *et al.* (2009) reported HpyVs and *M. smithii* each more human-specific than human *Bacteroidales*, and thus expected both to be useful additions to human MST.

Bacteroides spp. are Gram-negative, strictly anaerobic, nonspore-forming bacilli that exceed standard FIB, such as coliforms and enterococci in human and animal feces (McQuaig *et al.* 2012). They occur at concentrations of 10^9 to 10^{11} organisms per gram in feces and 10^9 *Bacteriodales* organisms per 100 mL in sewage (McQuaig *et al.* 2012). Meanwhile, traditional FIB concentrations in untreated sewage are approximately 10^7 CFU per 100 mL for total coliforms and 10^6 CFU per 100 mL for fecal coliforms and enterococci (McQuaig *et al.* 2012). The *Bacteroidales* are dominant members of the intestinal microflora in many animals, including humans, and were correlated with the presence of pathogenic enteric bacteria in environmental waters (Walters *et al.* 2007; Gordon *et al.* 2013). Human-associated *Bacteroidales* have been used as alternative indicators to fecal coliforms in previous studies to detect human fecal input in environmental waters (Bernhard & Field 2000; Bower *et al.* 2005). The human-associated *Bacteroidales* polymerase chain reaction (PCR) method has been found to detect sewage in more dilute samples than the HPyV or the *M. smithii* PCR methods and therefore has the most immediate relevance to regulatory limits based on indicator bacteria concentrations (Harwood *et al.* 2005; 2009).

The overall goal of this study was to investigate the correlations between OBs, mFC, *E. coli*, and PCR data in selected sites in Upper Bayou Lafourche and to determine if there is a correlation among these parameters to identify the source of fecal contamination in the water. The novelty of this study is it will accurately identify the problem of malfunctioning onsite sewage systems in the Bayou Lafourche watershed and also will demonstrate the use of OBs along with human fecal markers as a practical application tool to monitor watershed for fecal contamination from a human source.

MATERIALS AND METHODS

Study area

Bayou Lafourche, a branch of the Mississippi River, is located in the Barataria basin of southern Louisiana. Bayou Lafourche is a distributary of the Mississippi River, which starts at Donaldsonville and flows southeast for approximately 108 miles to the Gulf of Mexico. The 34 sites included bayou surface water at bridges and nonpoint sources such as culverts and ditches that directly led into Bayou Lafourche in Ascension (ASCE) and Assumption Parishes (ASSU). The general study site is given in Figure 1. Specific Site IDs were assigned based on parish location starting at the Mississippi River and numbered in sequential order as shown in Figure 2(a)–2(f) and Table 1 provides the GPS coordinates for these sites. Study sites began at the Mississippi River in Donaldsonville and continued through Ascension and Assumption parishes until the final site at Frank's Lane in Lafourche parish (LAFO). The land use is predominately rural and agricultural. Culverts and ditches draining homes and subdivisions with single dwelling package plants (PP) that emptied directly or indirectly into Bayou Lafourche were chosen for sampling based on a sanitary, microbiological, and OB survey of NPS inputs to the bayou from both LA 1 and HWY 308 during the summer of 2011.

Sample collection

Samples were divided into groups of three and collected once a week with alternating spatial and temporal morning and afternoon collections: group 1 had 11 sites, group 2 had

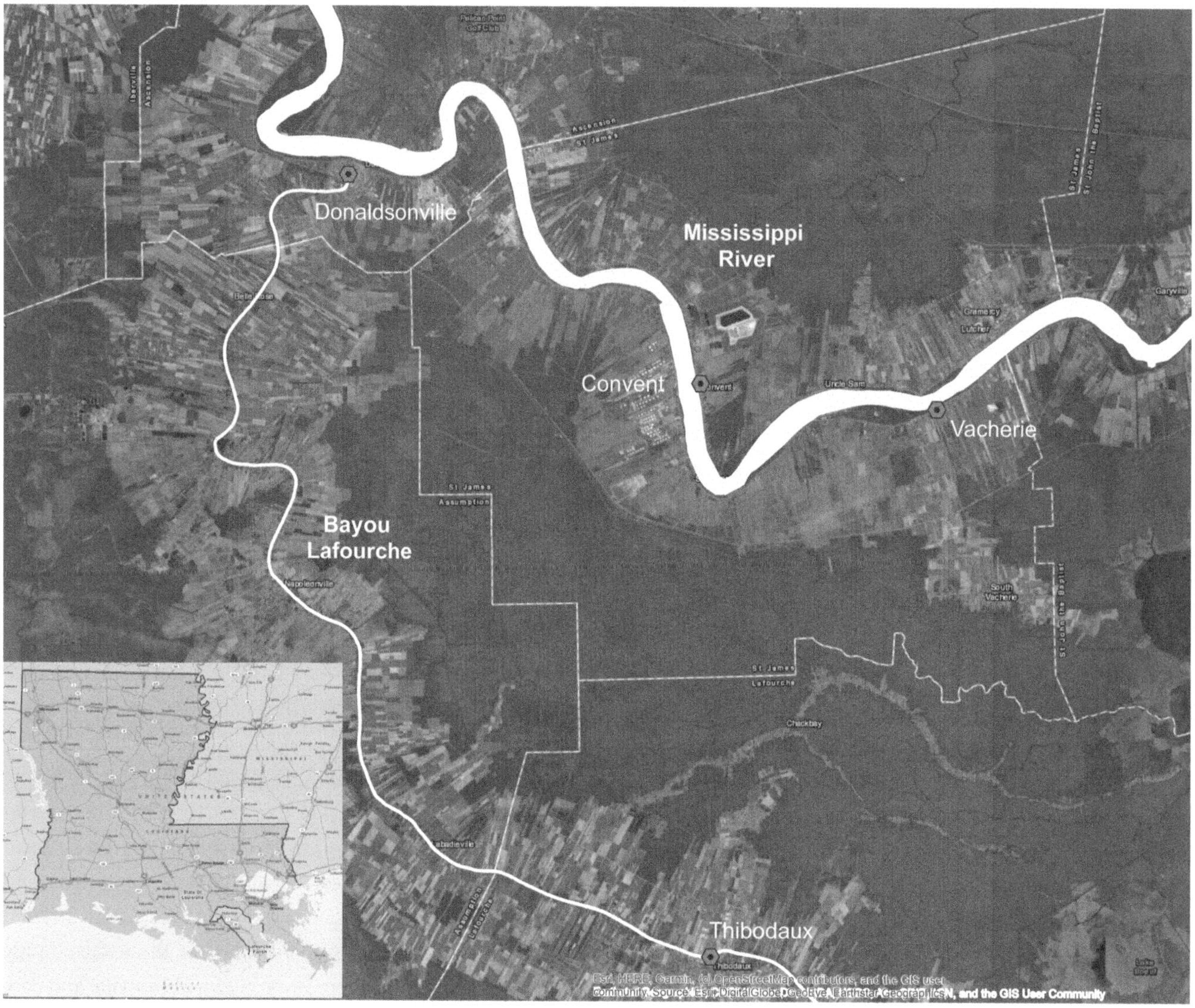

Figure 1 | A general view of the sampling sites from the Mississippi River to the Bayou Lafourche in the city of Thibodaux, Louisiana, USA with the state of Louisiana map inset.

11 sites, and group 3 had 12 sites for a total of 34 sites. Group 1 consisted of: ASCE-1–ASSU-4 (11 sites); group 2 consisted of: ASSU-5A–ASSU-14 (11 sites); and group 3 consisted of: ASSU-15–LAFO-1 (12 sites) as indicated in Figure 2.

An ice chest and thermometer were used to store the collected samples in an ice water bath at 4–10 °C. A blank consisting of DI water from our lab was always included for each sample collection. At each sampling site, a 1,000 mL whirl-pak bag (Nasco) was used to store at least 500 mL of water. The samples were put on ice and kept out of the sunlight by storing in an ice chest. All samples were processed within 6 h.

Fecal coliform detection

The detection of fecal coliforms is commonly used for analyzing environmental water quality. Samples were processed according to the membrane filtration technique as per the American Public Health Association's Standard Methods for the Examination of Water and Wastewater (APHA 1998).

Collected water samples were maintained on ice until processing. Each sample was inverted 25 times and then aliquoted into 250 mL Erlenmeyer flasks. Each sample was filtered through a separate membrane filtration unit

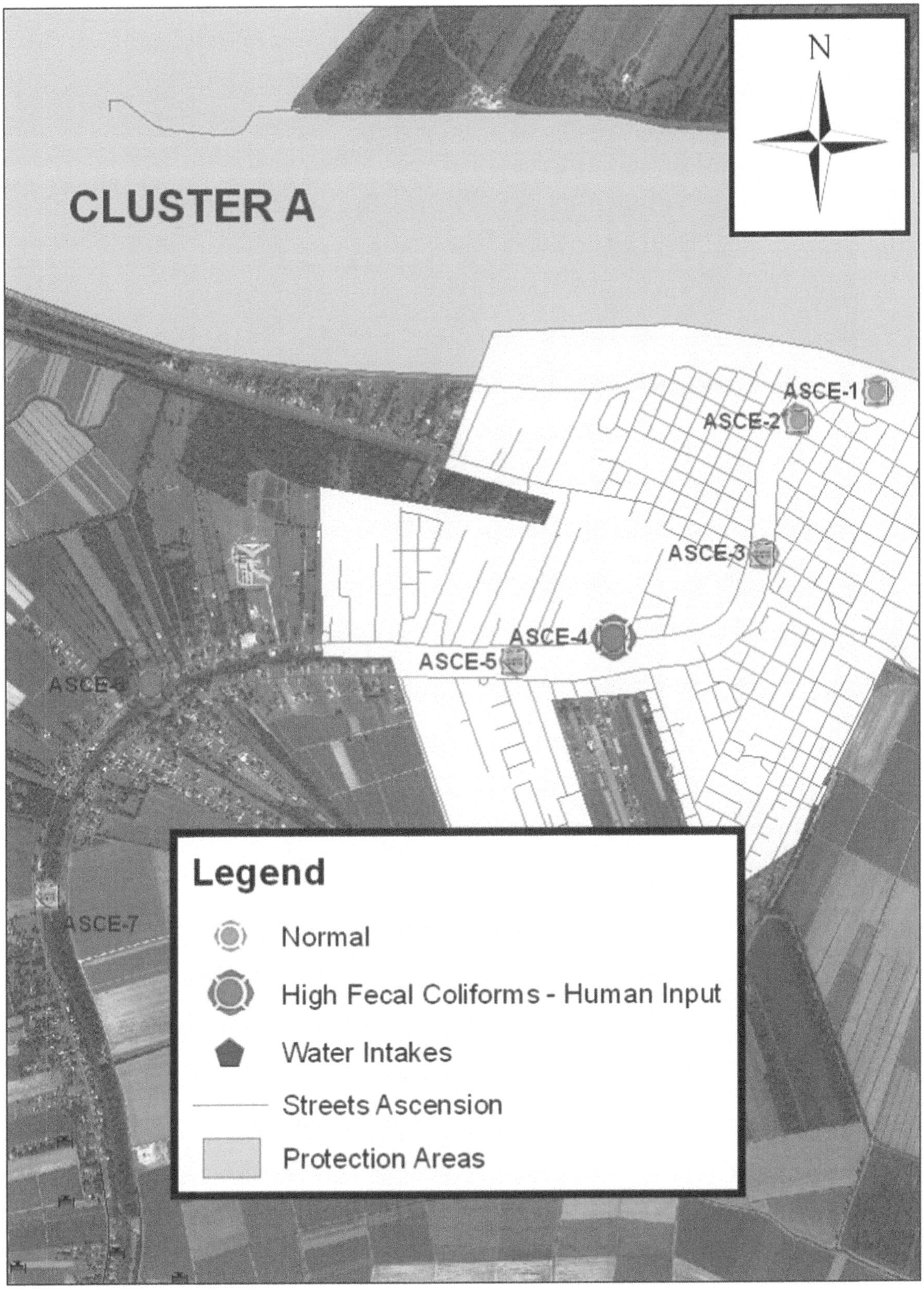

Figure 2 | Clusters A–F show the detailed site maps of Bayou Lafourche with specific sampling sites. Table 1 has more information on these sites including GPS coordinates. (*continued*.).

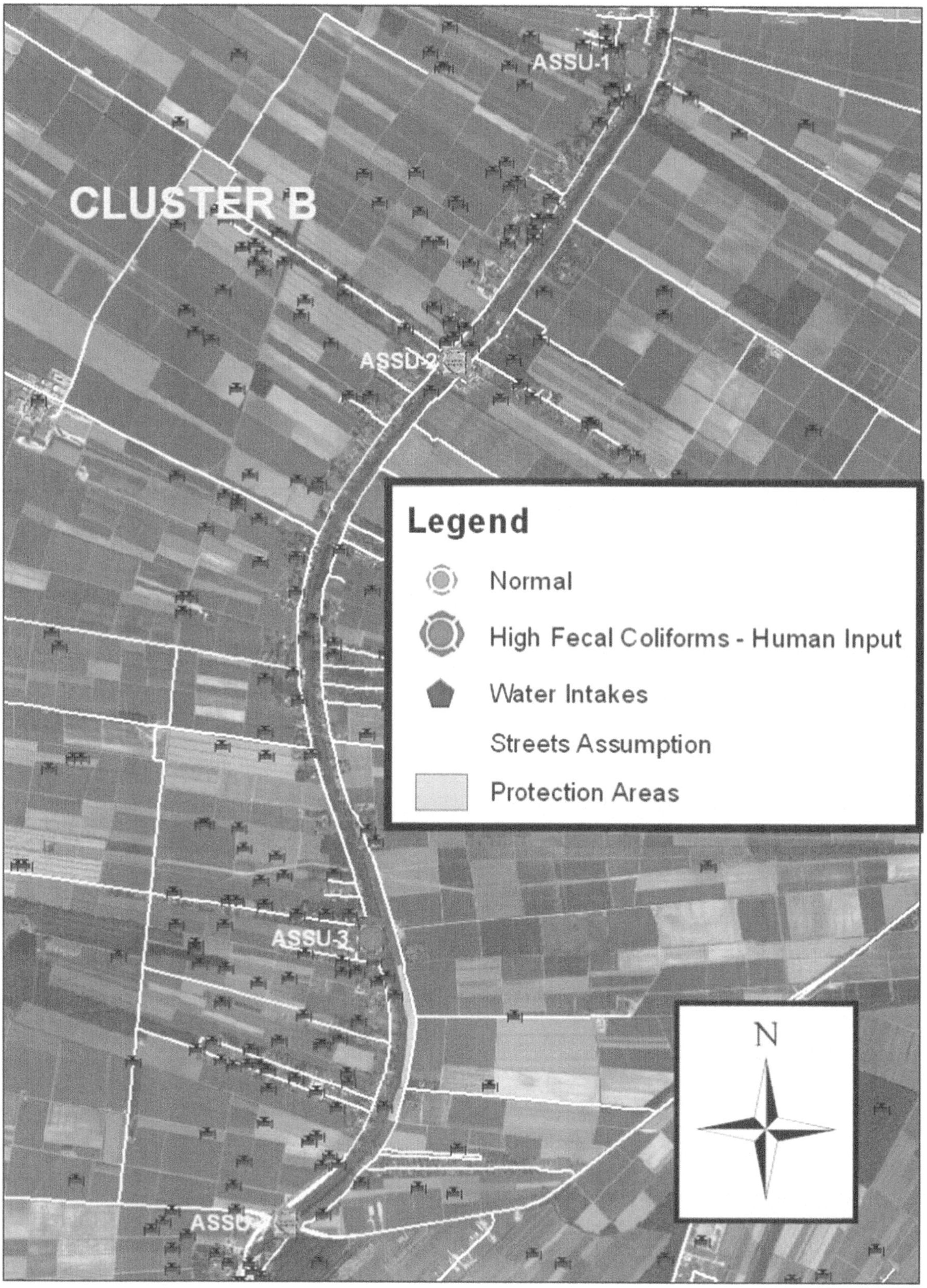

Figure 2 | Continued.

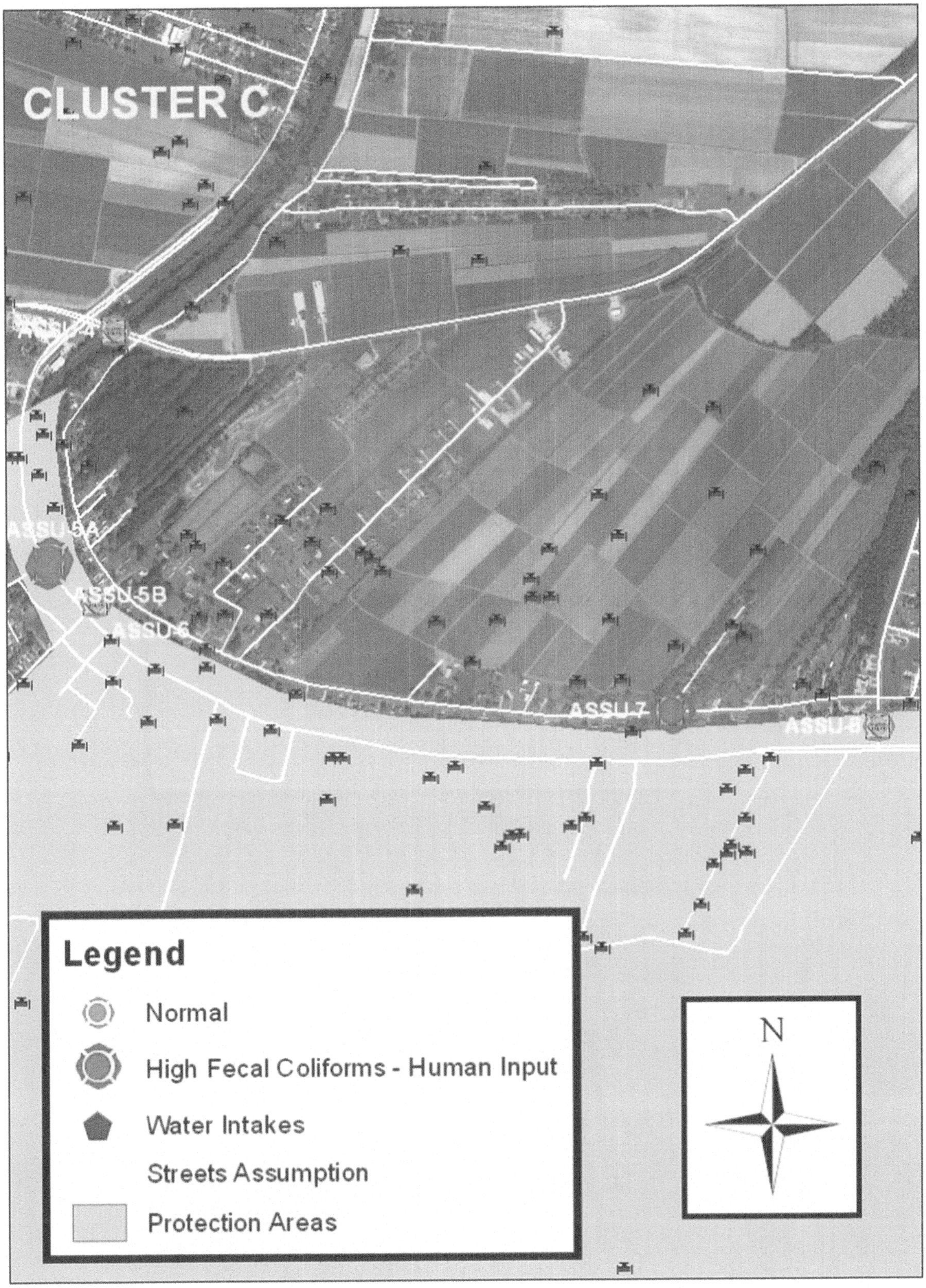

Figure 2 | Continued.

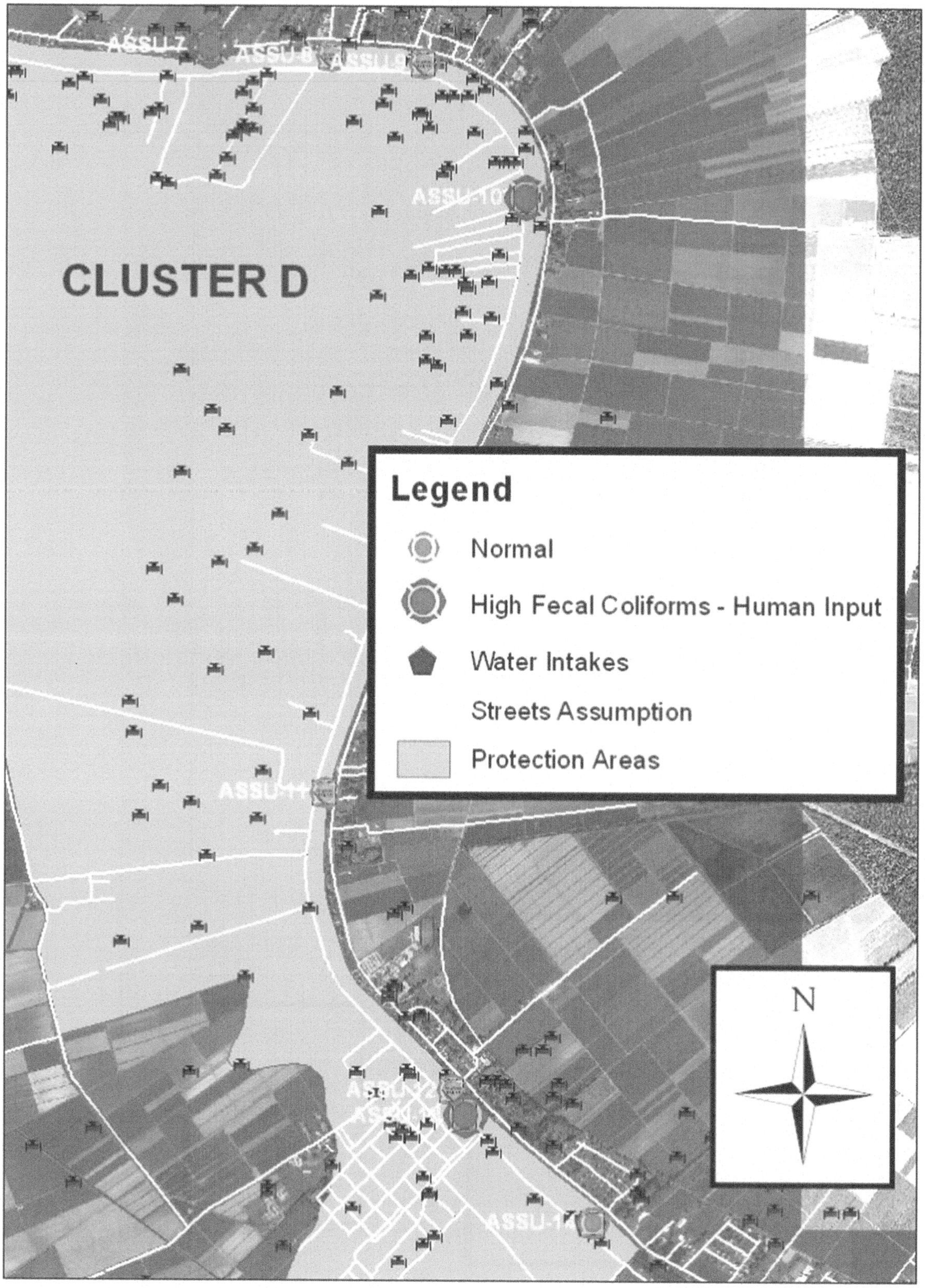

Figure 2 | Continued.

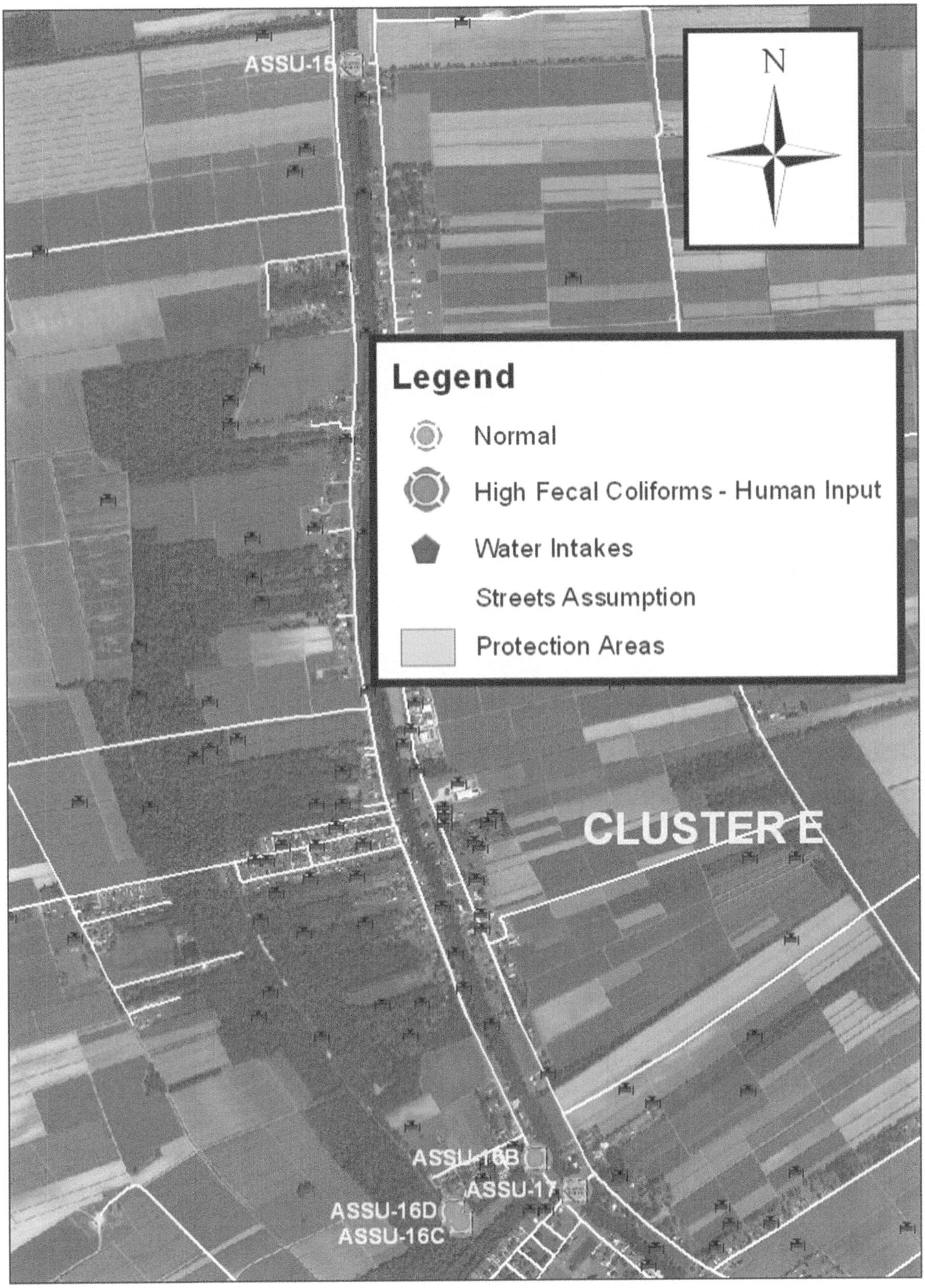

Figure 2 | Continued.

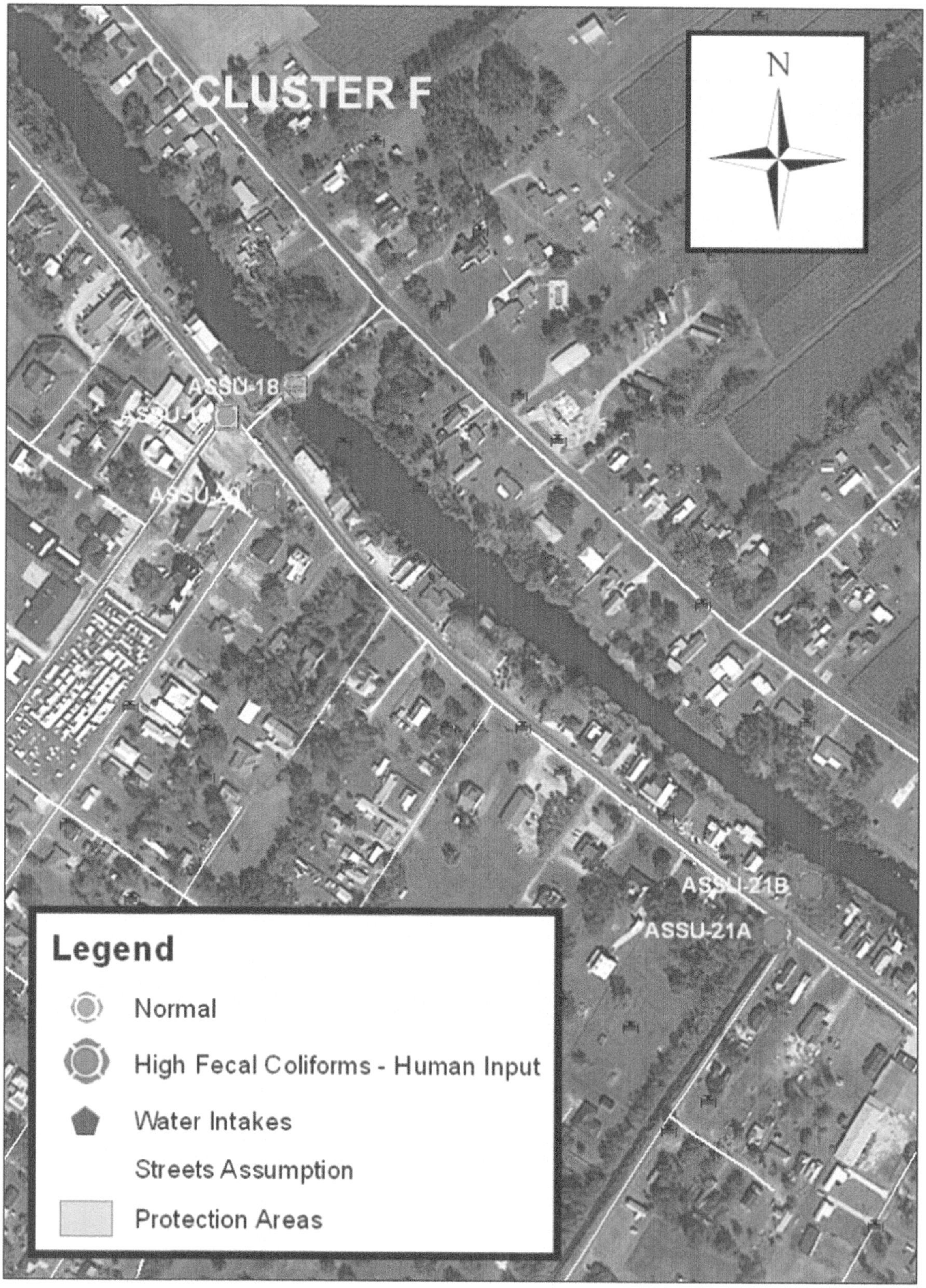

Figure 2 | Continued.

Table 1 | Site ID, description, latitude and longitude, wastewater management of study sites

Parish site ID	Sampling site brief description	Latitude	Longitude	Wastewater management
ASCE-1	Mississippi River	30.10763	−90.98754	River
ASCE-2	PEOPLES Water Intake	30.10647	−90.99157	WT
ASCE-3[a]	Albert (W. 10th) Street (Donaldsonville)	30.10088	−90.99348	[a]
ASCE-4	Popeyes/Stage/hwy LA 1/culvert	30.09747	−91.00112	WT
ASCE-5[a]	Rondinaud Lane (From Hwy 1)	30.09657	−91.00602	[a]
ASCE-6	2165 LA1 ditch	30.09593	−91.02417	PP
ASCE-7[a]	LA 943	30.08709	−91.02958	[a]
ASSU-1	7436LA1	30.06698	−91.02831	PP
ASSU-2[a]	HWY 998 (Belle Rose)	30.0499	−91.04082	[a]
ASSU-3	Tyler Lane/LA1	30.01685	−91.04706	PP
ASSU-4	Hwy 70 (Paincourtville)/LA1	30.00055	−91.05332	AG
ASSU-5A	Hwy 1005/LA1 culvert	29.99424	−91.0556	PP
ASSU-5B	LA1/bayouside culvert	29.9942	−91.0556	PP
ASSU-6[a]	St. Vincent Church	29.99323	−91.05408	[a]
ASSU-7	College Point/Canaan Baptist Church/Hwy 308/culvert	29.98994	−91.03596	PP
ASSU-8[a]	Spur 70 (Plattenville)	29.9894	−91.02932	[a]
ASSU-9[a]	Bridge St.	29.98899	−91.02415	[a]
ASSU-10	St. Benedict Church bayouside culvert	29.98272	−91.01869	PP
ASSU-11[a]	Hospital Road (Napoleonville)	29.95535	−91.03018	[a]
ASSU-12[a]	Hwy 1008	29.94149	−91.02329	[a]
ASSU-13	Jefferson St./Ace Hardware/LA1/culvert	29.94029	−91.02266	WT
ASSU-14	Assumption Water Plant Intake/LA1/bayou sample	29.93526	−91.01575	WT
ASSU-15[a]	HWY 1010	29.8986	−90.98867	[a]
ASSU-16A	Georgette Street ditch	29.87065	−90.99386	PP
ASSU-16B	Cancienne Rd Back canal	29.86096	−90.98226	PP
ASSU-16C	Cancienne Rd Levee Backend canal	29.85885	−90.98533	PP
ASSU-16D	Cancienne Road at LA1 in Bayou	29.85921	−90.98562	PP
ASSU-17[a]	HWY 1011	29.85985	−90.98073	[a]
ASSU-18[a]	Hwy 398/1247 (Labadieville)	29.84007	−90.95428	[a]
ASSU-19	Convent St./LA 1 culvert	29.83989	−90.95481	PP
ASSU-20	Brule Street/Labadieville/ LA 1 culvert	29.83935	−90.95454	PP
ASSU-21A	Pear St./Canal/LA1	29.99423	−90.05545	PP
ASSU-21B	Pear St./Bayou/LA1	29.83672	−90.95028	PP
LAFO-1[a]	Frank Lane (Supreme)	29.81632	−90.88052	[a]

[a]Bridges.
WT, Municipal Wastewater Treatment Plant; PP, Individual Package Plant(s); AG, Agricultural.

using a 0.45 μM filter (Fisher Scientific, Pittsburgh, PA). Four volumes of 0.01 mL, 0.1 mL, 1 mL, and 10 mL, respectively, were filtered and the filters placed on Chrom-ECC petri plates. Chrom-ECC is a differential medium recommended for presumptive identification of *E. coli* and other coliforms in food and environmental samples (CHROMagar, Paris, France). For all volumes less than 10 mL, 10 mL of PBS was added to the filter before the sample was added. All of the petri plates were then placed in the 44.5 °C incubator for 18–24 h. After the samples were incubated, the colonies were interpreted and counted. On Chrom-ECC agar, *E. coli* appears as blue colonies. A detailed procedure is given in the Supplemental section.

Optical brighteners

The OB decay was determined by exposing water samples to low-wave ultraviolet (UV) light using a handheld Turner fluorometer to differentiate OB fluorescence from FWA and natural organic matter fluorescence. The quality controls used for each sample collection consisted of a deionized water (DI) blank, a laboratory fortified blank, an OB standard (10% PTSA, 1,3,6,8-pyrene tetrasulfonic acid tetrasodium salt) (Turner Design, Sunnyville, CA), and a laboratory fortified blank of a 0.005% solution of laundry detergent (2× Tide liquid, original scent, Procter & Gamble, Cincinnati, OH).

An aliquot of each sample including quality controls was put into separate 100 mL test tubes. Each sample was centrifuged at 2000 RPM for 10 min in order to standardize the samples and remove the turbidity. Each centrifuged sample supernatant was micropipetted into polymethacrylate cuvettes (10 × 10 × 45 mm, Turner Designs). The cuvettes were filled with 2 mL sample. OB levels for each sample were initially read using a Turner Aquafluor Handheld Fluorometer. Each sample was then exposed to UV light (wavelength 365 nm) with a UV BLAK-RAY longwave ultraviolet bench lamp (UVP, Model B-100 A, San Gabriel, CA) at intervals of 5 and 10 min. After each exposure, OB levels were recorded. All measurements were taken in duplicate and reported as averages. The percent degradation was calculated according to the manufacturer's instruction.

PCR detection of human *Polyomaviruses* (HPyV-BK), human *Bacteroidales*, and *M. smithii* from environmental waters

After the data were analyzed, the samples that exceed secondary contact recreation limit mFC counts and had positive OB readings were selected to be examined by the PCR detection of HPyV-BK, human *Bacteroidales*, and *M. smithii*. For each selected sample, approximately 500 mL of the environmental water sample was placed into a sterile beaker containing a magnetic stir bar. The pH probe was initially sterilized, and then re-sterilized between samples by immersing it in 10% bleach solution (~30 s), and rinsing in sterile DI water. The probe was then immersed in 10% sodium thiosulfate solution and finally rinsed with sterile DI water. The probe was then ready for use in a water sample. The pH of the water sample was adjusted to 3.0–3.5 using 20% HCl while stirring because lowering the pH gives the viral capsids a net positive charge.

The sample was placed into the Beckman Coulter Avanti J-E centrifuge at 10,000 RPM at 10 °C for 20 min. The supernatant of the sample was then filtered through a 0.45 μM pore nitrocellulose filter. At least 300 mL were processed. Using sterile forceps, the 0.45 μM filter was folded and placed into a bead tube from the Mo Bio PowerSoil DNA extraction kit (Mo Bio Laboratories, Inc., Carlsbad, CA). If there was a pellet from the sample, then it was placed into a separate bead tube. The bead tubes containing the filters were stored at −80 °C until DNA extraction processing. The DNA of every sample was extracted using the manufacturer's instructions from the Mo Bio PowerSoil DNA kit. The extracted DNA was then stored at −80 °C until PCR processing. The primers and target DNA used (specific primer details are given in Supplemental Table 1) were human polyomavirus BK, total *Bacteroidales*, human *Bacteroidales*, and *M. smithii*.

To run the HPyV-BK, total *Bacteroidales*, human *Bacteroidales, and M. smithii* PCR, the reaction mix Go Taq Green (Promega, Madison, WI) was used. Each 25 μL reaction contained: GoTaq Green Master Mix 12.5 μL, forward primer (10 μM) 1.0 μL, reverse primer (10 μM) 1.0 μL, sterile PCR grade water 8.5 μL, and the template 2.0 μL. A thermocycler program was used for HPyV-BK, total *Bacteroidales*, human *Bacteroidales*, and *M. smithii*.

In order to visualize the reaction, 10 μL of each sample was loaded onto a 2% agarose gel for use with HPyV-BK. It was stained with 4 μL of ethidium bromide for each 100 mL, and the gel was run with 8 μL of a 100 bp DNA molecular weight marker. A 1% agarose gel was used for total *Bacteroidales*, human *Bacteroidales*, and *M. smithii* and it was stained with 4 µL of ethidium bromide for each 100 mL. Gels were loaded with 10 µL of each sample and run with 8 µL of a 100 bp DNA molecular weight marker (Fisher Scientific, Pittsburgh, PA).

Statistical analysis

Binary logistic regression was used to analyze the correlations between each method.

RESULTS AND DISCUSSION

Fecal coliform criteria

A total of 437 samples were collected for this study. Samples were collected starting in June 2011 and completed by August 2012. All sites were evaluated for FC counts, *E. coli* counts, OB reading, OB 5% reduction, and OB ratio. Thirty-four sites were evaluated and the geometric mean was calculated for FC counts and *E. coli* counts (Table 2).

The ingestion of water that is contaminated with human or animal feces poses the greatest risk for infections with enteric pathogens. Since enteric pathogens may not be detected when at low concentrations, the detection of indicator organisms is widely used to identify potential sources of fecal contamination. *E. coli*, *Enterococci*, and FC have been used for water quality monitoring due to low cost, ease of use, and as an indicator of the presence of pathogens in surface waters and the risk of disease (Sauer *et al.* 2011; Kishinhi *et al.* 2013). In Bayou Lafourche, the analysis of surface waters from the bayou and influent runoff sites were sampled monthly and were evaluated for two bacterial indicators: FC and *E. coli*. Influent sites in this study consisted of ditches and culverts that eventually flow to Bayou Lafourche.

Water samples were collected for a year alternating between rain and non-rain events and time of day. Surface waters from the bayou were collected on bridges that cross the waterway. As seen in Table 2, only water directly from the Mississippi River at the intake for Bayou Lafourche (ASCE-1) met the state's criteria for primary contact recreational use of water. All remaining sites, based on geometric mean calculations, did not meet the primary contact criteria, but all sites were within the state guidelines for secondary contact recreation/Drinking Water Intake (SCR/DWI). The evaluation of rain events shows an increase in the recovery of fecal coliforms and *E. coli*. These data represent the increased levels of potential anthropogenic contamination entering the waterway as NPS contamination. One site (ASCE-7), a rural area with older homes, exceeds the secondary recreation contact limits of the state during these events.

A total of 23 sites that drain directly into the bayou were selected based on earlier results. A total of 14 of these sites failed to meet the State on Louisiana criteria for secondary contact recreational based on fecal coliform (FC) levels. Two sites selected are drinking water intakes for Ascension Parish (ASCE-2) and Assumption Parish (ASSU-14). Each of these sites met the criteria for drinking water intake. Two of the cities along the bayou are incorporated: Donaldsonville (ASCE 1-4) and Napoleonville (ASSU 11-13). One of these sites (ASCE-4) failed to meet the secondary contact recreation limit. This area is a mixed commercial and residential area on both sides of the bayou. Of the remaining sites, 10 areas failed to meet the secondary contact and drinking water intake criteria. ASCE-6 in Ascension Parish is a rural/agricultural area, and ASSU-1, ASSU-3, ASSU-5A, ASSU-5B, ASSU-7, AASU-10, AASU-16A, AASU-16D, AASU 19, and AASU-21B in Assumption Parish are older residential areas on either side of the bayou. All of the waterway samples for fecal coliforms taken at bridges were elevated as compared to the original water entering the waterway (ASCE-1). However, all of the bridge sites remain within acceptable limits for secondary contact and drinking water intake but are not acceptable for primary contact recreation.

The influence of rain on areas of the study was noted at sites ASSU-10, ASSU-16A-C, and ASSU 21A and ASSU 21B. All sites were analyzed for the effect of rain and non-rain events on the recovery of fecal coliforms and *E. coli*. Though most of the bayou surface water sites showed some increase with rain events, there was no significant difference in the recovery of fecal coliforms. However, the recovery of *E. coli* in surface water showed a significant difference ($p < 0.05$). All of the influent sites failed to show any significant difference with rain events for either fecal coliforms or *E. coli*. Higher numbers of FIB and

Table 2 | Geometric mean of FC and *E. coli* counts from Bayou Lafourche study

Site	No. of times sampled	Geometric mean(range) (CFU[a] 100 mL^{-1}) FC	*E. coli*
ASCE-1	12	213(50–1,000)	33 (18–140)
ASCE-2	12	109 (20–1,532)	43 (10–145)
ASCE-3[a]	12	280(50–2,000)	46 (10–252)
ASCE-4	12	**152,106 (1,636–930,000)**	8,628 (718–160,000)
ASCE-5[a]	12	596 (155–9,091)	127 (10–636)
ASCE-6	13	**9,286 (685–310,000)**	664 (181–8,000)
ASCE-7[a]	14	1,084 (236–17,000)	239 (64–5,000)
ASSU-1	14	**267,954 (95,455–950,000)**	60,911 (270–530,000)
ASSU-2[a]	14	1,085 (218–8,739)	152 (36–4,363)
ASSU-3	12	**415,468 (36,364–980,000)**	85,070 (145–480,000)
ASSU-4	14	1,109 (82–480,000)	216 (20–51,181)
ASSU-5A	12	**11,098 (541–118,182)**	267 (40–2,000)
ASSU-5B	13	1,998 (20–85,455)	122 (10–2,636)
ASSU-6[a]	14	1,111 (198–6,667)	183 (20–420)
ASSU-7	12	**178,644 (2,216–970,000)**	42,673 (963–400,000)
ASSU-8	14	1,098 (218–5,315)	188 (90–390)
ASSU-9	13	1,029 (150–6,486)	209 (100–372)
ASSU-10	13	**4,792 (135–270,000)**	171 (20–11,818)
ASSU-11[a]	13	587 (140–2,273)	145 (70–380)
ASSU-12[a]	12	567 (73–4,144)	182 (45–1,636)
ASSU-13	12	**1,666 (255–40,000)**	94 (10–2,000)
ASSU-14	13	772 (20–26,364)	203 (70–763)
ASSU-15[a]	13	555 (90–3,423)	112 (60–818)
ASSU-16A	13	**2,273 (155–34,545)**	236 (40–3,727)
ASSU-16B	13	717 (40–10,455)	84 (10–5,272)
ASSU-16C	13	1,116 (70–11,636)	91 (20–5,818)
ASSU-16D	13	**3,477 (100–51,818)**	161 (30–4,454)
ASSU-17[a]	13	666 (55–7,545)	120 (30–2,400)
ASSU-18[a]	14	773 (60–12,973)	110 (20–4,414)
ASSU-19	13	**6,478 (391–63,000)**	90 (10–3,000)
ASSU-20	11	**144,909 (7,064–490,000)**	141 (27–2,972)
ASSU-21A	13	**2,007 (110–50,000)**	365 (20–29,090)
ASSU-21B	13	**3,104 (541–37,273)**	436 (10–19,090)
LAFO-1[a]	13	641 (73–5,946)	88 (20–2,600)

[a]Bridges.
Bold represents exceeding 2,000 secondary contact recreation and drinking water intake limit.

enteric pathogens have been previously reported following rain events (Englande *et al.* 2002; Sidhu *et al.* 2013). Effluent from poorly operating septic tank systems that are discharged to drainage ditches leading to waterways contribute to anthropogenic contamination (Englande *et al.* 2002). Leakages in aging sewage infrastructure and cross connections are under-recognized sources of sewage contamination in stormwater (Sidhu *et al.* 2013).

In this study, the FC input was somewhat diluted when it reached the bayou due to the larger volume of water as compared to a smaller volume of influent. In other studies of Mississippi waterways, high numbers of fecals were observed in Bayou Heron and Bayou Cumbest by Kishinhi *et al.* (2013). The area around Bayou Heron is surrounded by many trees that harbor different species of animals and birds, similar to Bayou Lafourche (Kishinhi *et al.* 2013). Studies using MST have revealed that wildlife and waterfowl make important contributions to fecal counts (Kishinhi *et al.* 2013). Monitoring for traditional indicators alone does not necessarily address the presence of human fecal pollution but is one tool among many in microbial water quality assessment. According to Hill *et al.* (2006) in Bayou Dorcheat (North Louisiana), NPS pollution that is carried by surface runoff has a significant effect on bacterial levels in water resources, and it was found that a significant increase in the FC numbers may be associated with average rainfall amounts. Fecal coliforms cannot be evaluated alone as an indicator of anthropogenic contamination in Bayou Lafourche.

Since fecal coliforms are found in both animal and human sources and vary greatly in their potential to predict human pathogens, measuring their levels contributes little to our determination of anthropogenic contamination. Testing for *E. coli* is more specific than FC because *E. coli* is only found in warm-blooded animals. According to Sidhu *et al.* (2013), high numbers of *E. coli* have been observed in the stormwater runoff and are likely due to the presence of fresh fecal contamination from sewage leakage and animal sources. In this study, the recovery of *E. coli* was used to evaluate samples for anthropogenic molecular markers. Samples with high FC concentration and *E. coli* data were used for analysis with PCR results for molecular markers associated with anthropogenic contamination.

Optical brightener criteria

The detection of OBs for anthropogenic contamination has been shown to better predict anthropogenic contamination (Cao *et al.* 2009). The data for OB are given in Figure 3.

All bridge sites had initial OB readings of less than 15 FU and was considered negative. During some of the rain events, 2 bridge sites (ASSU-18 and LAFO-1) had OB readings over 15 but did not have an 8% reduction with 5-min UV exposure indicating possible natural organic matter fluorescence. Eighteen of the influent sites were tested for the 8% reduction

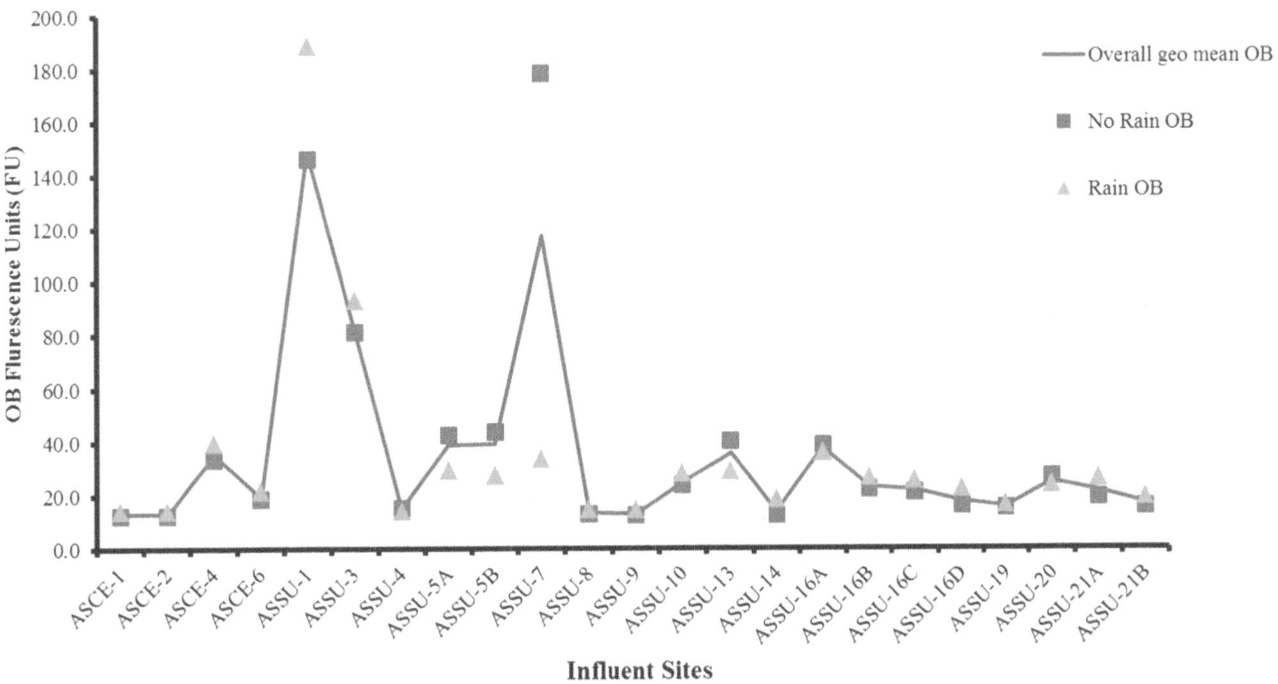

Figure 3 | Sites during rain and no rain events with OB reading of 15 FU or greater indicate potential anthropogenic contamination.

following UV exposure. Ten sites were then tested for the OB ratio according to procedures by Cao *et al.* (2009). The sites of ASCE-6 in Ascension Parish, ASSU-1, ASSU-3, ASSU-5B, ASSU-7, ASSU-10, ASSU-13, and ASSU-19 in Assumption Parish met all the OB ratio criteria for anthropogenic contamination when looking at the overall geometric mean. All sites tested positive for OB ratio are older rural locations on the east (HWY 308) and west (LA Hwy 1) bank of Bayou Lafourche.

The combination of FC and OB criteria showed that the sites of ASCE-6, ASSU-5B, ASSU-13, and ASSU-19 exceed the 400 FC/100 mL primary contact recreation limit and the <1.5 ratio. The sites of ASSU-1, ASSU-3, and ASSU-7 exceed the 2,000 FC/100 mL secondary contact recreation and drinking water intake limits.

Fluorometry-based OB detection is advantageous compared to other MST methods because it is rapid, simple, and inexpensive (Hartel *et al.* 2008). High initial fluorescence has been found to correspond with high levels of fecal bacteria in field studies (Hartel *et al.* 2008). In this study, initial OB readings of 15 fluorescence units (FU) or greater was the criterion for 'potential anthropogenic contamination.' Samples with initial OB reading of 15 or greater FU and a 5-min UV degradation of 8% or greater were exposed to UV light for an additional 5 min. A ratio of 1.5 or less for the 10-min to 5-min exposure is an indication of 'positive anthropogenic contamination'. All data were calculated as a geometric mean. Based on geometric means at the data, samples failing the criteria for any of the tests were excluded from the subsequent dataset. Of the samples tested for initial OB levels, 18 sites were tested for an 8% reduction. Of these 18 sites, only 10 met criteria or ratio of reduction testing. Only 7 sites of the 34 met criteria for anthropogenic contamination by CST.

A concern with using OB measurements in water from ditches and the bayou is that some organic matter and aromatic compounds absorb and emit light at wavelengths similar to that of OBs (Hartel *et al.* 2008). According to Cao *et al.* (2009), organic compounds degrade at a slower rate when exposed to UV light as compared to laundry detergent, which will degrade rapidly when exposed to UV light. Therefore, the sites that exceeded the first OB criteria for 'potential anthropogenic' were then exposed to UV light for 5 min in order to further determine whether the water at each site had anthropogenic or natural organic compounds. Sites that showed a degradation of 8% or greater indicated a 'probable anthropogenic contamination'.

Rain events showed site-specific changes in the detection of OBs. OB detection for sites ASSU-5A, ASSU-5B, and ASSU-10 during rain events met 'probable anthropogenic' criteria (initial reading of greater than 15 FU). OB detection for the influent sites of ASSU-3, ASSU-7, ASSU-13, ASSU-19, and ASSU-20 during no rain events met 'probable anthropogenic' criteria. None of the bridge sites met the ≥8% reduction criteria confirming that these sites were not anthropogenic. The samples that met the criteria of >8% UV reduction criteria for probable anthropogenic were further tested for % reduction.

Each 'probable anthropogenic' sample was exposed to an additional 5 min of UV exposure and calculated as a ratio of 10 min/5 min for % reduction ratio. A ratio of 1.5 or less for the 10-min to 5-min UV exposure was designated as an indication of positive 'anthropogenic contamination' for OBs. The sites ASCE-6 in Ascension Parish, ASSU-1, ASSU-3, ASSU-5A, ASSU-7, ASSU-13, and ASSU-19 in Assumption Parish met all of the OB ratio criteria for 'anthropogenic contamination' using geometric mean determinations. These sites are considered positive for CST. All of these sites are from rural areas along the east and west bank of Bayou Lafourche. These residents use either PP or septic tanks for the treatment of wastewater.

Two bridge sites, ASSU-18 and LAFO-1, met the criteria for potential anthropogenic contamination during single rain events, but these samples did not meet the criteria (8% reduction with 5 min UV exposure) for the presence of anthropogenic markers. All other bridge sites are not considered anthropogenic according to preliminary OB criteria. Both water intake sites, ASCE-2 (Assumption Parish) and ASSU-14 (Assumption Parish), did not meet the criteria for initial OB and are negative for CST.

However, a positive OB ratio alone was not used as the final indicator of anthropogenic contamination in Bayou Lafourche. They were used in combination with FC data to further confirm these sites as anthropogenic. The combination of FC and OB criteria showed that the sites of ASCE-6, ASSU-1, ASSU-3, ASSU-5B, ASSU-7, ASSU-13, and ASSU-19 exceeded the Department of Environmental Quality limit of 2,000 FC/100 mL for the drinking water intake and the secondary contact recreation limit and the ≤1.5 OB ratio. These sites were positive for both CST and MST. These sites were further evaluated using three anthropogenic molecular markers to confirm if the sites were human in origin.

Polymerase chain reaction

A total of 129 samples were selected to run for PCR in this study based on passing both the FC and OB criteria (Table 3). Of the 129 samples, 27/129 were found to be positive by PCR for HB, 45/129 were found to be positive for MS, and 25/129 were found to be positive for HPyV. Binary logistic regression was used to find a predictive relationship between FC, EC, OB, OB 5% reduction, OB ratio, and anthropogenic molecular markers during rain and no rain events.

Table 3 | Number of positive PCRs for human markers of fecal pollution from Bayou Lafourche Study

Site	No. of times sampled for PCR	No. of samples positive by PCR/no. of samples tested		
		HB	MS	HpyV
ASCE-1	1	1/1	0/1	0/1
ASCE-2	12	0/12	1/12	0/12
ASCE-3[a]	0	0/0	0/0	0/0
ASCE-4	12	0/12	5/12	1/12
ASCE-5[a]	1	1/1	0/1	0/1
ASCE-6	7	1/7	2/7	0/7
ASCE-7[a]	1	0/1	1/1	0/1
ASSU-1	14	8/14	6/14	7/14
ASSU-2[a]	3	0/3	1/3	0/3
ASSU-3	12	9/12	7/12	7/12
ASSU-4	1	0/1	0/1	0/1
ASSU-5A	6	1/6	1/6	1/6
ASSU-5B	10	1/10	2/10	2/10
ASSU-6[a]	1	0/1	0/1	1/1
ASSU-7	11	0/11	6/11	3/11
ASSU-8	1	0/1	1/1	0/1
ASSU-9	0	0/0	0/0	0/0
ASSU-10	5	2/5	1/5	0/5
ASSU-11[a]	0	0/0	0/0	0/0
ASSU-12[a]	0	0/0	0/0	0/0
ASSU-13	5	0/5	3/5	0/5
ASSU-14	5	0/5	2/5	0/5
ASSU-15[a]	1	0/1	1/1	0/1
ASSU-16A	4	0/4	1/4	1/4
ASSU-16B	0	0/0	0/0	0/0
ASSU-16C	0	0/0	0/0	0/0
ASSU-16D	1	0/1	0/1	0/1

[a]Bridges.

During no rain events, microbial indicators of EC vs HpyV ($R^2 = 0.1019$, $P = 0.0097$) and EC vs MS ($R^2 = 0.0562$, $P = 0.0269$) both showed a correlation. FC concentrations did not show a correlation with any of the anthropogenic markers for both no rain and rain events. During no rain events, OB indicators of OB ratio vs HpyV ($R^2 = 0.1994$, $P = 0.0022$), OB 5 min vs HpyV ($R^2 = 0.184$, $P = 0.0031$), OB 5 min vs HB ($R^2 = 0.0841$, $P = 0.0099$), and OB ratio vs HB ($R^2 = 0.0972$, $P = 0.0056$) each showed a correlation. All OB indicator criteria did not show any correlations with anthropogenic markers during rain events (statistical analysis is given in Supplemental Tables 2 and 3).

Anthropogenic markers

Based on FC, the absence of markers is at or below 10^4 for all markers (Figure 4). The presence of markers for this study was detected at 10^5 for HB and HPyV based on FC. MS can be found between 10^4 and 10^5. The percent occurrence of HB was 20.9%, the percent occurrence of HPyV was at 18.5%, and the percent occurrence for MS was at 33.06% (Figure 5). The sensitivity of HB was 59.26%, the sensitivity of HpyV was 62.96%, and the sensitivity of MS was 55.56% (Table 4). The specificity of HB was 77.78%, the specificity of MS was 88.89%, and the specificity of HpyV was 100%.

The human polyomavirus BK (HPyV-BK), human-associated anaerobic bacteria *Bacteroidales* H183(HB), and archaebacteria *Methanobrevibacter smithii* (MS) have been used to track anthropogenic waste (Bower *et al.* 2005; Fong *et al.* 2005). Molecular detection of viruses offers several advantages over traditional viral assays, such as cell culture. PCR viral detection is less laborious and time-consuming and is more specific and sensitive than cell culture (Fong *et al.* 2005). The PCR is capable of detecting viruses that are either difficult to grow in cultured cells or replicate without producing cytopathogenic effects in cells

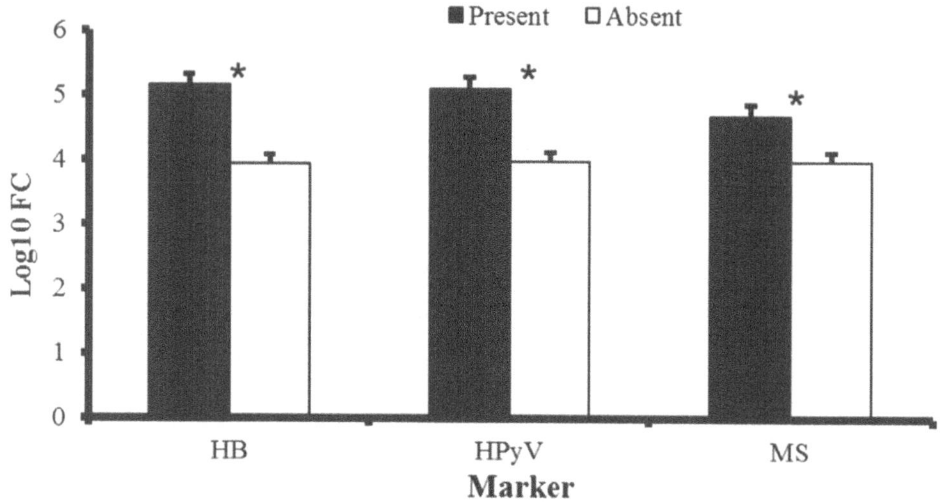

Figure 4 | The presence and absence of anthropogenic markers in relation to the total amount of FC present.

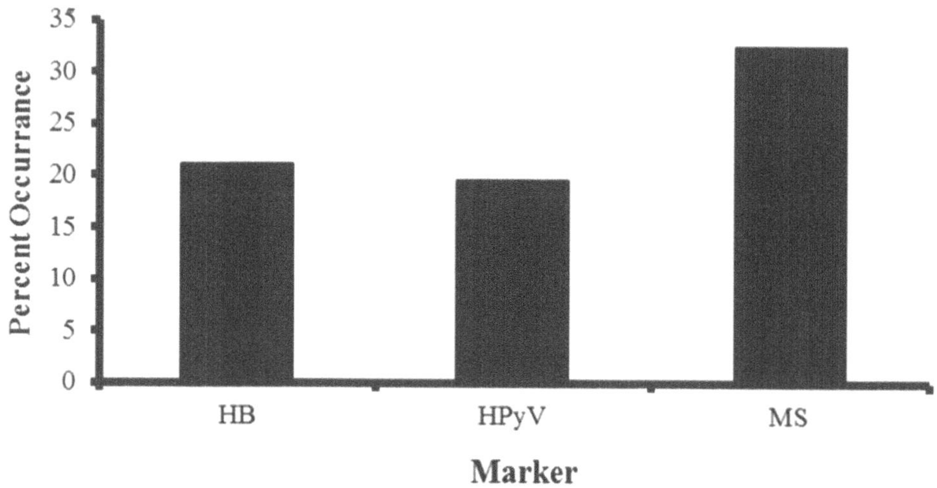

Figure 5 | The percent occurrence of each molecular marker.

Table 4 | Percent sensitivity and specificity of markers

Molecular markers	% Sensitivity	% Specificity
Human Bacteriodales HF 183	59.26	77.78
Methanobrevibacter smithii	55.56	88.89
Human polyomavirus BK	62.96	100

(Fong *et al.* 2005). The HPyV-BK is nonpathogenic, maintains lifelong productive infections in the kidneys, and is excreted in urine from human populations, which resulted in high titers in municipal sewage (McQuaig *et al.* 2006; Harwood *et al.* 2009). This marker is found in most of the population but not all. The HPyV-BK sensitivity, specificity, and correlation with other human-associated human markers imply that using HPyV-BK could be a useful predictor of human fecal pollution in environmental waters and a useful component of MST (McQuaig *et al.* 2006).

Methanobrevibacter smithii is an anaerobic archaeon and the most prominent methanogen in the human gastrointestinal tract, which has been found at concentrations of 10^7 to 10^{10} organisms per gram in feces (Ufnar *et al.* 2006; McQuaig *et al.* 2012). Fecal anaerobes make up the majority of fecal bacteria in the gastrointestinal tract of humans and may be present at 1,000-fold-higher densities than the FC group making these organisms highly sensitive indicators of fecal pollution (Bower *et al.* 2005). *Bacteroides* spp. are Gram-negative, strictly anaerobic, non-spore-forming bacilli that exceed standard FIB such as coliforms and enterococci in human and animal feces (McQuaig *et al.* 2012). They occur at concentrations of 10^9 to 10^{11} organisms per gram in feces and 10^9 *Bacteriodales* organisms per 100 mL in sewage (McQuaig *et al.* 2012). Human-associated *Bacteroidales* have been used as alternative indicators to fecal coliforms in previous studies to detect human fecal input in environmental waters (Bower *et al.* 2005).

In this study, 45/124 were found to be positive for MS and 25/124 were found to be positive for HpyV. The MS had the highest number of samples that were found to be positive, probably because it is a prominent marker in the human GI tract, but it is not the most human-specific. In other studies, HpyV has been shown to be the most human-specific (100%); however, the human-associated *Bacteroidaes* and *M. smithii* were 96% and 98% specific (Harwood *et al.* 2009). All three markers have been shown to be 100% sensitive in a previous study (Harwood *et al.* 2009). In this study, the HpyV showed 100% specificity and the HB and MS showed 77.78% and 88.89% specificity, respectively. The HpyV also had the highest sensitivity (62.96%) and the HB and MS showed 59.26% and 55.56% sensitivity, respectively. The absence of all molecular markers in this study is at or below 10^4. The markers were detected at 10^5 for HB and HPyV. The MS can be found between 10^4 and 10^5. Human *Bacteroides* is present at higher concentrations in sewage and therefore is a more sensitive marker in dilute samples (Gordon *et al.* 2013). Harwood *et al.* (2009) reported HpyVs and *M. smithii* each more human-specific than human *Bacteroidales*, thus expected both to be useful additions to human MST.

Binary logistic regression was used to find a predictive relationship between FC, EC, OB, OB 5% reduction, OB ratio, and anthropogenic molecular markers during rain and no rain events. During no rain events, microbial indicators of EC vs HpyV and EC vs MS showed a correlation. The FC concentrations did not show a correlation with any of the anthropogenic markers for both no rain and rain events. This indicated that FC was not a good indicator to determine if sites were anthropogenic for both rain and no rain events and that *E. coli* is a better indicator than FC. According to McQuaig *et al.* (2006), high FIB counts in the absence of marker detection were not unexpected because there are many sources of fecal bacteria other than human fecal pollution, including stormwater and agricultural runoff. Several studies have suggested that FC levels cannot be used to predict the occurrence of human viruses (Fong *et al.* 2005). The lack of correlation between indicator organisms and human-associated markers observed in this study is not an unusual outcome (McQuaig *et al.* 2006). The *Bacteroides* marker was detected by PCR during combined sewer overflow events with *E. coli* levels of <235 CFU and there were instances in which the human-specific *Bacteroides* marker was absent when *E. coli* levels were in excess of EPA-recommended levels (McQuaig *et al.* 2006).

In this study, during no rain events, the OB ratio vs HpyV, OB 5 min% reduction vs HpyV, OB 5 min% reduction vs HB, and OB ratio vs HB each showed a correlation. All OB indicator criteria did not show any correlations with anthropogenic markers during rain events. So, OB readings alone are not a good indicator but using the OB 5 min% reduction and the OB ratio during no rain events gave a predictive relationship when combined with HB or HpyV. In application to Bayou Lafourche, using more than one human sewage marker would be beneficial to exploit each marker. In this study, using HpyV or HB with OB ratio or OB 5 min% reduction during no rain events would be the best indicator to predict anthropogenic contamination in Bayou Lafourche. The MS marker should not be used in Bayou Lafourche as an indicator because it was the least sensitive and specific marker of this study along with having no correlation to any OB criteria. The absence of any of the three markers means that either there is no human input in that source or that the input was too diluted for the methods to detect. The elimination of anthropogenic contamination will help restore water quality to its designated uses for drinking water, and primary and secondary contact recreation.

The practical application of this study shows the use of anthropogenic sewage markers to positively identify the fecal origin in the watershed, so the corrective measures can be taken to rectify the malfunctioning sewage treatment systems. This method can be used easily in the developing countries where fecal contamination of watershed is a very common occurrence.

CONCLUSIONS

The standard fecal indicator bacteria such as FC are currently used for water quality monitoring in Louisiana, but the difficulty is that they are found in both animal and anthropogenic sources. These fecal coliforms can vary in their ability to carry human pathogens therefore measuring FC levels insufficiently contributes to knowledge of human contamination in Bayou Lafourche. In this study, initial OB readings, FC, *E. coli*, and *M. smithii* were found not to be good indicators of human fecal contamination. So, we recommend using the two anthropogenic sewage markers HB and HPyV in Bayou Lafourche because it would be beneficial to exploit each marker. We also suggest using HpyV or HB with OB ratio or OB 5 min% reduction during no rain events as the best indicator to predict anthropogenic contamination in Bayou Lafourche. In future studies, long-term monitoring of Bayou Lafourche is recommended to monitor the water quality using anthropogenic sewage markers.

SUPPLEMENTARY MATERIAL

The Supplementary Material for this paper is available online at https://dx.doi.org/10.2166/aqua.2019.063.

ACKNOWLEDGEMENT

This project was financially supported by a grant from the Louisiana Department of Environmental Quality.

REFERENCES

Anderson, K. L., Whitlock, J. E. & Harwood, V. J. 2005 Persistence and differential survival of fecal indicator bacteria in subtropical water and sediments. *Applied Environmental Microbiology* **71**, 3041–3048.

APHA 1998 *Standard Methods for the Examination of Water and Wastewater*, 20th edn. American Public Health Association, Washington, D.C.

Belding, C. & Boopathy, R. 2018 Presence of antibiotic-resistant bacteria and antibiotic resistance genes in coastal recreational waters of southeast Louisiana, USA. *Journal of Water Supply: Research and Technology-AQUA* **67**, 800–809.

Bergeron, S., Boopathy, R., Nathaniel, R., Crobin, A. & LaFleur, G. 2015 Presence of antibiotic resistant bacteria and antibiotic resistance genes in raw source water and treated drinking water. *International Biodeterioration & Biodegradation* **102**, 370–374.

Bergeron, S., Brown, R., Homer, J., Rehage, S. & Boopathy, R. 2016 Presence of antibiotic resistance genes in different salinity gradients of freshwater to saltwater marshes in southeast Louisiana, USA. *International Biodeterioration & Biodegradation* **113**, 80–87.

Bergeron, S., Boopathy, R., Nathaniel, R., Corbin, A. & LaFleur, G. 2017 Presence of antibiotic resistance genes in raw source water of a drinking water treatment plant in a rural community of USA. *International Biodeterioration & Biodegradation* **124**, 3–9.

Bernhard, A. E. & Field, K. G. 2000 A PCR assay to discriminate human and ruminant feces on the basis of host differences in Bacteroides-Prevotella genes encoding 16S rRNA. *Applied Environmental Microbiology* **66**, 4571–4574.

Bird, K., Boopathy, R., Nathaniel, R. & LaFleur, G. 2019 Water pollution and observation of acquired antibiotic resistance in Bayou Lafourche, a major drinking water source in southeast Louisiana, USA. *Environmental Science and Pollution Research*. https://doi.org/10.1007/s11356-018-4008-5

Bower, P. A., Scopel, C. O., Jensen, E. T., Depas, M. M. & McLellan, S. L. 2005 Detection of genetic markers of fecal indicator bacteria in Lake Michigan and determination of their relationship to *Escherichia coli* densities using standard microbiological methods. *Applied Environmental Microbiology* **71**, 8305–8313.

Cao, Y., Griffith, J. F. & Weisberg, S. B. 2009 Evaluation of optical brightener photodecay characteristics for detection of human fecal contamination. *Water Research* **43**, 2273–2279.

Englande Jr., A. J., Jin, G. & Dufrechou, C. 2002 Microbial contamination in Lake Ponchartrain basin and best management practices on microbial contamination reduction. *Journal of Environmental Science and Health, Part A* **37** (9), 1765–1779. doi:10.1081/ESE-120015435.

Fong, T. T., Griffen, D. W. & Lipp, E. K. 2005 Molecular assays for targeting human and bovine enteric viruses in coastal waters and their application for library independent source tracking. *Applied Environmental Microbiology* **71**, 2070–2078.

Gentry-Shields, J., Rowny, J. G. & Stewart, J. R. 2012 Hubac and nifH source tracking markers display a relationship to land use but not rainfall. *Water Research* **46**, 6162–6174.

Girones, R., Ferrus, M. A., Alonso, J. L., Rodriguez-Manzano, J., Calgua, B., CorreaA, d. A., Hundesa, A., Carratala, A. & Bofill-Mas, S. 2010 Molecular detection of pathogens in water – the pros and cons of molecular techniques. *Water Research* **44**, 4325–4339.

Gordon, K. V., Brownell, M., Wang, S. Y., Lepo, J. E., Mott, J., Nathaniel, R., Kilgen, M., Hellein, K. N., Kennedy, E. &

Harwood, V. J. 2013 Relationship of human-associated microbial source tracking markers with enterococci in Gulf of Mexico waters. *Water Research* **47**, 996–1004.

Hagedorn, C., Saluta, M., Hassall, A. & Dickerson, J. 2005 Fluorometric detection of optical brighteners as an indicator of human sources of water pollution. Part III. Development as a source tracking methodology for urban waters. *Crop and Soil Environmental News*. Available from: https://www.sites.ext.vt.edu/newsletter-archive/cses/2005-11/part1.html (accessed 23 August 2019).

Hartel, P. G., McDonald, J. L., Gentit, L. C., Hemmings, S. N. J., Rodgers, K., Smith, K. A., Belcher, C. N., Kuntz, R. L., Rivera-Torres, Y., Otero, E. & Schroder, E. C. 2007 Improving fluorometry as a source tracking method to detect human fecal contamination. *Estuaries and Coasts* **30**, 551–561.

Hartel, P. G., Rodgers, K., Moody, G. L., Hemmings, S. N. J., Fisher, J. A. & McDonald, J. L. 2008 Combined targeted sampling and fluorometry to identify human fecal contamination in a freshwater creek. *Journal of Water and Health* **6**, 105–116.

Harwood, V. J., Levine, A. D., Scott, T. M., Chivukula, V., Lukasik, J., Farrah, S. R. & Rose, J. B. 2005 Validity of the indicator organism paradigm for pathogen reduction in reclaimed water and public health protection. *Applied Environmental Microbiology* **71**, 3163–3170.

Harwood, V. J., Brownell, M., Wang, S., Lepo, J., Ellender, R. D., Ajidahunc, A., Hellein, K. N., Kennedy, E., Ye, X. & Flood, C. 2009 *Validation and Field Testing of Library-Independent Microbial Source Tracking Methods in the Gulf of Mexico*. U.S. EPA for Gulf of Mexico Alliance 403 Regional Partnership Projects MX-96478707-0, pp. 1–29.

Hill, D. D., Owens, W. E. & Tchounwou, P. B. 2006 The impact of rainfall on fecal coliform bacteria in Bayou Dorcheat (North Louisiana). *International Journal of Environmental Research and Public Health* **3**, 114–117.

Kishinhi, S. S., Tchounwou, P. B. & Farah, I. O. 2013 Molecular approach to microbiological examination of water quality in the Grand Bay National Estuarine Research Reserve (NERR) in Mississippi USA. *Environmental Health Insights* **7**, 33–41.

LDEQ 2012 State of Louisiana Water Quality Management Plan. ERC LAC Title 33 Environmental Quality Part IX. Water Quality (updated November 2012) Subpart 1. Water Pollution Control Chapter 11. Surface Water Quality Standards §1123.Numerical Criteria and Designated Uses. Available from: www.deq.louisiana.gov/portal/Portals/0/.../title33/33v09-201211.doc

Leskinen, S. D., Brownell, M., Lim, D. V. & Harwood, V. J. 2010 Hollow fiber ultrafiltration and PCR detection of human-associated genetic markers from various types of surface water in Florida. *Applied Environmental Microbiology* **76**, 4116–4117.

McQuaig, S. M., Scott, T. M., Harwood, V. J., Farrah, S. R. & Lukasik, J. O. 2006 Detection of human-derived fecal pollution in environmental waters by use of a PCR-based human polyomavirus assay. *Applied Environmental Microbiology* **72**, 7567–7574.

McQuaig, S. M., Scott, T. M., Lukasi, J. O., Paul, J. H. & Harwood, V. J. 2009 Quantification of human polyomaviruses JC virus and BK virus by TaqMan quantitative PCR and comparison to other water quality indicators in water and fecal samples. *Applied Environmental Microbiology* **75**, 3379–3388.

McQuaig, S., Griffith, J. & Harwood, V. J. 2012 Association of fecal indicator bacteria with human viruses and microbial source tracking markers at coastal beaches impacted by nonpoint source pollution. *Applied Environmental Biology* **78**, 6423–6432.

Roslev, P. & Bukh, A. S. 2011 State of the art molecular markers for fecal pollution source tracking in water. *Applied Microbiology Biotechnology* **89**, 1341–1355.

Sauer, E. P., VandeWalle, J. L., Bootsma, M. J. & McLellan, S. L. 2011 Detection of the human specific *Bacteroides* genetic marker provides evidence of widespread sewage contamination of stormwater in the urban environment. *Water Research* **45**, 4081–4091.

Sidhu, J. P. S., Ahmed, W., Gernjak, W., Arya, R. I., McCarthy, D., Palmer, A., Kolotelo, P. & Toze, S. 2013 Sewage pollution in urban stormwater runoff as evident from the widespread presence of multiple microbial and chemical source tracking markers. *Science of the Total Environment* **463–464**, 488–496.

Ufnar, J. A., Wang, S. Y., Christiansen, J. M., Yampara-Iquise, H., Carson, C. A. & Ellender, R. D. 2006 Detection of the nifH gene of *Methanobrevibacter smithii*: a potential tool to identify sewage pollution in recreational waters. *Journal of Applied Microbiology* **101**, 44–52.

Walters, S. P., Gannon, V. P. & Field, K. G. 2007 Detection of *Bacteroidales* fecal indicators and the zoonotic pathogens *E. coli* 0157:H7, *Salmonella*, and *Campylobacter* in river water. *Environmental Science and Technology* **41**, 1856–1862.

First received 18 April 2019; accepted in revised form 29 August 2019. Available online 23 October 2019

Assessment of drinking water quality in regional New South Wales, Australia

S. T. M. L. D. Senevirathna, Andrea M. Goncher and Aaron Hollier

ABSTRACT

There is a substantial research gap relating to the quality of drinking water in regional Australia and identifying possible improvements. In particular, the quality of water available in public places (washing and drinking), such as water bubblers installed in regional parks, schools, rest areas and railway stations, is poorly investigated. This paper discusses the primary and secondary water quality of eight water distribution networks in New South Wales (NSW) regional towns. An analysis of a large number of drinking water samples (more than 11,000) identified that maintaining microbial water quality and the required free chlorine level (>0.2 mg/L) are challenging issues for regional water distribution networks. Sixty-three per cent of the samples collected from the water outlets available in public places of a regional town showed free chlorine levels of <0.2 mg/L, and 30% of samples showed positive results for total coliform. All heavy metal levels of the samples were within the safe level. Water temperature was identified as the most problematic secondary water quality parameter in public water bubblers. Stainless steel was the common material used in bubblers where surface temperatures exceeded 50 °C during summer. This study identifies possible design and operational modifications to improve regional drinking water quality and make public water bubblers more usable.

Key words | drinking water, microbial water quality, regional Australia, water bubblers

S. T. M. L. D. Senevirathna (corresponding author)
Aaron Hollier
CSU Engineering, Faculty of Business, Justice and Behavioural Sciences,
Charles Sturt University,
Panorama Avenue, Bathurst, NSW,
Australia
E-mail: *lsenevirathna@csu.edu.au*

Andrea M. Goncher
Department of Engineering Education,
Herbert Wertheim College of Engineering,
University of Florida,
Gainesville, FL,
USA

INTRODUCTION

The provision of safe and continuously available drinking water is an essential requirement for human health and well-being and community development (Sagehashi & Akiba 2017). There are growing concerns about the quality of water distributed by comparatively smaller distribution networks (Gunnarsdottir *et al.* 2017). This is especially the case in regional towns where the increment of endemic cases of waterborne diseases may not be reported, and health surveillance systems may not be able to detect potential issues (Hrudey *et al.* 2006). Recent literature clearly identifies a correlation between water quality and the capacity of water distribution networks. For example, more than 99% of large water supply systems operated in 23 countries out of 27 European Union member countries meet microbiological compliance targets (Gunnarsdottir *et al.* 2017). However, for small water distribution networks, only three countries out of 27 meet microbiological compliance targets (Gunnarsdottir *et al.* 2017).

There is limited documented research in Australia to demonstrate the quality of water distribution networks that operate in regional towns (Cretikos *et al.* 2009). A few publications suggest the possibility of drinking water contamination with heavy metals, such as copper, nickel and lead in regional areas, particularly at the domestic level (Alam *et al.* 2008; Harvey *et al.* 2016). Several barriers have been identified to maintaining a safe and reliable drinking

doi: 10.2166/aqua.2019.103

water supply in rural Australia, including limited water supply infrastructure (water treatment plants and distribution networks), limited resources and skilled staff, and inadequate catchment monitoring (Whelan & Willis 2007). Mosley *et al.* (2012) have observed a correlation between river flow levels and some water quality parameters such as salinity, total nitrogen, total phosphorus, chlorophyll and turbidity in several regional locations. Several researchers have argued that the discharge of organic and inorganic pollutants through various land uses in regional Australia makes the provision of safe drinking water more challenging (Thornton 1996; McKay & Moeller 2001). For example, various chemicals and discharge related to agricultural industries, such as fertilizers, pesticides, insecticides and animal hormones, ultimately ended up in water environments and advanced processes are required to eliminate these pollutants from environmental water. Also, these chemicals make significant changes to the aquatic environment, such as causing excessive growth of algae. Uncontrolled and unintentional discharge from the mining industry have also been identified as a main water polluter in regional Australia (Hilson 2006; Wright *et al.* 2011).

The consumer perception of drinking water quality in regional Australia is also under-investigated. A community organization that works with secondary school students in a regional Australian town, identified that public drinking water outlets, such as school water bubblers, are rarely used by consumers. Several reasons are suggested for an unwillingness to use public water bubblers, including bad taste, odour and colour, and a high temperature during summer. A similar situation has been reported in California where more than 50% of middle school students were unhappy with the cleanliness of their water bubblers, 48% were dissatisfied with the taste of the water provided, and 33% had experienced illness due to drinking this water (Patel *et al.* 2014). People's use of public water outlets, such as water bubblers installed in parks and playgrounds, is influenced by several factors, including gender, location, age, household income, marital status and education levels (Park *et al.* 2011). However, the establishment of reliable drinking water outlets in public places, particularly in schools, is essential for public health, well-being, convenience and environmental sustainability. For example, a lack of desirable drinking water sources in schools may lead to students becoming dehydrated and their cognitive performance being weakened (D'Anci *et al.* 2006).

The aim of this paper is to identify drinking water quality issues in regional New South Wales (NSW). Analytical results of more than 11,000 water samples collected (2009–2018) from eight regional water distribution networks are summarized in this paper. Also, the quality of tap water (washing and drinking) available in public places, such as schools and playgrounds in a regional town, is discussed and possible improvements are identified.

METHODS

Water quality parameters

Chemical (aluminium, barium, boron, cadmium, copper, iron, manganese, mercury, molybdenum, nickel, selenium, silver, total hardness, free chlorine and pH), physical (turbidity and true colour) and microbial (*Escherichia coli* and total coliforms) parameters were used to evaluate the quality of drinking water. Water samples were collected, preserved and handled for chemical parameters according to Standards Australia (1998) AS/NZS 5667.1:1998. Sampling for microbial water quality parameters was carried out according to AS 2031:2012 (ISO 19458:2006, MOD). Briefly, sterile and leak-proof 250 ml PET bottles were used to collect samples. Sodium thiosulphate was used to neutralize free chlorine in the water samples. Checks were done to make sure that the tap was clean and in working order before 2–3 minutes of the first flush was discarded and samples were collected. Also, the mouths of the taps ware disinfected with ethanol (70%).

Public water taps

Water samples collected from first flushes of washing and drinking water taps (water outlets) available in parks, playgrounds, public places (a railway station, the city information centre, etc.) and education institutions (schools, university and technical colleges) of a NSW regional town were analysed at CSU Engineering for microbial water quality, free chlorine and total chlorine. Field blanks were used to conform the sample contamination that is negligible during the collection procedure. Prior to analysing the samples, all instruments were calibrated using the procedure given by the manufacturer of the instrument. The USEPA DPD (*N,N*-diethyl-*p*-phenylenediamine) method was used with a Hach DR 1900 Spectrophotometer to determine the total and free chlorine levels in the water samples. The Coliscan Easygel method (https://www.micrologylabs.com/page/93/Coliscan-Easygel) was used to determine the levels of total coliforms and *E. coli* in the collected samples. In this method, collected samples were immediately transported to CSU Engineering. Five millilitres of water was sterilely transferred from sample containers into the bottles of Coliscan Easygel. The bottles were swirled to distribute the inoculum and then the mixtures were gently poured into the labelled Petri dishes. The number of colonies was counted after 24 hours of incubation at 35 °C.

Data extraction from the NSW Health drinking water quality database

All water samples reported in this paper were analysed in the CSU Engineering water lab or laboratories accredited to perform such analyses. Procedures for water samples tested outside CSU Engineering used analytical methods approved by the EPA, and those results were directly extracted from the NSW Health (https://www.health.nsw.gov.au) drinking water quality database. Water distribution networks in the central tablelands of NSW were selected for the study (Figure 1). Ten years of water quality data for each network were extracted from the database (2009–2018). The locations of the targeted water distribution networks were within 200 km of the CSU Engineering School in Bathurst. These networks are designed to serve a total regional population of more than 150,000. The number of people who use each distribution network varies from about 5,000 to 38,000. Extracted data were carefully checked, and unrealistic water quality data were identified and eliminated from the analysis. All graphical and statistical computations were made using SPSS (version 25).

RESULTS AND DISCUSSION

Metal levels in drinking water

The metal levels measured in water samples collected from eight water distribution networks are summarized in Table 1. None of the water samples collected from small water distribution networks (<10,000 served population) exceeded any guideline values. All samples collected from other water distribution networks satisfied metal guidelines, except aluminium and iron. Two hundred and forty-four water samples were analysed from water distribution

Councils Map of New South Wales	Council
	Number of water distribution networks: 8 Largest distribution network: serves over 38,000 people Smallest distribution network: serves less than 6,000 people Location of distribution networks: within 200 km of CSU Engineering

Figure 1 | Water distribution networks investigated in this study.

Table 1 | Summary of maximum metal levels measured in samples collected from targeted water distribution networks for the last 10 years

Metal	Guideline value (mg/L)	Capacity (served population) 30,000–40,000	20,000–30,000	10,000–20,000	<10,000
Number of samples		244	124	234	14
Aluminium	0.2 (a)	1.16 (1%)	1.3 (18%)	0.59 (5%)	0.18
Barium	2 (h)	0.068	0.034	0.076	0.01
Boron	4 (h)	0.05	1.4	0.1	0.05
Cadmium	0.002 (h)	0.0014	0.00025	0.0005	0.00025
Copper	2 (h)	1.668	0.193	0.108	0.185
Iron	0.3 (a)	0.22	0.7 (4%)	0.16	0.02
Manganese	0.5 (h)	0.254	0.036	0.072	0.047
Mercury	0.001 (h)	0.0005	0.0005	0.0002	0.0001
Molybdenum	0.05 (h)	0.005	0.0025	0.005	0.0025
Nickel	0.02 (h)	0.01	0.02	0.01	0.005
Selenium	0.01 (h)	0.005	0.009	0.003	0.001
Silver	0.1 (h)	0.005	0.005	0.005	0.001

(a), aesthetic guideline value; (h), the health-based guideline value, the bracket percentage and the percentage exceeding guideline value.

networks with a served population of over 30,000 for metal parameters, and only two samples exceeded the aluminium aesthetic guideline value of 0.2 mg/L. One hundred and twenty-four samples were analysed from the distribution networks with a served population between 20,000 and 30,000 where 22 samples exceeded the aesthetic guideline value for aluminium (the maximum reported level was 1.3 mg/L) and five samples exceeded the aesthetic guideline value for iron (0.3 mg/L). Iron is often elevated in groundwater sources (mainly by weathering of iron-bearing minerals), and aluminium-based coagulant is most commonly used in drinking water treatment and may occasionally result in elevated aluminium concentrations. It is worth noting that the overall metal concentration of the water distribution networks, particularly health-related harmful heavy metal levels, such as arsenic, cadmium and lead, were well within the allowable ranges.

Previous studies have suggested the possible contamination of groundwater, river water and animal tissues with heavy metal by the mining industry in regional NSW. Researchers have measured the high level of cadmium (up to 0.016 mg/L), lead (up to 0.142 mg/L) and zinc (up to 0.025 mg/L) in samples collected from an NSW river (Norris 1986). The main sources of the metal contaminant of this river were seeps and previously deposited particulate material from mine workings. Low *et al.* (2005) have reported the mobility of zinc, cadmium and lead from anthropogenically contaminated soil to plants in western NSW.

Compliance of water quality parameters

The compliance results of selected water quality parameters for eight water distribution networks are shown in Table 2. Maintaining the required free chlorine level in the distribution network has been highlighted as a major operational issue in all water distribution networks. Also, the total hardness level measured in some samples was higher than the aesthetic guideline value. The hardness of water can be caused by dissolved metals, particularly divalent cations of calcium and magnesium, and hard water can lead to scaling in pipes. There is no health-based guidelines value for hardness. Highly increased intake of magnesium salts may cause a change in bowel habits (diarrhoea). Daily consumption of hard water may result in a comparatively higher level of magnesium in coronary arteries, bone and myocardial tissue. Several studies have reported the relationship between the hardness of drinking water and cardiovascular disease (Sengupta *et al.* 2013). Some

Table 2 | Commonly identified water quality parameters, which failed to meet guideline values on some occasions

Capacity (served population)	Characteristic	Guideline	Mean	SD	Min	Max	Number of samples	Meeting guideline value (%)
<10,000	Total hardness (mg/L)[a]	200	144	115	24	523	76	88
	True colour (HU)[a]	15	4	8	1	60	76	96
	E. coli (cfu/100 mL)[b]	0	0	0	0	1	1,757	100
	Free chlorine (mg/L)	0.2–5	0.6	0.5	0	7.2	1,515	76.2
	pH[a]	6.5–8.5	7.9	6.9	6.9	8.7	389	99.2
	Total coliforms (cfu/100 mL)[b]	0	4	22	0	202	1,757	92
	Turbidity (NTU)[a]	5	0.3	0.4	0	6	375	100
10,000–20,000	Total hardness (mg/L)[a]	200	110	32	38	257	124	99
	True colour (HU)[a]	15	1	1	1	7	124	100
	E. coli (cfu/100 mL)[b]	0	0	0	0	12	3,163	100
	Free chlorine (mg/L)	0.2–5	0.7	0.7	0	6	3,054	77
	pH[a]	6.5–8.5	7.7	0.3	5	8.7	2,938	99
	Total coliforms (cfu/100 mL)[b]	0	1	11	0	202	3,163	95
	Turbidity (NTU)[a]	5	0.6	1.9	0	12	2,891	99
20,000–30,000	Total hardness (mg/L)[a]	200	59	30	2	269	376	97
	True colour (HU)[a]	15	3	3	0	19	376	99
	E. coli (cfu/100 mL)[b]	0	0	0	0	12	3,896	99
	Free chlorine (mg/L)	0.2–5	0.8	1.1	0	5.5	3,553	67
	pH[a]	6.5–8.5	7.4	0.7	6.0	8.9	2,466	98
	Total coliforms (cfu/100 mL)[b]	0	4	23	0	202	3,896	90
	Turbidity (NTU)[a]	5	1	1	0	18	1,445	100
30,000–40,000	Total hardness (mg/L)[a]	200	97	37	43	230	153	99
	True colour (HU)[a]	15	2	4	1	41	153	99
	E. coli (cfu/100 mL)[b]	0	0	5	0	20	1,707	99
	Free chlorine (mg/L)	0.2–5	0.4	0.5	0	6.2	1,653	41
	pH[a]	6.5–8.5	7.9	0.4	6.7	9.1	1,534	95
	Temperature (°C)[a]	30	17	5	7	32	173	99
	Total coliforms (cfu/100 mL)[b]	0	1	7	0	200	1,707	93
	Turbidity (NTU)[a]	5	0.2	0.1	0	4	1,479	100

[a]Aesthetic guidelines.
[b]Health-based guidelines.

researchers have suggested a possible association between the risk of gastric cancer and the levels of calcium and magnesium (Yang *et al.* 1998). However, less than 5% of total samples failed to meet the required guideline, and there is insufficient evidence to say that there is hard water toxicity in the targeted areas.

To inactivate microorganisms, chlorination was used in all eight water supply systems investigated in this study. This process has proved its efficiency to destroy disease-causing pathogens, such as bacteria, viruses and protozoans. When chlorine is mixed with water, it is dissociated into hypochlorous acid and hydrochloric acid, which destroy microorganism by oxidating the organic molecules. Trihalomethanes and haloacetic acids are identified as common disinfection by-products, particularly, when the level of natural organic matter is comparatively high in the water.

This paper reports the free chlorine levels measured in more than 11,000 samples collected from various endpoints of eight water distribution networks for the past 10 years. Compared with other water quality parameters, free chlorine levels showed the lowest compliance with the minimum recommended free chlorine residuals. The Australian Drinking Water Guidelines (ADWG) level for free chlorine is 0.5 mg/L. However, a recommendation was made by the NSW Department of Primary Industries Office of Water in consultation with the NSW Health in 2014 to maintain the free chlorine residual level at more than 0.2 mg/L. Among 1,653 water samples collected from large water distribution networks (over 30,000 of the served population), only 678 (41%) samples showed more than 0.2 mg/L of free chlorine. Comparatively better free chlorine levels were reported from water distribution networks with a served population of less than 20,000 where more than 75% of the samples satisfied the minimum recommended free chlorine residual (>0.2 mg/L).

If the variation of free chlorine concentrations in water distribution systems is known, scheduling the re-chlorination is not a big challenge (Kim *et al.* 2014). The destruction of free chlorine in a distribution network is multi-factorial (Chowdhury *et al.* 2009). Previous studies have reported several important factors that impact free chlorine levels in a distribution network, such as temperature (Wang *et al.* 2019), water age (Xu *et al.* 2018), piping material, corrosion and pH (Zhang *et al.* 1992), hydraulic conditions and instantaneous mixing (Mompremier *et al.* 2015), dosage and microbial activity (Zhang *et al.* 2004). Bacteria that detach from pipe biofilm and enter treated water may be inactivated if the disinfectant residual is sufficient. In this case, there is no negative impact on the consumer even though bacterial regrowth has occurred on the pipe surface or recontamination has occurred. On the other hand, if the loss of the disinfectant residual in the treated water is large, as would be true for long water residence times, bacteria that are released from the pipe surface or by recontamination will not be inactivated quickly. In this case, indicator bacteria (e.g. coliforms, *E. coli* and enterococci) may be detected in the treated water and the consumer may be negatively impacted. Comparatively longer retention times and hydraulic conditions are identified as controlling factors for free chlorine levels in large water distribution networks operated in regional NSW.

Water quality in public water taps

Table 3 shows microbial water quality parameters and residual chorine levels measured in public water taps and bubblers installed in playgrounds, parks, public places and educational institutions in a regional town in NSW. These samples were collected from the first flush of each water outlet. A close positive correlation was observed between the free chlorine level and the detection of total coliforms. The safest microbial water quality was detected in parks and playgrounds where 87.5% of the samples showed safe free chlorine levels and only 10% of the samples showed positive results for total coliform. None of the water samples collected from public places, such as the railway station and city information centre, showed safe residual chorine levels, and 65% of the samples showed positive results for total coliforms. About 50% of public taps and water bubblers installed in educational institutions and investigated in this study showed positive results for total coliforms, and just 18% showed safe free chlorine levels. None of the samples collected from public water outlets indicated the potential contamination of *E. coli*.

The variation of the free chlorine level in different locations can be explained by the usage of water outlets. For example, public water outlets installed in parks and playgrounds are frequently used and water retention times in pipes are lower and free chlorine levels are comparatively higher.

It is worth noting that local authorities have already identified the issues with free chlorine levels in the town water distribution network and a significant improvement has been observed in free chlorine levels, particularly since 2017. Figure 2 shows the vitiation of town water free

Table 3 | Water quality parameters measured in public water taps and bubblers in a regional town in NSW

Parameter	Parks and playgrounds	Public places	Educational institutions
No. of water outlets	8	5	11
Total Cl mg/L (mean and SD)	0.90 (0.55)	0.17 (0.11)	0.17 (0.19)
Residual Cl mg/L (mean and SD)	0.71 (0.46)	0.03 (0.03)	0.09 (0.15)
Meeting guideline value (%) 0.2 mg/L of RCl (%)	87.5%	0%	18%
Detection of total coliforms (%)	10%	65%	50%
Detection of *E. coli* (%)	0	0	0

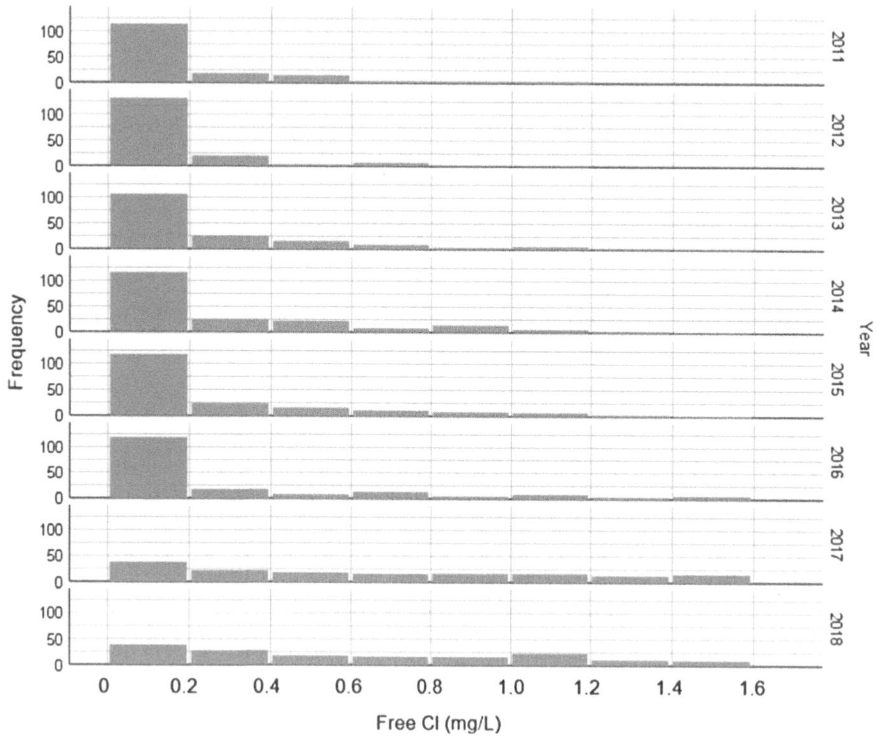

Figure 2 | Vitiation of free chlorine levels with time.

chlorine levels for the last eight years (from 2011 to 2018). Only 26.5% of samples analysed from 2011 to 2012 achieved the minimum recommended free chlorine residual of 0.2 mg/L. In 2014, the Department of Primary Industries Office of Water in consultation with the NSW Health recommended maintaining a free chlorine residual. Improvements were observed in 2015 and 2016 where more than 35% of the samples satisfied the guideline value. The free chlorine level improved significantly in 2017 and 2018 when more than 75% of analysed samples showed more than 0.2 mg/L of free chlorine. However, improvement is still needed and modelling of water distribution networks for free chlorine may be a promising tool to identify potential operational and maintenance measures to maintain free chlorine levels in entire distribution networks.

Temperature

Parents, teachers and community programme leaders of local schools have reported that school water bubblers are rarely used by students and the temperature has been identified as the main issue. Figure 3 shows the variation of daily average, atmospheric temperature, tap water temperature and steel temperature with time. Atmospheric temperature readings were extracted from the local airport meteorology station. Several days in summer 2018 reached more than 32 °C average atmospheric temperature.

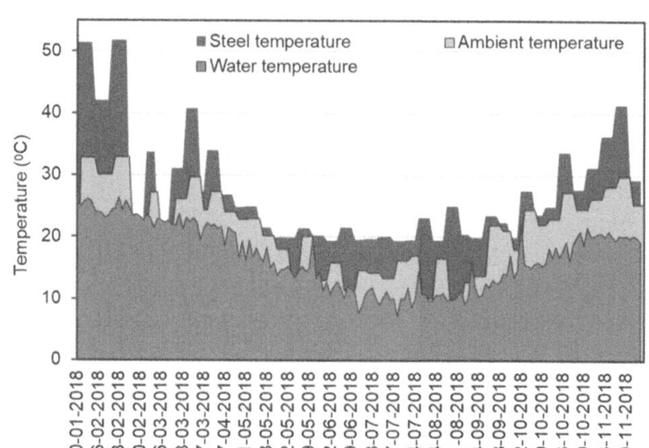

Figure 3 | Variation of atmospheric temperature, tap water temperature and steel temperature with time.

However, the tap water temperature was less than 30 °C throughout the year. Among 173 water temperature readings measured in the last 10 years, only one reading has exceeded the guideline value. The mean water temperature of the town water for the last 10 years is 16.9 °C (SD 5.2) with a minimum of 6.7 °C.

It was noticed that the frames, basin and casing of school bubblers were made with stainless steel and most of them were installed without a cover. Figure 3 indicates the temperature of uncovered stainless steel bubblers may exceed 50 °C in summer. This problem could be easily solved by shading uncovered bubblers or relocating them to more suitable places. Most of the bubblers investigated in this study were poorly maintained, and some of them were not in a usable condition. The selection of correct material (stainless steel, ceramic, carbon, etc.), the correct place of installation, correct design (robust, with treatment facility) and the implementation of proper operational and maintenance plans will make public water bubblers more usable.

Potential improvements

Major challenges identified in regional water distribution networks can be fixed by modifying the design and operational protocols. A calibrated hydraulic model of each water distribution network should be combined with available chlorine decay models to understand the variation of free chlorine levels in the entire water distribution network. This knowledge is helpful to identify problematic areas in the network and develop a re-chlorination schedule to overcome the issue. Modifications are needed to the guidelines relating to the design and installation of public water bubblers. In particular, serious consideration needs to be given to the bubbler basin material, and the place of installation and a regular maintenance programme should be seriously considered. Water outlets in public places for human consumption have been constructed to serve drinking and washing requirements. Design considerations include ergonomics for a range of ages, the angle of flow, the rate of flow, materials used and run-off disposal. Design solutions have been integrated to accommodate the demographics they are serving.

Role of education and public participation

Our review of the literature and study results support the importance of water quality, effects on consumption and the impacts on public health. Water outlets available for use in educational institutions, parks and playground and other public spaces can affect the health of consumers, as well as have secondary effects on consumers, e.g. lack of concentration, from dehydration. Educators, policy-makers, plant managers, researchers and citizens can all influence how water consumption is perceived and undertaken by the public.

Educators and administrators, or those who have authority in the development and enforcement of the policy, should consider the following points to encourage water consumption in educational institutions by students during school hours.

- Teachers, coaches and community programme leaders should encourage students to bring reusable water containers and fill up regularly from bubblers throughout the day; especially before and after activities, or conditions, that lead to dehydration.
- Linking water consumption via reusable containers and water bubblers to pro-environmental behaviours, e.g. recycling, reduced use of resources, provides students with a more holistic view, regarding their role in sustainability concepts.
- Steel bubblers should not be installed for public consumption without shade to help regulate exposure to high temperatures.

Individuals, or groups, who have influence or authority related to the water distribution networks and/or their regional councils should consider the following points to positively impact public water consumption and perception.

- Every water distribution network should have an accurate model to predict the free chlorine level at any point and any operational conditions.
- The accuracy of the model should be checked and improved frequently by incorporating new knowledge and field observations.
- A systematic programme should be implemented to maintain public water outlets.

Citizens in a general capacity, e.g. individuals, parents, should consider the following point before consumption of water from a public outlet.

- If you think a water outlet has not been used for some time, it is recommended to waste the first flush (1–2 minutes), before drinking, or filling up a reusable container.

There are a number of similarities between water distribution networks investigated in this study and operated in developing countries, such as chlorination to disinfect water, comparatively low connection density and limited skilled labours to operate networks. The results of this paper highlight the importance of free chlorine modelling in water distribution networks operated in developing countries. If the water leakages and possible contaminations are comparatively higher in a network, a safe free chlorine level throughout the network should be well maintained.

CONCLUSION

Analytical results of more than 11,000 water samples collected from various water distribution networks suggest that drinking water quality in regional NSW is generally good. Maintaining the required free chlorine level in entire distribution networks is identified as the major challenge in regional water supply systems. Water taps installed in public places are rarely used by the regional population, and most of them failed to meet drinking water quality guidelines. Almost all water bubblers found in this study were made of stainless steel. The surface temperature of steel bubblers in summer may reach more than 50 °C, which makes them unpopular among users, particularly among school children. Basic modifications in design and operational procedure could easily solve most issues in regional drinking water systems.

ACKNOWLEDGEMENT

The authors acknowledge the NSW Health for granting permission to use its NSW drinking water quality database and for helping to finalise this manuscript.

REFERENCES

Alam, N., Corbett, S. J. & Ptolemy, H. C. 2008 Environmental health risk assessment of nickel contamination of drinking water in a country town in NSW. *New South Wales Public Health Bulletin* **19** (10), 170–173.

Chowdhury, S., Champagne, P. & McLellan, P. J. 2009 Models for predicting disinfection byproduct (DBP) formation in drinking waters: a chronological review. *Science of The Total Environment* **407** (14), 4189–4206.

Cretikos, M., Byleveld, P., Durrheim, D. N., Porigneaux, P., Merritt, T. & Leask, S. 2009 Supply system factors associated with microbiological drinking water safety in regional New South Wales, Australia, 2001–2007. *Journal of Water and Health* **8** (2), 257–268.

D'Anci, K. E., Constant, F. & Rosenberg, I. H. 2006 Hydration and cognitive function in children. *Nutrition Reviews* **64** (10), 457–464.

Gunnarsdottir, M. J., Persson, K. M., Andradottir, H. O. & Gardarsson, S. M. 2017 Status of small water supplies in the Nordic countries: characteristics, water quality and challenges. *International Journal of Hygiene and Environmental Health* **220** (8), 1309–1317.

Harvey, P. J., Handley, H. K. & Taylor, M. P. 2016 Widespread copper and lead contamination of household drinking water, New South Wales, Australia. *Environmental Research* **151**, 275–285.

Hilson, G. 2006 Abatement of mercury pollution in the small-scale gold mining industry: restructuring the policy and research agendas. *Science of the Total Environment* **362** (1), 1–14.

Hrudey, S. E., Hrudey, E. J. & Pollard, S. J. 2006 Risk management for assuring safe drinking water. *Environment International* **32** (8), 948–957.

Kim, H., Koo, J. & Kim, S. 2014 A general framework of chlorine decay modeling at a pilot-scale water distribution system. *Journal of Water Supply: Research and Technology-AQUA* **64** (5), 543–557.

Low, C., Gore, D., Taylor, M. P. & Louie, H. 2005 Black bluebush and old man saltbush biogeochemistry on a contaminated site. In: *Regolith 2005 - Ten Years of CRC LEME*. I. Roach (ed.). Cooperative Research Centre for Landscape Environments and Mineral Exploration, Broken Hill, New South Wales, Australia, pp. 213–218.

McKay, J. & Moeller, A. 2001 Is risk associated with drinking water in Australia of significant concern to justify mandatory regulation? *Environmental Management* **28** (4), 469–481.

Mompremier, R., Pelletier, G., Fuentes Mariles, Ó. A. & Ghebremichael, K. 2015 Impact of incomplete mixing in the prediction of chlorine residuals in municipal water distribution systems. *Journal of Water Supply: Research and Technology-AQUA* **64** (8), 904–914.

Mosley, L. M., Zammit, B., Leyden, E., Heneker, T. M., Hipsey, M. R., Skinner, D. & Aldridge, K. T. 2012 The impact of extreme low flows on the water quality of the Lower Murray

River and Lakes (South Australia). *Water Resources Management* **26** (13), 3923–3946.

Norris, R. 1986 Mine waste pollution of the Molonglo River, New South Wales and the Australian Capital Territory: effectiveness of remedial works at Captains Flat mining area. *Marine and Freshwater Research* **37** (2), 147–157.

Park, S., Sherry, B., Wethington, H. & Pan, L. 2011 Use of parks or playgrounds: reported access to drinking water fountains among US adults, 2009. *Journal of Public Health* **34** (1), 65–72.

Patel, A. I., Bogart, L. M., Klein, D. J., Burt, C., Uyeda, K. E., Hawes-Dawson, J. & Schuster, M. A. 2014 Middle school student attitudes about school drinking fountains and water intake. *Academic Pediatrics* **14** (5), 471–477.

Sagehashi, M. & Akiba, M. 2017 Questionnaire survey on water consumption and preparedness for water outages at intensive care homes for the elderly in Japan. *Journal of Water Supply: Research and Technology-AQUA* **67** (2), 176–191.

Sengupta, P., Chaudhuri, P. & Bhattacharya, K. 2013 Screening obesity by direct and derived anthropometric indices with evaluation of physical efficiency among female college students of Kolkata. *Annual Medical Health Science Research* **3** (4), 517–522.

Standards Australia 1998 AS/NZS 5667.1:1998, Water quality: sampling. Guidance on the design of sampling programs, sampling techniques and the preservation and handling of samples, Standards Australia, Homebush, Australia.

Thornton, I. 1996 Impacts of mining on the environment; some local, regional and global issues. *Applied Geochemistry* **11** (1–2), 355–361.

Wang, Y.-H., Wu, Y.-H., Du, Y., Li, Q., Cong, Y., Huo, Z.-Y., Chen, Z., Yang, H.-W., Liu, S.-M. & Hu, H.-Y. 2019 Quantifying chlorine-reactive substances to establish a chlorine decay model of reclaimed water using chemical chlorine demands. *Chemical Engineering Journal* **356**, 791–798.

Whelan, J. & Willis, K. 2007 Problems with provision: barriers to drinking water quality and public health in rural Tasmania, Australia. *Rural and Remote Health* **7** (3), 1–9.

Wright, I. A., Wright, S., Graham, K. & Burgin, S. 2011 Environmental protection and management: a water pollution case study within the Greater Blue Mountains World Heritage Area, Australia. *Land Use Policy* **28** (1), 353–360.

Xu, J., Huang, C., Shi, X., Dong, S., Yuan, B. & Nguyen, T. H. 2018 Role of drinking water biofilms on residual chlorine decay and trihalomethane formation: an experimental and modeling study. *Science of The Total Environment* **642**, 516–525.

Yang, C. Y., Cheng, M. F., Tsai, S. S. & Hsieh, Y. L. 1998 Calcium, magnesium, and nitrate in drinking water and gastric cancer mortality. *Japanese Journal of Cancer Research* **89** (2), 124–130.

Zhang, G., Kiene, L., Wable, O., Chan, U. & Duguet, J. 1992 Modelling of chlorine residual in the water distribution network of Macao. *Environmental Technology* **13** (10), 937–946.

Zhang, W., Miller, C. T. & DiGiano, F. A. 2004 Bacterial regrowth model for water distribution systems incorporating alternating split-operator solution technique. *Journal of Environmental Engineering* **130** (9), 932–941.

First received 10 July 2019; accepted in revised form 6 October 2019. Available online 13 November 2019

Stress response and toxicity studies on zebrafish exposed to endosulfan and imidacloprid present in water

Barzah Muazzam, Kashif Munawar, Imtiaz Ahmad Khan, Sarwat Jahan, Mazhar Iqbal, Muhammad Rafique Asi, Abida Farooqi, Adila Nazli, Ijaz Hussain and Mazhar Iqbal Zafar IWA

ABSTRACT

Fish and other aquatic biota are hampered by mixtures of pesticides which pollute natural water through agricultural runoff and other sources. Toxicity of combined exposures of endosulfan and imidacloprid on zebrafish in terms of oxidative stress and deoxyribonucleic acid (DNA) damage in liver and histological alterations in gills and muscles was investigated. Zebrafish were exposed to three different sub-lethal concentrations of endosulfan and imidacloprid along with control selected for each treatment for 21 days: control treatment (CT), treatment 1 (T1), treatment 2 (T2) and treatment 3 (T3). T1, T2 and T3 groups were exposed to 0.1, 0.5 and 1 μg/L of endosulfan, respectively, while imidacloprid concentration was maintained at 1 ppm in all three treatments. Oxidative stress was evaluated by measuring levels of catalase (CAT), superoxide dismutase (SOD) and malondialdehyde (MDA). Comet assay was applied to measure degree of DNA damage. Dose- and time-dependent decrease in SOD and CAT activity was observed after 21 days of exposure while low concentrations of pesticides induced SOD and CAT activities after early exposure to reduce the oxidative stress. MDA content was found to be increased in T3 having high concentrations of pesticides. Substantial increase in DNA damage was noticed after 21 days' exposure to pesticides. Significant morphological changes were observed in gills relative to muscles.

Key words | antioxidant response, combined toxicity, *Denio rerio*, endosulfan, histopathological changes, imidacloprid

Barzah Muazzam
Mazhar Iqbal
Abida Farooqi
Mazhar Iqbal Zafar IWA (corresponding author)
Department of Environmental Sciences,
Faculty of Biological Sciences,
Quaid-i-Azam University,
Islamabad 45320, Pakistan
E-mail: *mzafar@qau.edu.pk*

Kashif Munawar
Plant Protection Department, College of Food and Agricultural Sciences,
King Saud University,
Riyadh, Kingdom of Saudi Arabia

Imtiaz Ahmad Khan
Department of Pathology, Faculty of Veterinary Sciences,
PMAS-Arid Agriculture University,
Rawalpindi, Pakistan

Sarwat Jahan
Department of Animal Sciences, Faculty of Biological Sciences,
Quaid-i-Azam University,
Islamabad, Pakistan

Muhammad Rafique Asi
Nuclear Institute for Agriculture and Biology (NIAB),
Faisalabad 38000, Pakistan

Adila Nazli
Department of Pharmacy, Faculty of Biological Sciences,
Quaid-I-Azam University,
Islamabad, Pakistan

Ijaz Hussain
Department of Statistics,
Quaid-i-Azam University,
Islamabad, Pakistan

INTRODUCTION

Water is an indispensable and precious component of human environments because it is used for many purposes such as drinking, irrigation, breeding of fish and other freshwater ecosystems and also as the main source of energy. Pesticide usage has become a necessary evil in developing countries and has increased several-fold where agriculture is anticipated to be the backbone of the economy. During the past few decades, agrochemicals have been widely used in most agricultural sectors for enhancing crop yield and improving the quality of the product (Doruchowski *et al.* 2017). Consequently, extensive application of pesticides poses potential risks to the biodiversity of freshwater aquatic

doi: 10.2166/aqua.2019.077

environments because of their bioaccumulation and intrinsic toxicity (Cui *et al.* 2015). Agricultural runoff, leaching to surface and groundwater and spray drift are assumed as the major entry routes which contribute to the addition of pesticides in water bodies (Zhang *et al.* 2018) that provide support to the biological integrity of these waters. Consumption of such water sources which are contaminated with pathogens and impermissible limits of chemical toxins cause acute and chronic diseases and are also a major cause of death (Elimelech 2006; Banihashemi *et al.* 2014). In developing countries, deterioration of water quality through different sources remains an important challenge (Sohel *et al.* 2003; Sadiq *et al.* 2010).

When pesticides are applied with the prevailing application methods used for crop protection, however, it is inevitable that a proportion of sprayed pesticides will reach such untargeted edge-of-field surface waters. Since aquatic ecosystems contain species (e.g., fish) related to the target organisms of pesticides, unintended repercussions are to be expected when these ecosystems become contaminated (Liess *et al.* 2005). Recently, evaluation of the potential adverse effects of pesticide stress on sensitive non-target aquatic organisms in aquatic ecosystems has been the subject of worldwide concern (Matozzo *et al.* 2018). One of the major dilemmas in environmental risk assessment is that pesticide contamination is frequently detected as a mixture of different substances rather than individual chemicals in the aquatic environment (Schreiner *et al.* 2016). The worldwide usage of pesticides is about 2 million tonnes/year; about 45% of pesticides are used in Europe, USA 25% and the rest of the world consumes 25% (De *et al.* 2014). Pakistan is the second largest consumer of pesticides among the south Asian countries.

Xenobiotics like pesticides can induce reactive oxygen species (ROS) in organisms. These excessive ROS may cause oxidative damage to deoxyribonucleic acid (DNA), i.e., breakage of DNA strands. Oxidative stress may cause modifications in antioxidants' enzyme systems. These enzymes include catalase (CAT), superoxide dismutase (SOD) and glutathione peroxidase (GPx) (Oruc *et al.* 2004). Excessive production of ROS causes lipid peroxidation that ultimately leads to malondialdehyde (MDA) formation, resulting in toxic stress to cells (Ge *et al.* 2015). Genotoxicity is investigated by measuring the alterations in DNA or mRNA that may be in the form of genome damage or genome mutations. For assessing genotoxicity, the most promising technique applied to examine DNA damage in a single cell is single-cell gel electrophoresis (SCGE) assay (known as comet assay) (Ritter & Knebel 2009).

Endosulfan (6,7,8,9,10,10-hexachloro-1,5,5a,6,9,9a-hexahydro-6,9-methano-2,4,3-benzo-dioxathiepin-3-oxide) is a non-systemic organochlorine insecticide, prevalently used as a broad-spectrum throughout the world. It is the most persistent and hazardous agent of all halogenated hydrocarbons, being highly resistant to microbial bioremediation and persisting as a xenobiotic. Oxidative stress is an important indicator of toxicity that is induced by endosulfan exposure to fish. Endosulfan can also affect many aquatic non-target organisms, particularly fish, for instance, modulation of antioxidant systems in fish liver can be induced (Shao *et al.* 2012). Owing to its toxic effects and persistence, endosulfan was considered as a new member of the persistent organic pollutants (POPs) group during the Stockholm Convention (POPRC 2010). Moreover, endosulfan is known as an endocrine disruptor and has the ability to bioaccumulate and biomagnify in food chains (US EPA 2002). Generally, it is found to be very toxic to fish as it is reported that concentrations of 0.50 to 0.75 μg/L have negative impacts on fish species (Alamdar *et al.* 2014) and cause chronic effects including genotoxic and developmental effects.

The insecticide imidacloprid [1-(6-chloro-3-pyridylmethyl)-N-nitroimidazolidin-2-ylideneaimine] is a chloronicotinyl nitroguanidine and due to its systemic nature, it is ranked as the second most widely used pesticide globally (c. 70 crops over 100 countries) (Bonmatin *et al.* 2003). Environmental studies carried out on imidacloprid have indicated that it can be detected in the soil and can be carried by storm rain and runoff through which it can move into irrigation canals, streams, rivers, lakes and can leach into the groundwater, thus becoming a potential risk to aquatic life. US EPA (United States Environmental Protection Agency) has classified imidacloprid as highly toxic to aquatic invertebrates. Imidacloprid is considered as a potential groundwater and surface water contaminant, because of its leaching potential (PAN Pesticides database 2006). Acute exposure of imidacloprid and curzate on *Labeo rohita* induced increased mortality, quick opercular and fin movement, loss of equilibrium and lethargy (Desai & Parikh 2014). Pesticides may cause acute and chronic poisoning of fish and damage their dynamic organs, causing various biochemical alterations, reproductive impairment and morphological deformities (Velasco-Santamaría *et al.* 2011; Wu *et al.* 2018).

For the present study, zebrafish (*Denio rerio*) were chosen because they are prominent biological model organisms for ecotoxicological research and are recommended by the Organization for Economic Co-operation and Development (OECD 2013) and International Organization for Standardization (ISO) as a standard test organism (ISO 1996). Zebrafish are well-accepted aquatic vertebrate models for investigating the toxicity of chemicals in aquatic environments and associated health risks. Hence, they are preferred for acute and chronic toxicity tests in the laboratory owing to many favourable characteristics, such as small body size, easy husbandry, morphology, short test cycle and rapid development. Due to these unique advantages, zebrafish can be used as a bioindicator to assess the pesticidal pollution in aquatic environments.

The liver is the organ mainly affected by contaminants as it is concerned with detoxification and biotransformation (Camargo & Martinez 2007). The gills are important organs for fish and also the first target organ to be in contact with pollutants after exposure. Therefore, pesticides accumulate in gills and their concentrations reflect the pesticide level in water where fish species reside (Napit 2013). Gills are respiration sites that are involved in osmoregulation (Fernandez & Mazon 2003). Due to their large surface area they are susceptible to pollutants in water, and are therefore considered as an appropriate water contamination indicator. Histopathology provides knowledge about the functionality of different organs. For this purpose, histological examination is a sensitive parameter and is critical for describing alterations in cells of target organs, like the liver and gills (Dutta 1996). The zebrafish is an important bioindicator organism for monitoring the contamination of freshwater resources since it has a genetic similarity to humans. As a result of illegal use of endosulfan and excessive use of imidacloprid in agricultural fields entering aquatic systems above maximum permissible limits, it is presumed that the zebrafish population is declining in developing countries (e.g., Pakistan) and may result in its extinction.

Literature shows that, according to the OECD test guidelines, zebrafish is found to be one of the most widely used model species. Therefore, the purpose of the present study is to assess the potential sublethal effects of endosulfan and imidacloprid on levels of antioxidant enzymes, MDA, and DNA damage and histopathological alterations in gills and muscle tissues of zebrafish.

MATERIALS AND METHODS

Materials and reagents

Two pesticides were evaluated in the present study, one a neonicotinoid insecticide (imidacloprid) and the other an organochlorine insecticide (endosulfan), as they are usually detected in surface waters worldwide, and known to be highly toxic to aquatic fauna. Imidacloprid (IMI; CAS: 138261-41-3; purity: 96.40% technical product) was purchased from the Chem Service, USA. Endosulfan (END; CAS: 115-29-7; purity: 99.0%) was purchased from the local distributor of Sigma-Aldrich, USA. All other chemicals and solvents used were of analytical purity and purchased from Sigma Chemical Co. (USA).

Procurement and maintenance of zebrafish

In the present study, about 360 mature adult zebrafish (*D. rerio*) in equal ratio of male and female were procured from a commercial local fish supplier (Fish Harbour) and transported to the Chemical Stress Ecology Laboratory, Department of Environmental Sciences, Quaid-i-Azam University. The fish were transferred from the polyethylene bag to a glass aquarium with continuous aeration. The size (dimensions) of the aquarium was $30 \times 30 \times 30$ cm ($L \times W \times H$) with a capacity of 25 L of water. Prior to the experiment, the fish were acclimatized to laboratory conditions for 2 weeks using dechlorinated tap water . During acclimatization, the water was changed on alternate days. The fish were maintained under specified conditions according to Diekmann *et al.* (2004), with photoperiod 12-hour light, 12-hour dark, water temperature adjusted to 26 ± 1 °C, oxygen saturation above 70% and pH fluctuating from 7.4 to 8.1. The acclimatized healthy fish of mixed sex were selected randomly for the experiment.

Experimental design and pesticide exposure concentrations

The experiment was performed in a static closed test system. Healthy and uniform sized fish (mean body weight: 0.38 ± 0.01 g; mean body length: 3.85 ± 0.03 cm) used in this study were selected and evenly distributed into 12 glass aquariums. The acclimatized fish were divided into four different experimental groups as follows: one control group and three treatment groups (30 individuals were placed in each aquarium). Three different sub-lethal concentrations of endosulfan and imidacloprid along with control were selected in triplicate for each treatment, namely, control treatment (CT), treatment 1 (T1), treatment 2 (T2) and treatment 3 (T3). In this experiment, three different concentrations of endosulfan and imidacloprid in the following treatments, i.e., T1, T2 and T3, were administered to zebrafish over a period of 21 days. T1 was exposed to 0.1 µg/L of endosulfan, T2 to 0.5 µg/L and T3 to 1 µg/L of endosulfan while the imidacloprid concentration of 1 mg/L was maintained in all these three treatments. Control treatment (CT) was treated without any treatment of agro-chemicals (exposed to acetone only). For each tested concentration, the experiment was performed in triplicate.

Water quality parameters (pH, temperature, dissolved oxygen and ammonia) were regularly observed and maintained within a fixed range. The water in aquariums was changed every alternate day and pesticide concentration level was maintained by re-dosing. The fish were fed once per day until satiation with commercially available dry flakes (2% of their body weight). The feed was not given to fish a day before sampling in order to avoid any trusion with faecal material. The experiment was conducted for a period of 21 days and behaviour of fish monitored regularly. No mortality was observed either in control individuals or in any of the treatment groups. After 7, 14 and 21 days of the experiment, fish were sampled, underwent biochemical analysis, estimation of DNA damage in liver tissues and histopathological alterations were examined in gills and muscles of zebrafish in order to estimate the effects of pesticide on the given species. Thirty fish at each sampling occasion were sacrificed from each treatment and control group. Before dissection, the fish were anaesthetized on ice. Liver tissues were obtained by dissecting the fish on an ice box. Tissue samples were kept on ice during preparation and later kept at −80 °C until further analysis. The whole experimental trial was subject to approval by Bioethics Committee, Quaid-i-Azam University.

Preparation of stock and treatment solutions

Stock solutions were prepared from technical-grade endosulfan (analytical standard α and β isomers) and imidacloprid and then stored at 4 °C in a refrigerator prior to further analyses. For endosulfan stock solution, 25 mg endosulfan was dissolved in 50 mL of acetone analytical reagent. From stock solution, three dosing/working solutions of concentrations, 10,000 µg/L, 5,000 µg/L and 1,000 µg/L, were made. From each dosing solution, 1.5 mL was dissolved in 14,998.5 mL of water to make three treatment levels of 0.1, 0.5 and 1 µg/L. The concentration of acetone was kept at <0.05% in all pesticide solutions used. For imidacloprid stock preparation, 100 mg of imidacloprid was dissolved in 1 L of distilled water then 150 mL of stock was added in each treatment (T1, T2, T3) level to maintain the concentration of 1 ppm in the three treatments. Concentrations of both insectcides were conserved by changing the water every alternate day.

Enzyme extraction and protein determination

Liver tissues were homogenized in 50 mM ice-cold potassium phosphate buffer (pH 7.0) in a teflon tissue homogenizer. Centrifugation of homogenate was done at 10,000 rpm at 4 °C for 10 minutes. Supernatant was instantly used for analysing enzyme activity and protein determination. Protein contents of liver tissues were estimated by Sigma Bradford method using bovine serum albumin as a standard.

Measurement of enzyme activities (CAT and SOD) and MDA contents

CAT was measured according to the method described by Claiborne (1985). For estimation of SOD activity, the protocol of Shao *et al.* (2012) was followed in this study, while MDA content was determined following the methods used by Zhang *et al.* (2013), with the help of thiobarbituric acid (TBA) assay.

Measurement of DNA damage

SCGE was used to assess the DNA damage. Comet assay was executed to investigate the DNA strand breaks in zebrafish liver. Cell suspensions were prepared using the method described by Ge *et al.* (2015). The extent of DNA damage was estimated by Comet Assay Software Project 1.2.3.b software.

Histopathological examinations

For histology analysis, gill and muscle tissues of zebrafish were dissected and fixed in 4% paraformaldehyde for 24 hr at 4 °C. Afterwards, fixation tissues were dehydrated and further rinsed and processed in an ascending sequence of alcohol (70%, 80%, 90% and 100%) and then cleared in xylene. Tissues were embedded in paraffin and mounted on wooden blocks with the help of melted wax. Thin sections of 7-μm thickness were cut using microtome. Tissue ribbons were stretched by fixation on albumenized glass slides. Slides were hydrated using a sequence of reducing ethanol and run through a series of histological stains, Mayer's haematoxylin and eosin (H&E). Afterwards, slides were examined under light microscope (Olympus-CX41) at 400× and photographs of gill and muscle tissue were obtained using a Tucsen digital camera (Model: ISH500).

Data analysis

All statistical analyses were performed using the IBM SPSS Statistics version 21 for Windows 7 (SPSS Inc., USA). All the results were expressed as mean ± SD (standard deviation), unless specified otherwise. Quantitative data were subjected to one-way analysis of variance followed by Duncan's multiple range test (post hoc test) to estimate the significance of difference among different treatments at the same sampling point. The criterion for significance that was used in all statistical tests was set as $p < 0.05$.

RESULTS AND DISCUSSION

Combined effects of endosulfan and imidacloprid on antioxidant enzymes

Catalase activity

Changes in zebrafish CAT activities after exposure to endosulfan and imidacloprid are depicted in Figure 1.

Significant increase of CAT activity was observed in the second week of exposure in T1 as compared to the control. Endosulfan and imidacloprid induced CAT synthesis to maintain the oxygen free radical equilibrium by scavenging H_2O_2 into H_2O and O_2. CAT activity was significantly decreased in T3 after 7, 14 and 21 days of exposure, which shows excessive oxidative stress, so the antioxidant enzymes were unable to eliminate that stress and were exhausted. Control treatment showed enzyme activity in the range of 20.5–22.3 U/mg Pr (protein), indicating the normal level of catalase in zebrafish in given environmental conditions, as shown in Figure 1. There was no significant difference in CAT activity of T1 on day 7, just a slight increase of 5.3% as compared to the control was observed while CAT activity significantly decreased by 30.9 and 62.1% as compared to the control on day 7 of sampling in T2 and T3. After 14 days of exposure, a 19.9% decrease in T3 was observed while a

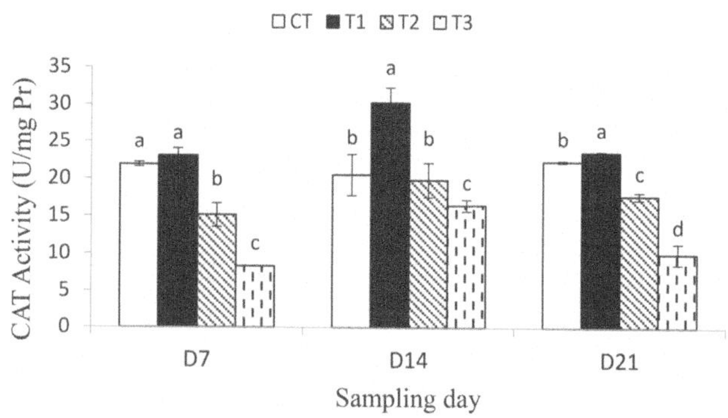

Figure 1 | Combined effects of endosulfan and imidacloprid on catalase activity in liver tissues of zebrafish: CT represents control treatment without exposure to pesticides; T1 represents treatment 1 having low concentration of endosulfan (0.1 μg/L) and 1 ppm imidacloprid; T2 represents treatment 2 having endosulfan concentration (0.5 μg/L) and 1 ppm imidacloprid; T3 represents treatment 3 having high concentration of endosulfan (1 μg/L) and 1 ppm imidacloprid. Each bar represents the mean of three replicates ± standard deviation (SD). Bars denoted with different lowercase letters indicate significant differences between treatments (Duncan's post hoc test, $p \leq 0.05$).

significant decline of 22.7 and 55.3% in T2 and T3 was observed as compared to the control after 21 days of exposure.

Superoxide dismutase activity

The effects of endosulfan and imidacloprid on SOD activity in zebrafish are illustrated in Figure 2.

A significant increase was observed in SOD activity after mild exposure to pesticides on day 7. After 21 days of exposure, a significant decrease in SOD activity was observed at the highest concentration (1 μg/L endosulfan and 1 mg/L imidacloprid) which suggests the inhibition of SOD enzyme. Fish in the control treatment (without pesticide exposure) demonstrated constant SOD activity of 12 U/mg Pr (protein) exhibiting consistent enzyme activity without toxicant, as shown in Figure 2. Compared to the control, a substantial increase of SOD activity was observed in treatments T1, T2 and T3 on day 7, i.e., 98.6, 70.2 and 22.6%, respectively. Less SOD activity was noticed in T2 and T3 as compared to T1 on day 7, but this difference was insignificant. However, SOD activity was greater in T1 and T2, by 19.5% and 9.3%, respectively, as compared to the control on day 14, but there was a insignificant difference among all three treatments in comparison to the control. On day 21, SOD activity was inhibited in all three treatments as compared to the control, but no significant

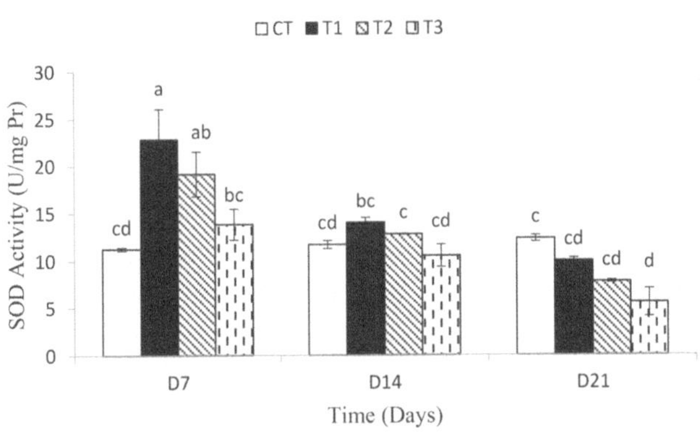

Figure 2 | Combined effects of endosulfan and imidacloprid on superoxide dismutase activity in liver tissues of zebrafish: CT represents control treatment without exposure to pesticides; T1 represents treatment 1 having low concentration of endosulfan (0.1 μg/L) and 1 ppm imidacloprid; T2 represents treatment 2 having endosulfan concentration (0.5 μg/L) and 1 ppm imidacloprid; T3 represents treatment 3 having high concentration of endosulfan (1 μg/L) and 1 ppm imidacloprid. Each bar represents the mean of three replicates ± standard deviation (SD). Bars denoted with different lowercase letters indicate significant differences between treatments (Duncan's post hoc test, $p \leq 0.05$).

decrease was observed in T1 and T2 from the control, whereas T3 activity showed a significant decline of 55.1% with respect to CT.

For monitoring the health of aquatic environments, fish can be used as an indicator to assess the quality of aquatic ecosystems. Oxidative damage is a toxicity mechanism induced by pollutants in aquatic organisms (Santos *et al.* 2004). SOD and CAT are considered as the two most important enzymes of antioxidant defensive systems. They have the ability to capture hydrogen peroxide and superoxide anions and protect organisms from oxidative stress (Han *et al.* 2016). In the present study, significant oxidative stress induced by endosulfan and imidacloprid in the livers of zebrafish was observed, which is likely to be due to the synergistic effects of both insecticides. Superoxide dismutase (SOD) activity was greatly inhibited by the increased pesticide exposure. Highest pesticide concentration and increased exposure time showed significant inhibition in SOD activity (day 21). Earlier exposures showed increased SOD activity (day 7 and 14) as compared to the control. During early exposures, SOD activity may increase to resist the oxidative stress caused by the pesticide exposure. Decreased SOD activity in all treatments after 21 days of exposure to pesticide mixtures may be explained by the fact that oxidative stress increased so much that the SOD level was not sufficient to resist that stress. These results are in agreement with the outcomes published by Ge *et al.* (2015) and Shao *et al.* (2012). A slight increase in CAT activity at low concentration (T1) of pesticide after 7 days of exposure accompanied by a significant increase after 14 days of exposure occurred. It was followed by a decrease in T1 after 21 days of exposure as compared to T1 on day 14. This may show that early exposure to pesticides induces CAT activity to reduce or balance the excessive ROS production by converting hydrogen peroxide into water and molecular oxygen. On the other hand, prolonged exposure to and higher concentration of pesticides can induce oxidative damage (Shao *et al.* 2012). Similar results were reported in a study performed by Han *et al.* (2016), in which, after exposure of azoxystrobin to zebrafish, SOD activities decreased significantly in liver as compared to control. This can happen because the increase in oxygen-free radicals due to constant exposure to pesticides for long durations reduced SOD and the active sites of this antioxidant enzyme became inactivated, leading to enzyme dysfunction (Butterfield *et al.* 1998). Reduced SOD activity was observed in brains and kidneys of zebrafish exposed to atrazine (ATR) and chlorpyrifos (CPF). As the concentration of ATR, CPF and ATR/CPF combination increased, SOD activity reached its lowest levels. This may show the negative impact of pesticides either alone or in a mixture on zebrafish, and these observations in the form of synergistic effects are found to be in accordance with our study findings. CAT activity also followed the same pattern of decrease in activity as the pesticide concentration increased along with exposure time (Xing *et al.* 2012). Increase in CAT activity at low doses of pesticides may represent a compensatory mechanism against oxidative stress. The findings of CAT activity are consistent with the results of Rosety *et al.* (2005), where CAT activity was decreased by the exposure of insecticide malathion.

Generally, the presence of pesticide mixture is likely to be expected in aquatic ecosystems and, consequently, fish can be affected due to synergistic effects. In response to oxidative stress, highly ROS are produced that may cause oxidative damage to the cellular bodies. These highly reactive intermediates are scavenged by the natural defensive systems in the form of antioxidant enzymes, These antioxidant enzymes do not allow excessive ROS production since their basic function is to maintain a balance between production and elimination of ROS, but if the antioxidant enzymes fail to maintain this balance oxidative stress occurs (Ge *et al.* 2015; Han *et al.* 2016).

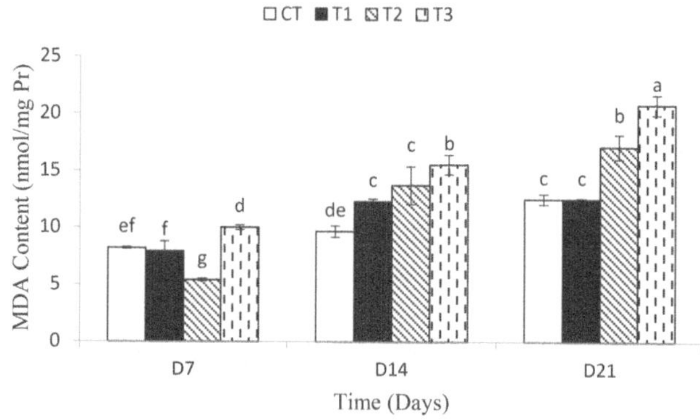

Figure 3 | Combined effects of endosulfan and imidacloprid on MDA content in liver tissues of zebrafish: CT represents control treatment without exposure to pesticides; T1 represents treatment 1 having low concentration of endosulfan (0.1 μg/L) and 1 ppm imidacloprid; T2 represents treatment 2 having endosulfan concentration (0.5 μg/L) and 1 ppm imidacloprid; T3 represents treatment 3 having high concentration of endosulfan (1 μg/L) and 1 ppm imidacloprid. Each bar represents the mean of three replicates ± standard deviation (SD). Bars denoted with different lowercase letters indicate significant differences between treatments (Duncan's post hoc test, $p \leq 0.05$).

Combined effect of endosulfan and imidacloprid on MDA levels

Zebrafish MDA level was used to assess the level of lipid peroxidation exposed to endosulfan and imidacloprid, and the results are depicted in Figure 3.

For the evaluation of lipid peroxidation, MDA content was measured. No significant increase in MDA content was observed in groups having low concentrations of pesticides after day 7. On day 14, MDA level increased in T1, T2 and T3 by 55.9, 154.3 and 55.2%, respectively, as compared to similar treatments on day 7 of sampling. On day 21, a 36.8 and 66.2% increase in MDA content was observed as compared to the control. Significant increase in MDA content was observed in T3 after 7, 14 and 21 days of exposure, which shows that higher concentrations of pesticides and increased exposure durations can induce lipid peroxidation in cell membranes.

Environmental pollution can be measured by employing MDA as a biomarker of pollution. It is reported that dose-dependent increase in MDA level was observed in female zebrafish livers when exposed to atrazine (Jin *et al.* 2010) and these findings are in accordance with our results where the highest increase in MDA content was observed in groups having the highest concentration of pesticides; or, this significant increase could also be due to the synergistic effects of pesticide mixtures. Excessive ROS causes inhibition of antioxidant enzymes that leads to lipid peroxidation and MDA is the end product which is measured to evaluate the extent of oxidative stress. Similar results were observed in a study performed by Han *et al.* (2016), where increased MDA content was observed in both male and female zebrafish with increased concentration of fungicide azoxystrobin. Pesticide mixtures may also induce the same kind of changes in MDA levels as the trends observed in the case of exposure to pesticides alone. It has already been revealed that a high concentration of pesticides (atrazine and chlorpyrifos alone and in mixture) increased MDA levels of the brain and kidney in common carp (Xing *et al.* 2012). These results are in accordance with our study findings and synergistic effects are observed.

Effects on DNA damage in zebrafish

Results of DNA damage in the form of olive tail moments (OTMs) in zebrafish treated with different concentrations of endosulfan and imidacloprid are portrayed in Figure 4.

Single cell gel electrophoresis assay (comet assay) was used to evaluate DNA damage in zebrafish livers. Dose- and time-dependent increases in DNA damage were observed in zebrafish livers. Prolonged exposure increased the OTM, with the highest OTM being observed in T3 at day 21. Lipid peroxidation products and ROS produced as a result of oxidative stress can induce DNA damage. Olive tail moments in zebrafish liver tissues exposed to different treatments are shown in Figure 4. Control groups of all three sampling intervals showed the same value of per cent DNA in the head, i.e., 99.9% which shows intact DNA. DNA content in the tail increased as the pesticide concentration and exposure time period increased and this could be attributed to the factor of synergistic effects of both insecticide mixtures.

DNA damage is quite a suitable biomarker in toxicological studies assessed through comet assay. Oxidative stress and lipid peroxidation lead to the production of alkyl-free radicals which are also accountable for DNA damage. Tail length and per cent DNA in the tail gradually increased as the concentration of pesticides increased. Increase in OTM was also observed after 21 days' exposure to high concentrations of pesticides (endosulfan and imidacloprid). These results are inconsistent with the findings reported by Ge *et al.* (2015).

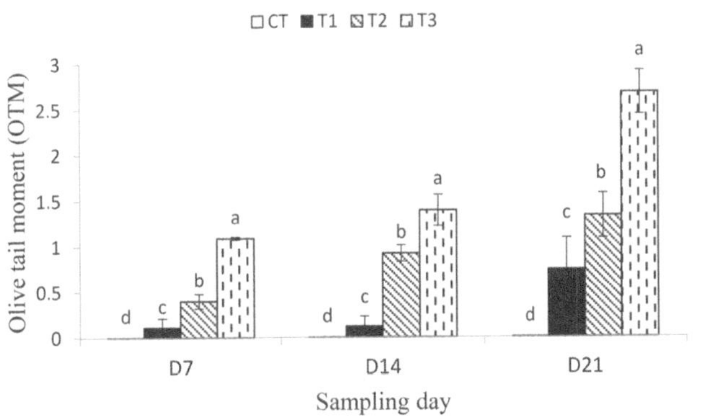

Figure 4 | Combined effects of endosulfan and imidacloprid on olive tail moments in liver tissues of zebrafish: CT represents control treatment without exposure to pesticides; T1 represents treatment 1 having low concentration of endosulfan (0.1 μg/L) and 1 ppm imidacloprid; T2 represents treatment 2 having endosulfan concentration (0.5 μg/L) and 1 ppm imidacloprid; T3 represents treatment 3 having high concentration of endosulfan (1 μg/L) and 1 ppm imidacloprid. Each bar represents the mean of three replicates ± standard deviation (SD). Bars denoted with different lowercase letters indicate significant differences between treatments (Duncan's post hoc test, $p \leq 0.05$).

DNA damage may be induced by biochemical processes, lipid peroxidation products, pesticide concentration and oxygen-free radicals. It has already been reported that significant increase in OTM was observed with increasing concentration of endosulfan in both male and female zebrafish (Shao *et al.* 2012), which is in accordance with the results of the current research. It has been revealed that endosulfan can induce genotoxicity in cultured hepatic cells by forming DNA adducts and high endosulfan-mediated genotoxicity was observed in rats and human cells (Dubois *et al.* 1996). Therefore, we may infer that at high endosulfan and imidacloprid concentrations, formation of DNA adducts could be the cause of DNA damage during the undertaken research, which can be an indication of combined toxicity showing synergistic effects. Dose-dependent response in DNA damage was observed in zebrafish hepatopancreas that was exposed to atrazine (Zhu *et al.* 2011). These results show similarity with the data obtained from the current study.

Tail length is an important parameter to investigate the magnitude of DNA damage as the damaged DNA migrates from the head towards the tail, and consequently tail length increases. In the present study, owing to synergistic effects, length increased because of increased migration of damaged DNA as the concentration of pesticide mixture and exposure time increased.

Effects on histopathological changes in gills and muscles

Hyperplasia, lifting of epithelia, fusion of secondary lamellae and narrowed water channels were observed as a result of pesticide exposure as shown in Figure 5. Moreover, marked lesions were observed in high concentration groups after prolonged exposure. After 21 days of exposure, a significant reduction in primary and secondary lamellae occurred and a higher degree of sloughing was observed, as shown in Figure 5. The findings illustrate that a constant exposure to pesticides for a longer duration leads to the deleterious effects.

Muscles of control groups at all sampling intervals showed normal morphology with the nucleus at the

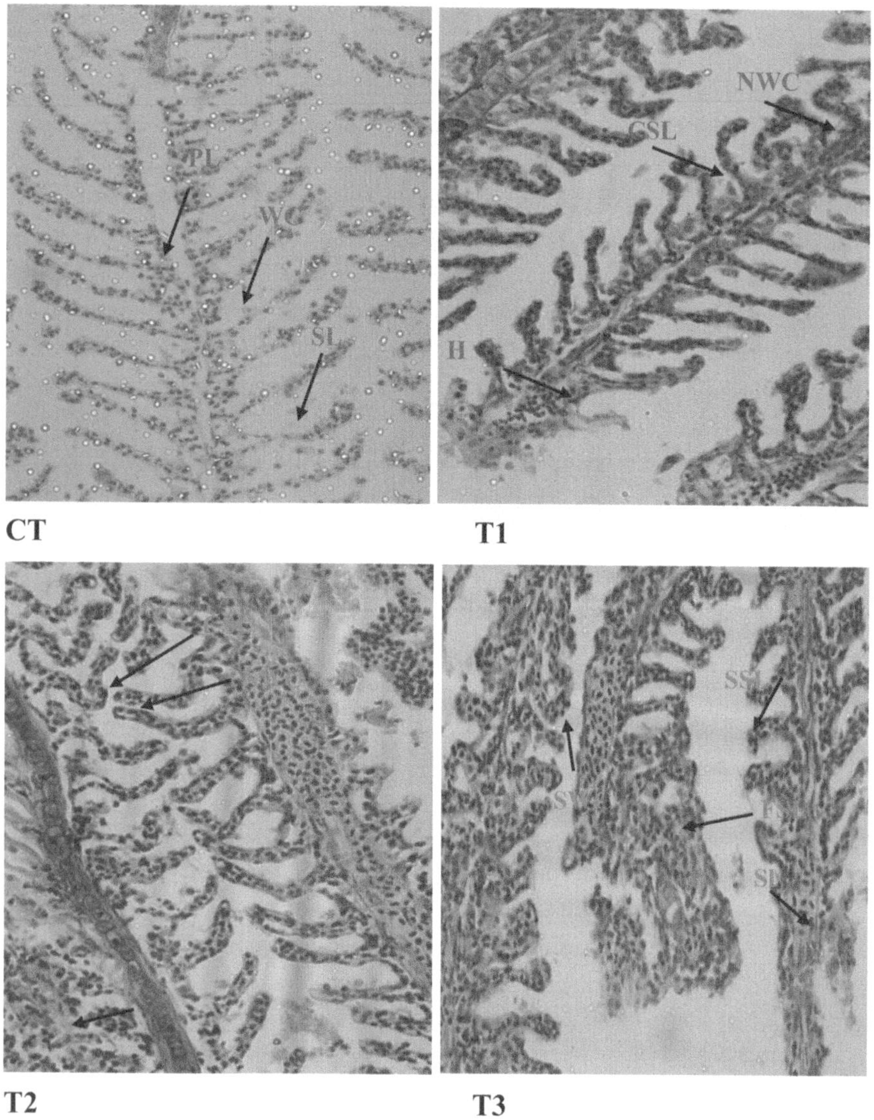

Figure 5 | Photomicrographs of gill sections of zebrafish H&E stained, (400× magnification) after 21 days of exposure. CT (control treatment): SL, secondary lamellae; PL, primary lamellae; WC, water channel. T1 (treatment group having lower concentration (endosulfan 0.1 μg/L and 1 ppm imidacloprid)): CSL, curling of secondary lamellae; NWC, narrowed water channels; H, hyperplasia. T2 (treatment group having medium concentration (endosulfan 0.5 μg/L and 1 ppm imidacloprid)): H, hyperplasia; SWC, shortening of water channels; CSL, curling of secondary lamellae. T3 (treatment group having high concentration (endosulfan 1 μg/L and 1 ppm imidacloprid)): Sl, sloughing of secondary lamellae; SSL, shortening of secondary lamellae; SWC, shortening of water channels; FSL, fusion of secondary lamellae.

periphery of fibres while mild splitting and degeneration of muscle bundles was observed in T2 and T3 after 7 days of exposure, as shown in Figure 6. After 14 and 21 days of exposure, quite significant changes were observed in muscles as can be seen from Figure 6. Severe muscle damage was noticed in the form of atrophy, necrosis, degeneration and splitting of muscle bundles.

Pathological changes in gills can be used as an indicator of water pollution. Hyperplasia, lifting of epithelia, fusion of secondary lamellae and narrowed water channels were observed as results of pesticide exposure due to synergistic effects. Sloughing was also observed in high-dose groups. This uplifting of lamellar epithelium and fusion of secondary lamellae could be the adaptations of gill tissues in order to reduce the contact between pollutants by reducing the surface area of gills. Results of the current research are in accordance with the study presented by Nowak (1992), where hyperplasia, lifting of lamellar epithelium and lamellar hypertrophy were

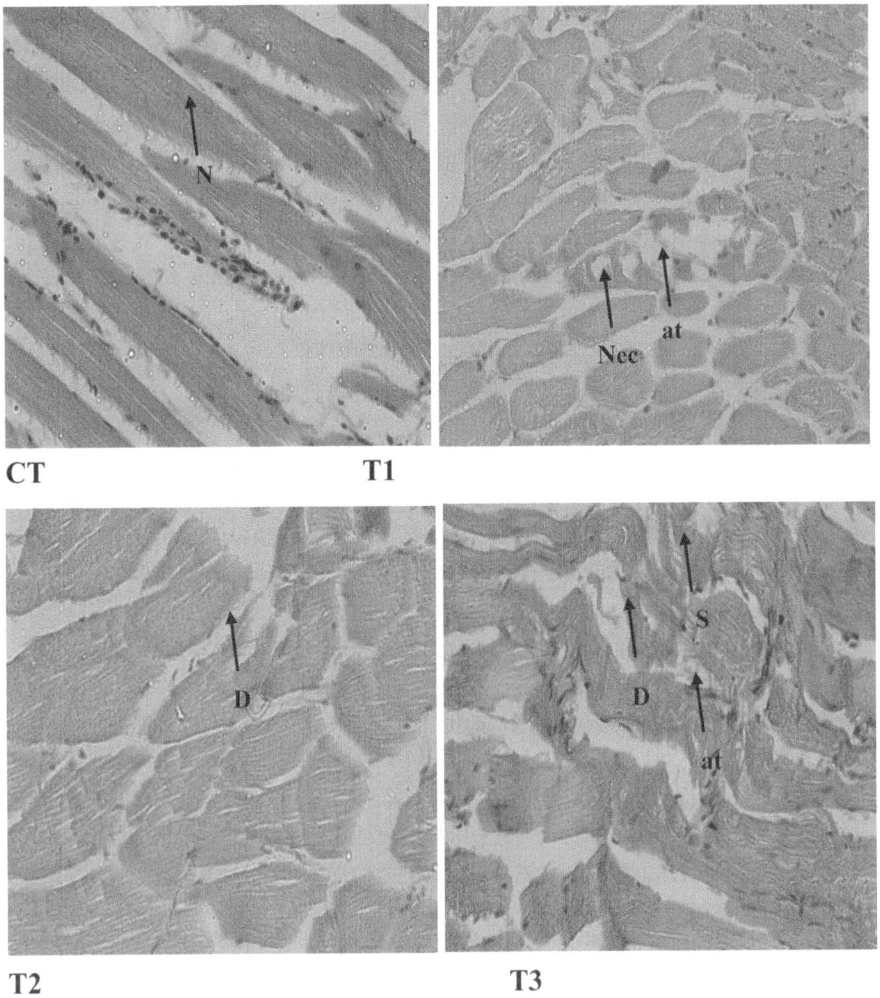

Figure 6 | Photomicrographs of zebrafish muscle sections H&E staining, (400×) after 21 days of exposure. CT (represents the muscles of zebrafish from the control group); T1 (muscles of zebrafish in treatment 1); T2 (muscles of zebrafish in treatment 2); T3 (muscles of zebrafish in treatment 3): N represents the nucleus located at the periphery of muscle fibres; S represents the splitting of muscle fibres; D represents the degeneration of muscle bundles; Nec represents necrosis and at represents atrophy.

observed in the gills of catfish having residues of endosulfan. Results presented by Bhuvaneshwari *et al.* (2015) are in accordance with our findings. Histopathological damage previously noticed in the gills of rainbow trout is in accordance with our findings and was induced by chronic exposure to diazinon at concentrations of 0.1 mg/L and 0.2 mg/L (Banaee *et al.* 2013). Low doses of pesticides did not induce any significant alterations in muscles while high doses and prolonged exposures induced splitting and degeneration of muscle bundles; slight atrophy was also observed in muscle bundles and this could be attributed to the combined effects of both insecticides in terms of synergistic impacts. Results reported by Bhuvaneshwari *et al.* (2015) are in accordance with our findings where the same alterations were produced when zebrafish were exposed to a mixture of organochlorine pesticides. Splitting of muscle fibres, disintegration of muscle bundles, focal area of necrosis, vacuolar degeneration in muscle bundles were the histopathological alterations observed in tilapia (*Orechromis mossambicus*) exposed to textile dyes (0.5, 1 and 1.5 ppm) for 21 days (Sripriya *et al.* 2014), and these findings are found to be similar to our observations.

Practical applications

Fish are an essential component of the aquatic environment and affected by pesticides which pollute natural water through agricultural runoff and various other means. The polluted water can result in the death of fish or decreasing their numbers and other sensitive aquatic fauna in the food web resulting in an ecological imbalance. With the help of this study, we will be able to identify the oxidative stress and antioxidant levels which are supposed to be good indicators of aquatic pollution. The present work will offer valuable information related to the possible impact on biochemical parameters of zebrafish. These findings can also contribute to understanding the effect of mixture toxicity on the growth rate, feeding efficiency, survival and production of fish. Moreover, our study can also provide valuable support to fishery managers and aquaculturists in order to improve the water quality for culturing or growing fish. However, the improper use of pesticides can result in adverse harmful effects on fish; thus this study could be used for creating awareness among local farmer communities so that the indiscriminate use of these pesticides could be reduced. This information can also be useful for policy makers, the agro-chemical and fish industry, academia and the public at large.

CONCLUSIONS AND FUTURE RESEARCH PROSPECTS

Tainted water is responsible for the weak growth, disease outbreak, reproduction efficiency and also for survival and death of fish. To our knowledge, this is the first study to report the combined effects on oxidative stress and histopathology in zebrafish exposed to endosulfan and imidacloprid present in water. Successively, it is inferred that both pesticides induced oxidative stress in the liver of zebrafish which leads to lipid peroxidation resulting in DNA damage. The present study revealed that the pesticides stimulated the antioxidant enzyme activities (CAT and SOD) at low concentrations, but high concentrations of pesticides and prolonged exposure duration decreased the activity of both enzymes. As the antioxidant enzyme activity decreased, MDA content increased, which enhanced the DNA damage because of the excessive ROS. MDA content increased with increased exposure time and pesticide concentration, which demonstrated increased lipid peroxidation in cellular bodies. Similarly, a substantial increase in DNA damage was noticed after 21 days' exposure to pesticides. Based on the present study, it may be concluded that environmental pollution can be evaluated by measuring biochemical responses in organisms against stress. In summary, three biomarkers, oxidative stress, DNA damage and pathological changes may be helpful to understand the mechanism of the combined toxicity of endosulfan and imidacloprid. All the negative effects of pesticides must be taken into account while using these agrochemicals for insect-pest control in agricultural fields surrounding freshwater ecosystems. Thus, it is recommended that proper maintenance and measures are necessary in order to keep these water bodies clean and safe. Since many pesticides coexist and mixtures of pesticides are usually detected in the natural environment, additional work is required to understand the research into pesticide interaction towards non-target organisms. In the future, higher concentrations of pesticides may be expected due to their increasing use and a higher risk to aquatic organisms is anticipated. In the case that pesticides are accidentally spilled into small water bodies, their predicted environmental concentrations would affect less sensitive organisms (chronically) and more sensitive aquatic invertebrates (acutely). Henceforth, this work suggests that further (eco) toxicological studies with a broader spectrum of aquatic organisms belonging to different taxonomic groups and trophic levels need to be performed.

ACKNOWLEDGEMENTS

The present study was financially supported by the specialized University Research Funds (URF) of Quaid-i-Azam University (QAU), Islamabad. Moreover, the authors wish to express their thanks to Dept. Animal Sciences, QAU for providing technical expertise and assistance in the comet assay. We also thank the Pharmacology Dept. at Arid Agriculture University, Rawalpindi for the support in identification and interpretation of histopathology slides. The authors declare that there is no conflict of interest regarding this article.

REFERENCES

Alamdar, A., Syed, J. H., Malik, R. N., Katsoyiannis, A., Liu, J., Li, J., Zhang, G. & Jones, K. C. 2014 Organochlorine pesticides in surface soils from obsolete pesticide dumping ground in Hyderabad City, Pakistan: contamination levels and their potential for air–soil exchange. *Sci. Total. Environ.* **470**, 733–741.

Banaee, M., Sureda, A., Mirvagefei, A. & Ahmadi, K. 2013 Histopathological alterations induced by diazinon in rainbow trout (*Oncorhynchus mykiss*). *Int. J. Environ. Res.* **7**, 735–744.

Banihashemi, A., Dyke, M. I. V. & Huck, P. M. 2014 Detection of viable bacterial pathogens in a drinking water source using propidium monoazide-quantitative PCR. *J. Water Supply Res. Technol.-Aqua* **64** (2), 139–148.

Bhuvaneshwari, R., Padmanaban, K. & Babu Rajendran, R. 2015 Histopathological alterations in muscle, liver and gill tissues of zebrafish *Danio rerio* due to environmentally relevant concentrations of organochlorine pesticides (OCPs) and heavy metals. *Int. J. Environ. Res.* **9**, 1365–1372.

Bonmatin, J. M., Moineau, I., Charvet, R. C., Fleche, C., & Bengsch, M. E. & R, E. 2003 A LC/APCI-MS/MS method for analysis of imidacloprid in soils, in plants, and in pollens. *Anal. Chem.* **75**, 2027–2033.

Butterfield, D. A., Koppal, T., Howard, B., Subramaniam, R., Hall, N., Hensley, K., Yatin, S., Allen, K., Aksenov, M. & Aksenova, M. 1998 Structural and functional changes in proteins induced by free radical-mediated oxidative stress and protective action of the antioxidants N-tert-butyl-α-phenylnitrone and vitamin Ea. *Ann. New York Acad. Sci.* **854**, 448–462.

Camargo, M. M. & Martinez, C. B. 2007 Histopathology of gills, kidney and liver of a neotropical fish caged in an urban stream. *Neotropical Ichthyology* **5**, 327–336.

Claiborne, A. 1985 Catalase activity. In: *Handbook of Methods for Oxygene Radical Research* (R. A. Greenwald, ed.). CRC Press, Boca Raton, FL, USA, pp. 283–284.

Cui, L. L., Ge, J., Zhu, Y. D., Yang, Y. Y. & Wang, J. 2015 Concentrations, bioaccumulation, and human health risk assessment of organochlorine pesticides and heavy metals in edible fish from Wuhan. *China. Environ. Sci. Pollut. Res.* **22**, 15866–15879.

De, A., Bose, R., Kumar, A. & Mozumdar, S. 2014 Worldwide pesticide use. In: *Targeted Delivery of Pesticides Using Biodegradable Polymeric Nanoparticles* (A. De, R. Bose, A. Kumar & S. Mozumdar, eds). Springer, New Delhi, pp. 5–6.

Desai, B. & Parikh, P. 2014 Behavioural responses to acute exposure of imidacloprid and curzate on *Labeo rohita*. *Int. J. Open Sci. Res.* **2** (1), 1–12.

Diekmann, M., Hultsch, V. & Nagel, R. 2004 On the relevance of genotoxicity for fish populations I: effects of a model genotoxicant on zebrafish (*Danio rerio*) in a complete life-cycle test. *Aquat. Toxicol.* **68**, 13–26.

Doruchowski, G., Swiechowski, W., Masny, S., Maciesiak, A., Tartanus, M., Bryk, H. & Hołownicki, R. 2017 Low-drift nozzles vs. standard nozzles for pesticide application in the biological efficacy trials of pesticides in apple pest and disease control. *Sci. Total Environ.* **575**, 1239–1246.

Dubois, M., Pfohl-Leszkowicz, A., De Waziers, I. & Kremers, P. 1996 Selective induction of the CYP3A family by endosulfan and DNA-adduct formation in different hepatic and hepatoma cells. *Environ. Toxicol. Pharmacol.* **1**, 249–256.

Dutta, H. M. 1996 A composite approach for evaluation of the effects of pesticides on fish. In: *Fish Morphology* (H. M. Dutta & J. S. Datta Munshi, eds). CRC Press, Boca Raton, FL, USA, pp. 249–269.

Elimelech, M. 2006 The global challenge for adequate and safe water. *J. Water Supply Res. T-Aqua* **55** (1), 3–10.

Fernandez, M. & Mazon, A. 2003 Environmental pollution and fish gill morphology. In: *Fish Adaptations* (A. L. Val & B. G. Kapoor, eds). Science Publishers, Enfield, NH, USA, pp. 203–231.

Ge, W., Yan, S., Wang, J., Zhu, L., Chen, A. & Wang, J. 2015 Oxidative stress and DNA damage induced by imidacloprid in zebrafish (*Danio rerio*). *J. Agr. Food. Chem.* **63**, 1856–1862.

Han, Y., Liu, T., Wang, J., Wang, J., Zhang, C. & Zhu, L. 2016 Genotoxicity and oxidative stress induced by the fungicide azoxystrobin in zebrafish (*Danio rerio*) livers. *Pest. Biochem. Physiol.* **133**, 13–19.

ISO 1996 Water quality – determination of the acute lethal toxicity of substances to a freshwater fish [*Brachydanio rerio* Hamilton Buchanan (Teleostei, Cyprinidae)] – part 3: flow-through method. ISO 7346-7343.

Jin, Y., Zhang, X., Shu, L., Chen, L., Sun, L., Qian, H., Liu, W. & Fu, Z. 2010 Oxidative stress response and gene expression with atrazine exposure in adult female zebrafish (*Danio rerio*). *Chemosphere* **78**, 846–852.

Liess, M., Brown, C., Dohmen, P., Duquesne, S., Hart, A., Heimbach, F., Kreuger, J., Lagadic, L., Maund, S., Reinert, W., Streloke, M. & Tarazona, J. V. 2005 *Effects of Pesticides in the Field-EPIF, European Union and SETAC Europe Workshop, 2003 LeCroic, France*. SETAC, Brussels, Belgium.

Matozzo, V., Fabrello, J., Masiero, L., Ferraccioli, F., Finos, L., Pastore, P., Gangi, I. M. D. & Bogialli, S. 2018 Ecotoxicological risk assessment for the herbicide glyphosate to non-target aquatic species: a case study with the mussel Mytilus galloprovincialis. *Environ. Pollut.* **233**, 623–632.

Napit, M. K. 2013 The effect of pesticides on fish fauna of Bhopal lower lake (MP). *Afr. J. Environ. Sci. Technol.* **7**, 725–727.

Nowak, B. 1992 Histological changes in gills induced by residues of endosulfan. *Aquat. Toxicol* **23** (1), 65–83.

OECD 2013 *OECD Guidelines for the Testing of Chemicals, Fish Embryo Acute Toxicity (FET) Test*. OECD, Paris, France, No. 236.

Oruc, E. O., Sevgiler, Y. & Uner, N. 2004 Tissue-specific oxidative stress responses in fish exposed to 2, 4-D and azinphosmethyl. *Comp. Biochem. Physiol. C Toxicol. Pharmacol.* **137**, 43–51.

PAN Pesticides database 2006 http://www.pesticideinfo.org.

POPRC 2010 *Recommendation of the Persistent Organic Pollutants Review Committee (POPRC) on Endosulfan*. Stockholm Convention on Persistent Organic Pollutants, Geneva, Switzerland. http://www.pops.int/poprc.

Ritter, D. & Knebel, J. 2009 Genotoxicity testing in vitro – development of a higher throughput analysis method based on the comet assay. *Toxicol. in Vitro* **23**, 1570–1575.

Rosety, M., Rosety-Rodríguez, M., Ordonez, F. & Rosety, I. 2005 Time course variations of antioxidant enzyme activities and histopathology of gilthead seabream gills exposed to malathion. *Histol. Histopathol.* **20**, 1017–1021.

Sadiq, R., Kleiner, Y. & Rajani, B. 2010 Modelling the potential for water quality failures in distribution networks: framework (I). *J. Water Supply Res. Technol.-Aqua* **59** (4), 255–276.

Santos, M., Pacheco, M. & Ahmad, I. 2004 *Anguilla anguilla* L. antioxidants responses to in situ bleached kraft pulp mill effluent outlet exposure. *Environ. Int.* **30**, 301–308.

Schreiner, V. C., Szocs, E., Bhowmik, A. K., Vijver, M. G. & Schafer, R. B. 2016 Pesticide mixtures in streams of several European countries and the USA. *Sci. Total. Environ.* **573**, 680–689.

Shao, B., Zhu, L., Dong, M., Wang, J., Wang, J., Xie, H., Zhang, Q., Du, Z. & Zhu, S. 2012 DNA damage and oxidative stress induced by endosulfan exposure in zebrafish (*Danio rerio*). *Ecotoxicology* **21**, 1533–1540.

Sohel, K. M. A., Chowdhury, M. A. I. & Ahmed, M. F. 2003 Surface water quality in and around Dhaka City. *J. Water Supply Res. Technol.-Aqua* **52** (2), 141–153.

Sripriya, L., Vijayalakshmi, M., Sumathy, R. & Sharmila, J. 2014 The impact of textile dyes on the biochemistry and histology of liver, a fresh water fish, tilapia, *Oreochromis mossambicus* (peters.). *Int. J. Pharma. Bio Sci.* **5**, 271–298.

United States Environmental Protection Agency (US EPA) 2002 *Re-registration Eligibility Decision for Endosulfan. EPA 738-R-02-013. Pollution, Pesticides and Toxic Substances (7508C).* United States Environmental Protection Agency, Washington, DC, USA.

Velasco-Santamaría, Y. M., Handy, R. D. & Sloman, K. A. 2011 Endosulfan affects health variables in adult zebrafish (*Danio rerio*) and induces alterations in larvae development. *Comp. Biochem. Physiol. C* **153**, 372–380.

Wu, S., Li, X., Liu, X., Yang, G., An, X., Wang, Q. & Wang, Y. 2018 Joint toxic effects of triazophos and imidacloprid on zebrafish (*Danio rerio*). *Environ. Pollut.* **235**, 470–481.

Xing, H., Li, S., Wang, Z., Gao, X., Xu, S. & Wang, X. 2012 Histopathological changes and antioxidant response in brain and kidney of common carp exposed to atrazine and chlorpyrifos. *Chemosphere* **88**, 377–383.

Zhang, Q., Zhu, L., Wang, J., Xie, H., Wang, J., Han, Y. & Yang, J. 2013 Oxidative stress and lipid peroxidation in the earthworm *Eisenia fetida* induced by low doses of fomesafen. *Environ. Sci. Pollut. Res.* **20**, 201–208.

Zhang, D., Zheng, G., Zheng, S., Guan, W., Zhao, W. & Jia, X. 2018 Assessing water quality of Nen River, the neighboring section of three provinces, using multivariate statistical analysis. *J. Water Supply Res. Technol.-Aqua* **67** (8), 779–789.

Zhu, L., Dong, X., Xie, H., Wang, J., Wang, J., Su, J. & Yu, C. 2011 DNA damage and effects on glutathione-S-transferase activity induced by atrazine exposure in zebrafish (*Danio rerio*). *Environ. Toxicol.* **26**, 480–488.

First received 31 May 2019; accepted in revised form 23 July 2019. Available online 4 September 2019

Phytoremediation of nitrate contaminated water using ornamental plants

S. Shyamala, N. Arul Manikandan, Kannan Pakshirajan, Van Tai Tang, Eldon R. Rene, Hung-Suck Park and Shishir Kumar Behera

ABSTRACT

This work aims at evaluating the potential of two ornamental plant species, i.e., money plant (*Epiprennum aureum*) and arrowhead plant (*Syngonium podophyllum*), to treat nitrate containing wastewater. Statistically designed experiments were performed to ascertain the effect of initial nitrate concentration (40–120 mg/L), growth period (1–12 days) and plant density (20–80 g/L) on nitrate removal. Based on the results of analysis of variance (ANOVA), it was observed that the individual effects ($F = 78.04$ and $P = 0.013$) of process parameters influenced the nitrate removal efficiency by money plant stronger than the 2-way ($F = 0.2$ and $P = 0.89$) and 3-way interaction effects ($F = 0.46$ and $P = 0.569$). In the case of the arrowhead plant, the individual effects significantly affected the nitrate removal efficiency than the 2-way and 3-way interaction effects. Low nitrate concentrations (40 mg/L) and high plant density (80 g/L), showed ~88% nitrate removal by arrowhead plant, during a growth period of 6 d. On the contrary, under similar conditions, the money plant showed a nitrate removal efficiency of ~93% during a growth period of 12 d. Concerning the removal kinetics, an increase in the growth period increased the nitrate removal rate for both the plants.

Key words | arrowhead plant, growth period, initial nitrate concentration, money plant, plant density, phytoremediation

S. Shyamala
Shishir Kumar Behera (corresponding author)
Department of Chemical Engineering,
GMR Institute of Technology,
Rajam 532 127, Srikakulam District, Andhra Pradesh,
India
E-mail: *shishir.kb@gmail.com*

N. Arul Manikandan
Department of Chemical Engineering,
IIT Guwahati,
Guwahati 781039, Assam,
India

Kannan Pakshirajan
Department of Biosciences & Bioengineering,
IIT Guwahati,
Guwahati 781039, Assam,
India

Van Tai Tang
Faculty of Environment and Labour Safety,
Ton Duc Thang University,
Ho Chi Minh City,
Vietnam

Eldon R. Rene
Department of Environmental Engineering and Water Technology,
UNESCO-IHE Institute of Water Education,
P. O. Box 3015, 2601 DA, Delft,
The Netherlands

Hung-Suck Park
Center for Clean Technology and Resource Recycling,
University of Ulsan,
Ulsan,
South Korea

Shishir Kumar Behera
Industrial Ecology Research Group, School of Chemical Engineering,
Vellore Institute of Technology,
Vellore 632014,
India

INTRODUCTION

As a consequence of rapid urbanization, industrialization and the excessive use of fertilizers, the levels of nitrate in ground and surface water has increased several-fold in both developing and rapidly developing countries (Weldeslassie *et al.* 2018). The presence of high nitrate concentrations in drinking water results in health issues such as blue baby syndrome (methemoglobinemia), stomach cancer, high blood pressure, thyroiditis, cytogenetic

doi: 10.2166/aqua.2019.111

malfunction and several other defects in newborn babies (Sharma *et al.* 2017; Soomro *et al.* 2017; Nieder *et al.* 2018). According to the World Health Organization (WHO), the maximum level of nitrate in drinking water has been fixed at ~11.0 mg N/L. The maximum nitrate level stipulated by US Environmental Protection Agency (USEPA) and China Environmental Agency is 10 mg N/L in drinking water (Liu *et al.* 2015; Pennino *et al.* 2017). According to the prevailing Indian situation, Rai (2003) reported that ~118 million people are drinking water with nitrate levels in the range of 45–100 mg N/L and more than 108 million people with levels >100 mg N/L (e.g., the Bureau of Indian Standards desirable limit is <45 mg N/L).

The main anthropogenic source for the release of nitrate into groundwater and surface water includes a number of issues such as improper agricultural practices, excessive use of synthetic fertilizers, fossil fuel combustion, among others (Chen *et al.* 2017; Wollheim *et al.* 2017). Quintessentially, in the developing nations, groundwater is the real source of drinking water as the risk of microbial contamination is much less compared to other drinking water sources (Fout *et al.* 2017; Li *et al.* 2018). Therefore, the removal of nitrate from both groundwater and surface water is essential before it is used for human consumption.

Reverse osmosis, adsorption, ion exchange, chemical denitrification, photocatalytic reduction and electro-dialysis are conventional physico-chemical processes for the remediation of water containing excessive levels of nitrate (Bulgariu *et al.* 2012; Arun *et al.* 2017; Challagulla *et al.* 2017; Tyagi *et al.* 2018). Some of these physico-chemical processes are rather expensive, energy intensive, and they also generate secondary pollutants during their operation (Sinha *et al.* 2017; Wang *et al.* 2018). Nevertheless, for the treatment of low nitrate concentrations in ground and surface waters, biological processes are often considered as the best suitable technological option from a techno-economic viewpoint.

In this line of continuous research that targets biotechniques for water treatment, phytoremediation is considered to be a versatile and sustainable natural treatment technique. Phytoremediation is an inexpensive, non-invasive and versatile treatment technology for removing nutrients and trace contaminants from wastewater and surface water sources (Sinha *et al.* 2018). The inherent capacity of certain (native) plant species in sequestering nitrate from water for their growth is used in this technique for the removal of nitrate from contaminated water environments (Li *et al.* 2016). More specifically, macrophytes are extensively employed in the field of phytoremediation because of their ability to sequester high amounts of pollutant, in a shorter span of time, thereby demonstrating high pollutant removal rates (Bartucca *et al.* 2016). Besides, these plants promote the growth of microbial communities around their root systems, thus having a cumulative (synergistic) effect in removing the pollutants from water (Lingua *et al.* 2015; Xu *et al.* 2018).

According to the literature, a variety of macrophytes has been employed for the removal of pollutants present in water (Liu *et al.* 2017; Sinha *et al.* 2018). These include N, P and K, heavy metals, organic and inorganic forms of nitrogen present in water bodies, micropollutants, fluoride and other trace metallic species (Pavlineri *et al.* 2017). For example, *Pistia stratiotes* and *Spirodela polyrhiza* were used to remove fluoride and heavy metal ions from water (Volf *et al.* 2015; Karmakar *et al.* 2016). *Vallisneria natans* showed good potential for the treatment of wastewater contaminated with excessive levels of arsenic (Chen *et al.* 2015). In another recent study, the Italian ryegrass grown in a vegetated floating system was tested for its potential to treat nitrate-contaminated water (Bartucca *et al.* 2016). Rye-grass (*Lolium perenne*) has proven to be effective in removing inorganic nitrogen from water, in both hydroponic systems and in sand filters (Liu *et al.* 2017; Escobar-Alvarado *et al.* 2018; Radziemska *et al.* 2019). However, due to their aesthetic looks, commercial value and easy adaptability to diverse environments, ornamental plants seem to be a promising alternative to fast-growing plants/weeds for treating nitrogen-contaminated surface/groundwater in households. In this study, as a proof-of-concept, and to demonstrate the application of ornamental plants for the treatment of nitrate-containing water, laboratory-scale experiments were performed with the following objectives: (i) to test the nitrate removal efficiency by two ornamental plants, i.e., money plant and arrowhead plant; (ii) to elucidate the main and interaction effects of process factors such as initial nitrate concentration, growth period and plant density on the nitrate removal efficiency; and (iii) to perform a kinetic analysis on the nitrate uptake capacity by the two plants.

MATERIALS AND METHODS

Growth conditions of plants

Money plants (*Epiprennum aureum*) and arrowhead plants (*Syngonium podophyllum*) were collected from a local nursery, at the GMRIT campus, India. These two plants were selected based on their wide occurrence as well as their ability to grow under harsh conditions, with minimal nutrient requirements. Prior to the regular experiments, all the selected plants were allowed to grow only with tap water for a total period of 2 weeks. As per the experimental requirement, in order to study the effect of plant density, varying sizes of well-grown and healthy plants were chosen. Prior to use, these plants were rinsed thoroughly in tap water in order to eliminate sediments and other dirt elements attached to them. Finally, the washed plants were placed in clean, transparent cylindrical plastic containers for regular experiments (Figure 1). Different experimental setups were used for money plants and arrowhead plants, respectively. To each of these containers was added 2 L of 50% (*v/v*) Hoagland's solution (HS) of composition (in mM): KNO_3 (3.0), $Ca(NO_3)_2$ (2.4), $MgSO_4$ (1.0), KH_2PO_4 (1.0) and NaCl (0.5), and Fe-EDTA (44.8), H_3BO_3 (23.1), $MnCl_2$ (4.6), $ZnSO_4$ (0.38), $CuSO_4$ (0.16) and H_2MoO_4 (0.052) complex prepared using tap water (Sinha *et al.* 2017).

The plants were grown under ambient environmental conditions, with a temperature range of 25–30 °C. All the chemicals used in the study were obtained from Desai Chemical Company (Visakhapatnam, India) and Lotus Granges India Limited (Visakhapatnam, India), respectively. All the chemicals used in this study were of analytical grade.

Statistical design of experiments

The factorial design of experiments is a statistically significant technique used widely to infer maximum information by performing fewer actual experiments (Sinha *et al.* 2017). In this design, the effects of various processes or operational factors on a specific response variable, at different levels, can be measured simultaneously. Both the individual and interaction effects induced by each factor can be ascertained in a statistically significant manner (Hazarika *et al.* 2015). A 2^3 full-factorial experimental design, comprising ten runs, i.e., eight actual runs and two additional runs, was used in this study. The initial nitrate concentration (X_1), growth period or time (X_2) and plant density (X_3) were chosen as the three important factors that affect the nitrate removal efficiency in plants. Table 1 presents the three different factors, along with the three different levels chosen in the present study. The low (−1), medium (0)

Figure 1 | Photograph of the experimental setup used in the study: (a) money plant and (b) arrowhead plant.

Table 1 | Experimental factors and levels investigated in this work

Levels	Factors: Initial nitrate concentration (mg/L) Money plant and arrowhead plant	Growth period (d): Money plant	Growth period (d): Arrowhead plant	Plant density (g/L) Money plant and arrowhead plant
Low	40	3	1	20
Centre	80	7.5	3.5	50
High	120	12	6	80

and high (+1) levels of each factor were chosen based on previous literature reports (Khajuria & Kanae 2013; Sinha *et al.* 2017) and based on preliminary experiments conducted prior to this study. The responses of nitrate removal by money plant and arrowhead plant were expressed in terms of the nitrate removal efficiency (%). Thereafter, a linear polynomial model was proposed to correlate the individual and interaction effects of these factors with the response variable. The equation takes the form as shown in Equation (1):

$$Y = \beta_0 + \beta_1 X_1 + \beta_2 X_2 + \beta_3 X_3 + \beta_{12} X_1 \cdot X_2 + \beta_{13} X_1 \cdot X_3 + \beta_{23} X_2 \cdot X_3 + \beta_{123} X_1 \cdot X_2 \cdot X_3 \quad (1)$$

where Y is the nitrate removal efficiency as observed in two different plants, β_0 is a constant term, β_1, β_2 and β_3 are the coefficients that describe the individual effects of initial nitrate concentration (X_1), growth period (X_2) and plant density (X_3), respectively. β_{12}, β_{13} and β_{23} are the coefficients that reveal information on the interaction effects between initial nitrate concentration × growth period, initial nitrate concentration × plant density and growth period × plant density, respectively. The coefficient β_{123} indicates the interaction effect of initial nitrate concentration × growth period × plant density, while X_1, X_2 and X_3 are the independent variables. The difference observed in the response of any factor amid its variation from low to high levels is defined as the effect and the value of the effect is divided by two to obtain the regression model coefficients. Thereafter, the regression model coefficients are divided by the standard error coefficient to get the value of standardized effects (T) (Villa-Gomez *et al.* 2015).

The statistical significance of any factors in the regression model equation was determined based on its probability (*P*) value (Sinha *et al.* 2017). The coefficient values of the individual and interaction effects of each factor observed for a specific response showed its effect on a particular experimental condition. The individual and interaction effects of the factors which cross the reference line (vertical line at 4.30) in a Pareto plot (Figure 2(a) and 2(b)) are considered to be potentially important at the 95% significance level (Roy *et al.* 2015). In the present study, Minitab™ Release 16, 2006 software was used to analyse the experimental results and the same software was also used to determine the statistical indices, namely, the regression coefficient, the probability value (*P*) and Fischer's value (*F*).

Experimental methodology

1.631 g of potassium nitrate was dissolved in 1 L of double distilled water to prepare a nitrate stock solution of 1,000 mg/L (Sinha *et al.* 2017). Thereafter, the stock solution was stored under refrigerated conditions. The desired concentration of nitrate for each experimental run (Table 2) was obtained by diluting the stock solution with 50% (*v/v*) Hoagland's solution. Control experiments in the present study refer to the experiments that were performed in the absence of plants.

Analytical method and kinetics

In order to determine the nitrate removal efficiency by the plants, samples were collected at different time intervals as shown in Table 2. For this analysis, 2 mL of working solution was collected in a beaker and 0.1 mL of concentrated HCl was added. The samples were agitated thoroughly for 5 min and allowed to remain in suspension for 10 min. Thereafter, the samples were examined using a UV-Vis spectrophotometer at a wavelength (λ_{max}) of 220 nm. The nitrate removal efficiency was calculated as follows:

$$\textit{Nitrate removal efficiency}\ (\%) = \frac{C_0 - C_e}{C_0} \times 100 \quad (2)$$

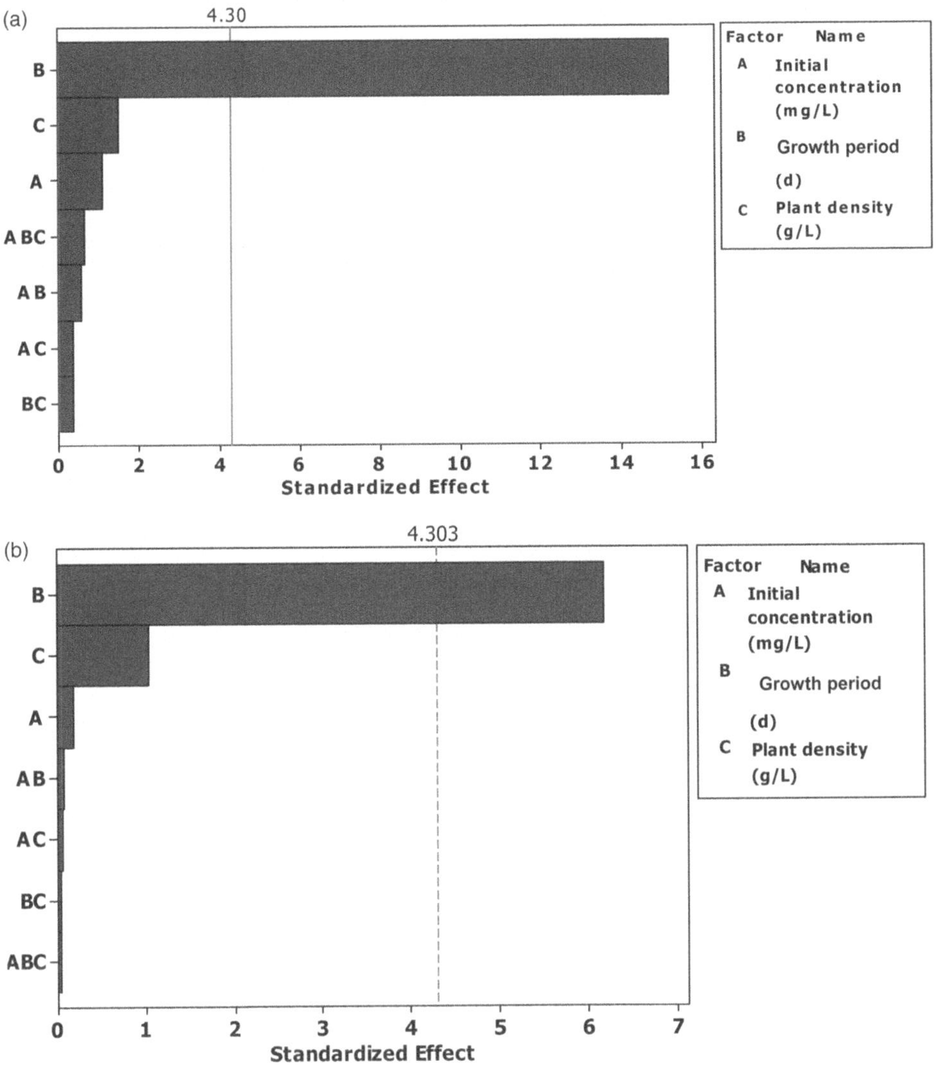

Figure 2 | Pareto chart showing the standardized effects of various factors on nitrate removal efficiency (%) by (a) money plant and (b) arrowhead plant.

where C_0 is the initial nitrate concentration (mg/L) and C_e is the equilibrium nitrate concentration (mg/L).

The kinetics of nitrate uptake by money plant and arrowhead plant was ascertained by fitting the nitrate removal results obtained in each experimental run at different intervals of time to different kinetic models reported in the literature. However, only the Lagergren pseudo first-order gave an accurate fit. The pseudo first-order rate expression can be defined as shown in Equation (3) (Sinha *et al.* 2017):

$$\ln (C_e - C_t) = -kt + \ln C_e \tag{3}$$

where C_e and C_t, both expressed in mg/g, are the uptake capacities at equilibrium and at time t (min), respectively, and k (min^{-1}) is the pseudo first-order rate constant. The values of k for both the ornamental plants were calculated from the slope of a linear plot of $\ln (C_e - C_t)$ versus t.

RESULTS AND DISCUSSION

Model fitting and statistical analysis

Table 2 presents the nitrate removal efficiency observed at three levels of various factors tested for money plant and

Table 2 | Full factorial design for determining the nitrate removal by money plant and arrowhead plant

	Factors				Nitrate removal efficiency (%)	
Run no.	Initial nitrate concentration (mg/L), X_1	Money plant Growth period (d) (d), X_2	Arrowhead plant Growth period (d) (d), X_2	Plant density (g/L), X_3	Money plant	Arrowhead plant
1	40	1	3	20	9	15
2	120	1	3	20	5	7
3	40	6	12	20	79	77
4	120	6	12	20	76	71
5	40	1	3	80	13	24
6	120	1	3	80	12	18
7	40	6	12	80	93	88
8	120	6	12	80	80	82
9	80	3.5	7.5	50	37	32
10	80	3.5	7.5	50	40	34

arrowhead plant, respectively. The highest nitrate removal efficiency for both the plants occurred when experiments were performed under the conditions described for run no. 7 where a lower nitrate concentration, a higher growth period and a higher plant density were used. Under this condition, the maximum nitrate removal efficiency was found to be 93% and 88% for money and arrowhead plants, respectively.

The lowest value of nitrate removal efficiency, i.e., 5% and 7% was obtained in run no. 2 for both money and arrowhead plants, respectively. The increase of the plant growth period and plant density showed a steady increase in nitrate removal efficiency for all the experimental runs. These observations were similar to those reported previously by Bartucca *et al.* (2016) and Lingua *et al.* (2015). On the contrary, increasing the initial nitrate concentration from low to high levels decreased the nitrate removal efficiency (Iamchaturapatr *et al.* 2007; Ayyasamy *et al.* 2009).

The individual (main) and interaction effects of all the factors and the coefficients of regression, total sum of squares (TSS), probability value (*P*) and Fischer's value (*F*) for each response (i.e., the nitrate removal by money plant and arrowhead plant) are shown in Table 3. The coefficients (β) shown in Table 4 can be substituted in the regression model Equations (4) and (5), and the polynomial equations developed in the present study can be used further for evaluating the nitrate removal efficiency under different values of process parameters. The polynomial equations are as follows:

$$\begin{aligned} &\textit{Nitrate removal efficiency in money plant} \\ &\quad = -12.7376 - 0.083\,X_1 + 7.12\,X_2 - 0.045\,X_3 \\ &\qquad + 0.0073\,X_1X_2 + 0.0015\,X_1\,X_3 + 0.031\,X_2\,X_3 \end{aligned} \tag{4}$$

$$\begin{aligned} &\textit{Nitrate removal efficiency in arrowhead plant} \\ &\quad = 1.26 - 0.114\,X_1 + 12\,X_2 - 0.132\,X_3 \\ &\qquad + 0.01\,X_1\,X_2 + 0.0004\,X_1\,X_3 + 0.009\,X_2\,X_3 \end{aligned} \tag{5}$$

Based on the results shown in Figure 2(a) and 2(b), it can be ascertained that the plant growth period posed the maximum effect on nitrate removal efficiency, for both the ornamental plants. Figure 3 depicts the degree of fitness of the proposed model with the experimental results. The proposed model for estimating the nitrate removal efficiency by both the ornamental plants had R^2 values >0.95.

Student's *t*-test and analysis of variance

The Student's *t*-test and analysis of variance (ANOVA) was carried out to get a better insight into the experimental results. In general, regression models are said to be statistically significant if the Fischer's '*F*' value is high and the probability '*P*' value is low. Furthermore, the Student's

Table 3 | ANOVA for nitrate removal by (a) money plant and (b) arrowhead plant

	DF	Seq SS	Adj MS	F	P
(a) Source					
Main effects	3	10,596.2	3,532.1	78.04	0.013
X_1	1	55.5	55.5	1.23	0.383
X_2	1	10,434.5	10,434.5	230.53	0.004
X_3	1	106.3	106.3	2.35	0.265
2-way interactions	3	26.9	9	0.2	0.89
$X_1 \times X_2$	1	14.7	14.7	0.32	0.626
$X_1 \times X_3$	1	6.3	6.3	0.14	0.746
$X_2 \times X_3$	1	6	6	0.13	0.751
3-way interactions	1	20.6	20.6	0.46	0.569
$X_1 \times X_2 \times X_3$	1	20.6	20.6	0.46	0.569
Error	1	4.2	4.2	4.2	
Total	9	10,734.3			
(b) Source					
Main effects	3	8,316.27	2,772.09	13.11	0.072
X_1	1	6.93	6.93	0.03	0.873
X_2	1	8,083.56	8,083.56	38.22	0.025
X_3	1	225.78	225.78	1.07	0.41
2-way interactions	3	1.86	0.62	1	1
$X_1 \times X_2$	1	1.05	1.05	0	1.95
$X_1 \times X_3$	1	0.45	0.45	0	0.967
$X_2 \times X_3$	1	0.36	0.36	0	0.971
3-way interactions	1	0.36	0.36	0	0.971
$X_1 \times X_2 \times X_3$	1	0.36	0.36	0	0.971
Error	1	2.51	2.51		
Total	9	8,741.45			

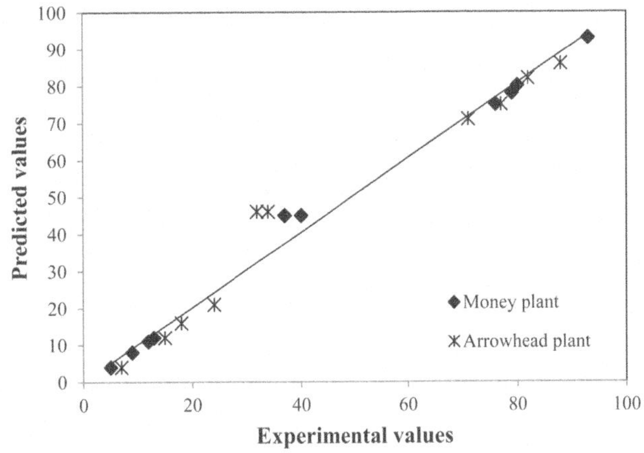

Figure 3 | Linear plot showing the distribution of experimental and predicted values of nitrate removal efficiency by money plant and arrowhead plant.

t-test was conducted to identify the 'course of the effect (positive or negative)', posed by each individual factor and its significance on the nitrate removal efficiency. The Student's *t*-test results were additionally interpreted for their interaction effects and their statistical index (*t*, *F* and *P*) values are shown in Table 4 for money plant and arrowhead plant, respectively.

The corresponding coefficient terms can be considered to be highly significant when the magnitude of '*t*' is larger and smaller for '*P*'. Generally, the Student's *t*-test and ANOVA conducted in the present study show that the plant growth period had a positive effect on the nitrate removal efficiency, while there were no significant effects by other factors in both the plants. The growth period showed the highest positive effect ($t = 15.18$ and

Table 4 | Student's *t*-test of the model coefficients for nitrate removal by money plant and arrowhead plants

	Nitrate removal efficiency (%)							
	Money plant				Arrowhead plant			
Term	Effect	Coefficient	*t*-value	*P*-value	Effect	Coefficient	*t*-value	*P*-value
Constant		−12.74	20.88	0.002		1.26	9.62	0.011
X_1	−0.166	−0.083	−1.11	0.383	−0.228	−0.114	−0.18	0.873
X_2	14.24	7.12	15.18	0.004	24	12	6.18	0.025
X_3	−0.09	−0.045	1.53	0.265	0.264	0.132	1.03	0.41
$X_1 \times X_2$	0.0146	0.0073	−0.57	0.626	0.02	0.01	0.07	0.95
$X_1 \times X_3$	0.003	0.0015	−0.37	0.746	0.0008	0.0004	0.05	0.967
$X_2 \times X_3$	0.062	0.031	0.36	0.751	0.018	0.009	0.04	0.971

$P = 0.004$) for money plant. Similarly, the growth period has shown the highest positive effect ($t = 6.18$ and $P = 0.025$) for arrowhead plant. Thus, it was observed that the individual effects had greater influence on the nitrate removal efficiency when compared to the interaction effects (Table 3).

Individual (main) effects plot

The individual effects or the main effects plot for money plant and arrowhead plant are shown in Figure 4(a) and 4(b), respectively. The individual effects demonstrate the relative quality of the effects of different factors in an

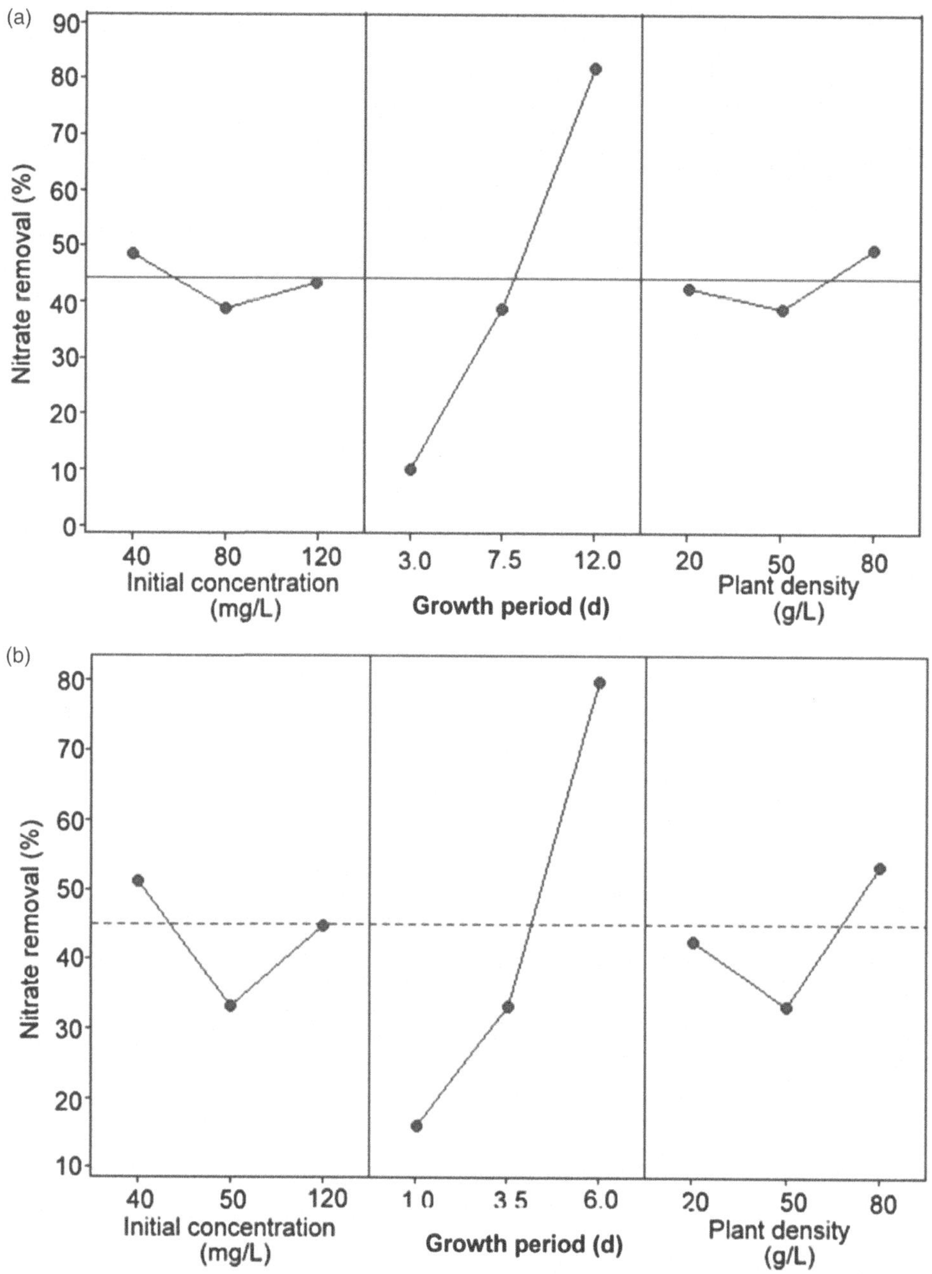

Figure 4 | Individual (main) effects plot for nitrate removal efficiency (%) by (a) money plant and (b) arrowhead plant.

experimental design. A factor is said to affect the process individually without the interaction of other factors when the mean responses of the individual factor exceed the other levels of the same factor (Singh *et al.* 2016). Further, the sign associated with the individual effects of the factors clearly depicts its course of action in removing nitrate from the water.

The increase in the nitrate removal profiles with an increase in the growth period and plant density at different levels reveals that both these factors positively affected the nitrate removal efficiency in money plant (Figure 4(a)). For money plant, an increase in the growth period from low to high level resulted in an increase in nitrate removal efficiency from 10% to 92%. Similarly, in the case of arrowhead plant, the nitrate removal efficiency increased from 18% to 80%. Enhancement in nitrate removal efficiency along with an increase in the growth period may be correlated with the fact that longer retention of the pollutants in the system will result in longer contact time between the plant surface and the pollutants (Xu & Shen 2011; Shu *et al.* 2017).

However, it was observed that the nitrate removal efficiency always decreased with an increase in the initial nitrate concentration, irrespective of the growth period and plant density factors being at any level. For instance, in the case of money plant, an increase in the initial nitrate concentration from low level to high level resulted in a decrease in the nitrate removal efficiency from 49% to 44%. Similarly, for the arrowhead plant, the nitrate removal efficiency decreased from 52% to 45%. This was also supported by the negative Student's *t*-test value ($t = -1.11$) observed in the present experiment. In conventional phytoremediation systems, the uptake capacity of plants reduces with an increase in initial nitrate concentration due to changes in the osmotic pressure (Ayyasamy *et al.* 2009; Ng & Chan 2017). Moreover, high nitrate concentration reduces the regeneration rate of active sites in the plant tissues which would eventually end up in slow metabolic nitrate uptake (Saber *et al.* 2018). On the other hand, at low nitrate concentrations, the plant accumulation capacity is usually high. Similar results were also observed in the case of arrowhead plant, i.e., the growth period and plant density showed a positive effect on the nitrate removal efficiency, while the initial nitrate concentration had a negative effect. This observation was also supported by the negative Student's *t*-test values ($t = -0.18$) for arrowhead plant. An increase in the plant density from low to high levels, both in money plant as well as the arrowhead plant, resulted in an increase in the nitrate removal efficiency from 44% to 50% and 43% and 55%, respectively. In the case of arrowhead plant, plant density was found to play a significant role in nitrate removal. It is because of the plant's wider leaves, with a well-developed root system and high growth rate (Sinha *et al.* 2017). Thus, an increase in the growth period in both the cases resulted in an increase in nitrate removal efficiency. However, in both the plants, nitrate removal efficiency was found to decrease with an increase in initial nitrate concentration, at some growth period and plant density conditions.

Interaction effects plot

The mean responses of any two factors, in all possible combinations, are usually plotted in the form of interaction effects plot. Non-parallel lines in the interaction effect plot indicate there is an interaction between the two tested factors. Nevertheless, the appearance of parallel lines in the plot shows that there is no interaction between the two factors (Singh *et al.* 2016). Figure 5(a) and 5(b) show the interaction effect plots for money plant and arrowhead plant, respectively. In the case of money plant (Figure 5(a)), positive interaction between the growth period × plant density was observed. On the contrary, a negative interaction effect was observed between the growth period × initial nitrate concentration and the plant density × initial nitrate concentration. An increase in the growth period from 3 to 12 d increased the nitrate removal efficiency by 72% (from 11 to 83%), at a plant density value of 80 g/L. Increasing the initial nitrate concentration from 40 to 120 g/L caused the nitrate removal efficiency to reduce by 5% (from 82 to 77%), at a growth period value of 12 d. Likewise, an increase in the initial nitrate concentration from low to high level decreased the nitrate removal efficiency value by 3% (from 45 to 42%), at a plant density value of 80 g/L.

A similar trend as observed in money plant was also observed in the case of arrowhead plant. From Figure 5(b), it is evident that an increase in the growth period from 1 to 6 d increased the nitrate removal efficiency by 63%

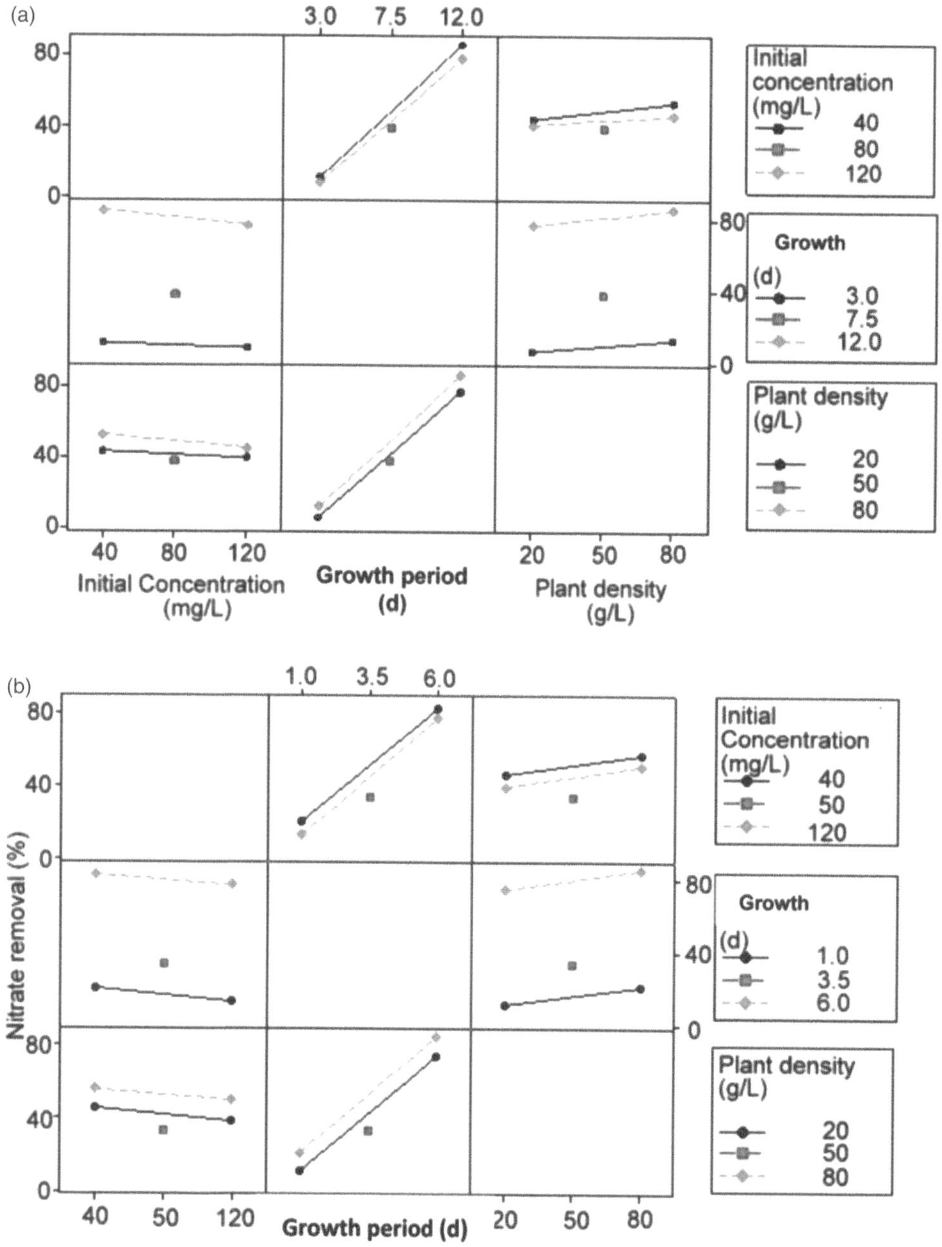

Figure 5 | Interaction effect plot for nitrate removal efficiency (%) by: (a) money plant and (b) arrowhead plant.

(from 20 to 83%), at an initial nitrate concentration of 40 g/L. Increasing the initial nitrate concentration from 40 to 120 g/L decreased the nitrate removal efficiency by 3% (from 82 to 79%), at a growth period of 6 d. Likewise, an increase in the initial nitrate concentration from low to a high level slightly reduced the nitrate removal efficiency by 2% (from 45 to 43%) at a plant density value of 80 g/L. The cumulative effects of different factors on the nitrate removal efficiency can be visualized in the form of 3D surface plots (Supplementary Material, Figures S1 and S2), for both money plant and arrowhead plant, respectively.

Kinetic modelling of nitrate removal by two plants

The values of the estimated kinetic model parameters, obtained using actively growing money and arrowhead

Table 5 | Estimated pseudo first-order kinetic parameters of nitrate uptake by (a) money plant and (b) arrowhead plant

Factors				Response
Run no.	Initial nitrate concentration (mg/L)	Plant density (g/L)	Growth period (d)	Kinetic constant (d^{-1})
(a) Money plant				
1	40	20	3	0.031
2	120	20	3	0.017
3	40	80	3	0.046
4	120	80	3	0.042
5	40	20	12	0.13
6	120	20	12	0.118
7	40	80	12	0.221
8	120	80	12	0.134
(b) Arrowhead plant				
1	40	20	1	0.162
2	120	20	1	0.072
3	40	80	1	0.278
4	120	80	1	0.199
5	40	20	6	0.244
6	120	20	6	0.208
7	40	80	6	0.353
8	120	80	6	0.289

plants, are presented in Table 5. The results reveal that nitrate removal by both the ornamental plants followed the pseudo first-order kinetics well with an R^2 value close to 1, which is in agreement with the literature report on pollutant removal in various other plant species (Li *et al.* 2002; Lv *et al.* 2016). For both the plants, the maximum estimated nitrate removal rate (Table 5(a) and 5(b)) obtained in run 7 using this model also matched well with the maximum nitrate removal efficiency (84.3%) obtained in run no. 7 (Table 1). From Table 5, it can be observed that at similar initial nitrate concentration and plant density, the nitrate removal rate constant ($k = 0.353\ d^{-1}$) for arrowhead plant was superior to money plant ($k = 0.221\ d^{-1}$), irrespective of the different growth period values (Shyamala *et al.* 2017). The maximum nitrate removal rate observed in run no. 7 may be attributed to the higher metabolic rate of these plants, which in turn, can be attributed to the higher plant density, i.e., the presence of a greater number of active sites in the plants to draw nitrate from the system (Saber *et al.* 2018). Further, low concentration of nitrate in this run would have kept the regeneration rate of these active sites intact, thereby leading to an unhindered nitrate uptake. The results obtained from this study clearly show that both these ornamental plants may be used in constructed wetland systems as a cheap, natural, environmentally and technically viable treatment technique for nitrate removal from water.

Future perspectives

Since the money plant used in the present study offered a greater nitrate removal efficiency and the arrowhead plant presented a maximum nitrate uptake rate, further investigation by combining both these plants in a single system may result in the development of a faster and efficient nitrate removal system. Moreover, it would be fascinating to carry out optimization studies by involving various other process parameters like pH, dissolved oxygen and wastewater volume, etc. However, in order to realize the full potential of the ornamental plants used in this study (money plant and arrowhead plant) for treatment of nitrate-containing wastewater, a detailed investigation by adopting the pilot-scale wetland system is needed instead of the lab-scale hydroponic system. As well, investigation on the effect of co-ions on nitrate removal and mimicking the current system with that of the real-world scenario by subjecting it to shock loads of different volumes of wastewater is warranted. A cost–benefit analysis on such a system would unleash the use of ornamental plants in phytoremediation. Thus, the results of this study would be valuable to establish the use of these ornamental plants for nitrate removal from water in field-scale systems.

CONCLUSIONS

The individual (main) effect of the process variables, i.e., the initial nitrate concentration, growth period, plant density, was found to play a major role in affecting the nitrate removal process. Among the individual effects, growth period appeared to have a significant influence on nitrate removal for both money plant ($t = 15.18$, $P = 0.004$) and arrowhead plant ($t = 6.18$, $P = 0.025$), respectively. At their highest growth period, the arrowhead plant demonstrated better nitrate removal kinetics when compared to money plant. For example, at an initial nitrate concentration of 40 mg/L and a plant density of 20 g/L, the rate constant for nitrate removal was found to be 0.13 d^{-1} and 0.244 d^{-1}, respectively, at the growth periods of 12 and 6 d for money plant and arrowhead plant.

ACKNOWLEDGEMENTS

The authors acknowledge GMR Institute of Technology, Rajam, Andhra Pradesh (India), India for the infrastructural support provided to carry out this research work. ERR thanks IHE-Delft, Institute for Water Education, Delft (The Netherlands) for providing staff time support to collaborate with researchers from India.

SUPPLEMENTARY MATERIAL

The Supplementary Material for this paper is available online at https://dx.doi.org/10.2166/aqua.2019.111.

REFERENCES

Arun, S., Manikandan, N. A., Pakshirajan, K., Pugazhenthi, G. & Syiem, M. B. 2017 Cu (II) removal by *Nostoc muscorum* and its effect on biomass growth and nitrate uptake: a photobioreactor study. *Int. Biodeterior. Biodegrad.* **119**, 111–117.

Ayyasamy, P. M., Rajakumar, S., Sathishkumar, M., Swaminathan, K., Shanthi, K., Lakshmanaperumalsamy, P. & Lee, S. 2009 Nitrate removal from synthetic medium and groundwater with aquatic macrophytes. *Desalination* **242** (1–3), 286–296.

Bartucca, M. L., Mimmo, T., Cesco, S. & Del Buono, D. 2016 Nitrate removal from polluted water by using a vegetated floating system. *Sci. Total Environ.* **542**, 803–808.

Bulgariu, L., Ceica, A., Lazar, L., Creţescu, I. & Balasanian, I. 2012 Equilibrium and kinetics study of nitrate removal from water by Purolite a520-E resin. *Environ. Eng. Manage. J. (EEMJ)* **11** (1), 37–45.

Challagulla, S., Tarafder, K., Ganesan, R. & Roy, S. 2017 All that glitters is not gold: a probe into photocatalytic nitrate reduction mechanism over noble metal doped and undoped TiO_2. *J. Phys. Chem. C* **121** (49), 27406–27416.

Chen, G., Liu, X., Brookes, P. C. & Xu, J. 2015 Opportunities for phytoremediation and bioindication of arsenic contaminated water using a submerged aquatic plant: *Vallisneria natans* (lour.) Hara. *Int. J. Phytoremed.* **17** (3), 249–255.

Chen, J., Wu, H., Qian, H. & Gao, Y. 2017 Assessing nitrate and fluoride contaminants in drinking water and their health risk of rural residents living in a semiarid region of northwest China. *Expos. Health* **9** (3), 183–195.

Escobar-Alvarado, L. F., Vaca-Mier, M., López-Callejas, R. & Rojas-Valencia, M. N. 2018 Efficiency of *Opuntia ficus* in the phytoremediation of a soil contaminated with used motor oil and lead, compared to that of *Lolium perenne* and *Aloe barbadensis*. *Int. J. Phytoremed.* **20** (2), 184–189.

Fout, G. S., Borchardt, M. A., Kieke, B. A. & Karim, M. R. 2017 Human virus and microbial indicator occurrence in public-supply groundwater systems: meta-analysis of 12 international studies. *Hydrogeol. J.* **25** (4), 903–919.

Hazarika, J., Pakshirajan, K., Sinharoy, A. & Syiem, M. B. 2015 Bioremoval of Cu (II), Zn (II), Pb (II) and Cd (II) by *Nostoc muscorum* isolated from a coal mining site. *J. Appl. Phycol.* **27** (4), 1525–1534.

Iamchaturapatr, J., Yi, S. W. & Rhee, J. S. 2007 Nutrient removals by 21 aquatic plants for vertical free surface-flow (VFS) constructed wetland. *Ecol. Eng.* **29** (3), 287–293.

Karmakar, S., Mukherjee, J. & Mukherjee, S. 2016 Removal of fluoride contamination in water by three aquatic plants. *Int. J. Phytoremed.* **18** (3), 222–227.

Khajuria, A. & Kanae, S. 2013 Potential and use of nitrate in agricultural purposes. *J. Water Resour. Prot.* **5** (5), 529.

Li, H., Sheng, G., Sheng, W. & Xu, O. 2002 Uptake of trifluralin and lindane from water by ryegrass. *Chemosphere* **48** (3), 335–341.

Li, K., Liu, L., Yang, H., Zhang, C., Xie, H. & Li, C. 2016 Phytoremediation potential of three species of macrophytes for nitrate in contaminated water. *Am. J. Plant Sci.* **7** (08), 1259.

Li, Q., Yang, J., Fan, W., Zhou, D., Wang, X., Zhang, L., Huo, M. & Crittenden, J. C. 2018 Different transport behaviors of *Bacillus subtilis* cells and spores in saturated porous media: implications for contamination risks associated with bacterial sporulation in aquifer. *Colloids Surf. B. Biointerfaces* **162**, 35–42.

Lingua, G., Copetta, A., Musso, D., Aimo, S., Ranzenigo, A., Buico, A., Gianotti, V., Osella, D. & Berta, G. 2015 Effect of arbuscular mycorrhizal and bacterial inocula on nitrate concentration in mesocosms simulating a wastewater treatment system relying on phytodepuration. *Environ. Sci. Pollut. Res.* **22** (23), 18616–18625.

Liu, H., Tong, S., Chen, N., Liu, Y., Feng, C. & Hu, Q. 2015 Effect of electro-stimulation on activity of heterotrophic denitrifying bacteria and denitrification performance. *Bioresour. Technol.* **196**, 123–128.

Liu, J., Xin, X. & Zhou, Q. 2017 Phytoremediation of contaminated soils using ornamental plants. *Environ. Rev.* **999**, 1–12.

Lv, T., Zhang, Y., Casas, M. E., Carvalho, P. N., Arias, C. A., Bester, K. & Brix, H. 2016 Phytoremediation of imazalil and tebuconazole by four emergent wetland plant species in hydroponic medium. *Chemosphere* **148**, 459–466.

Ng, Y. S. & Chan, D. J. C. 2017 Wastewater phytoremediation by *Salvinia molesta*. *J. Water Process Eng.* **15**, 107–115.

Nieder, R., Benbi, D. K. & Reichl, F. X. (eds) 2018 Reactive water-soluble forms of nitrogen and phosphorus and their impacts on environment and human health. In: *Soil Components and Human Health*. Springer, Dordrecht, The Netherlands, pp. 223–255.

Pavlineri, N., Skoulikidis, N. T. & Tsihrintzis, V. A. 2017 Constructed floating wetlands: a review of research, design, operation and management aspects, and data meta-analysis. *Chem. Eng. J.* **308**, 1120–1132.

Pennino, M. J., Compton, J. E. & Leibowitz, S. G. 2017 Trends in drinking water nitrate violations across the United States. *Environ. Sci. Technol.* **51** (22), 13450–13460.

Radziemska, M., Vaverková, M. D. & Mazur, Z. 2019 Pilot scale use of compost combined with sorbents to phytostabilize ni-contaminated soil using *Lolium perenne* L. *Waste Biomass Valori.* **10** (6), 1585–1595.

Rai, S. N. 2003 Groundwater pollution in India – an overview. In: *Groundwater Pollution* (V. P. Singh & R. N. Yadava, eds). Allied Publishers Pvt. Ltd, New Delhi, India, pp. 419–436.

Roy, A. S., Hazarika, J., Manikandan, N. A., Pakshirajan, K. & Syiem, M. B. 2015 Heavy metal removal from multicomponent system by the cyanobacterium *Nostoc muscorum*: kinetics and interaction study. *Appl. Biochem. Biotechnol.* **175** (8), 3863–3874.

Saber, A., Tafazzoli, M., Mortazavian, S. & James, D. E. 2018 Investigation of kinetics and absorption isotherm models for hydroponic phytoremediation of waters contaminated with sulfate. *J. Environ. Manage.* **207** (1), 276–291.

Sharma, B., Parul, A. K., Jain, U., Yadav, J. K., Singh, R. & Mishra, R. 2017 Occurrence of multidrug resistant *Escherichia coli* in groundwater of Brij region (Uttar Pradesh) and its public health implications. *Vet. World* **10** (3), 293.

Shu, L., Jegatheesan, V., Keir, G. & Keir, M. 2017 *CESE 2017 – Abstract Book: The Tenth Annual Conference on the Challenges in Environmental Science and Engineering*. LJS Environment, Highton, Australia.

Shyamala, S., Manikandan, N. A., Pakshirajan, K., Rene, E. R. & Behera, S. K. 2017 Phytoremediation potential of two ornamental plants treating nitrate contaminated water. In: *CESE 2017 – Abstract Book: The Tenth Annual Conference on the Challenges in Environmental Science and Engineering* (L. Shu, V. Jegatheesan, G. Keir & M. Keir, eds). LJS Environment, Highton, Australia.

Singh, M., Pakshirajan, K. & Trivedi, V. 2016 Photo-inactivation of *Escherichia coli* and *Enterococcus hirae* using methylene blue and sodium anthraquinone-2-sulphonate: effect of process parameters. *3 Biotech* **6** (2), 176.

Sinha, V., Pakshirajan, K., Manikandan, N. A. & Chaturvedi, R. 2017 Kinetics, biochemical and factorial analysis of chromium uptake in a multi-ion system by *Tradescantia pallida* (Rose) DR Hunt. *Int. J. Phytoremed.* **19** (11), 1007–1016.

Sinha, V., Pakshirajan, K. & Chaturvedi, R. 2018 Chromium tolerance, bioaccumulation and localization in plants: an overview. *J. Environ. Manage.* **206**, 715–730.

Soomro, F., Rafique, T., Michalski, G., Ali, S. A., Naseem, S. & Khan, M. U. 2017 Occurrence and delineation of high nitrate contamination in the groundwater of Mithi sub-district, Thar Desert, Pakistan. *Environ. Earth Sci.* **76** (10), 355.

Tyagi, S., Rawtani, D., Khatri, N. & Tharmavaram, M. 2018 Strategies for nitrate removal from aqueous environment using nanotechnology: a review. *J. Water Process Eng.* **21**, 84–95.

Villa-Gomez, D. K., Pakshirajan, K., Maestro, R., Mushi, S. & Lens, P. N. L. 2015 Effect of process variables on the sulfate reduction process in bioreactors treating metal-containing wastewaters: factorial design and response surface analyses. *Biodegradation* **26** (4), 299–311.

Volf, I., Rakoto, N. G. & Bulgariu, L. 2015 Valorization of *Pistia stratiotes* biomass as biosorbent for lead (II) ions removal from aqueous media. *Sep. Sci. Technol.* **50** (10), 1577–1586.

Wang, Z., Jiang, Y., Awasthi, M. K., Wang, J., Yang, X., Amjad, A., Wang, Q., Lahori, A. H. & Zhang, Z. 2018 Nitrate removal by combined heterotrophic and autotrophic denitrification processes: impact of coexistent ions. *Bioresour. Technol.* **250**, 838–845.

Weldeslassie, T., Naz, H., Singh, B. & Oves, M. 2018 Chemical contaminants for soil, air and aquatic ecosystem. In: *Modern Age Environmental Problems and Their Remediation* (M. Oves, M. Z. Khan & I. M. Ismail, eds). Springer, Cham, pp. 1–22.

Wollheim, W. M., Mulukutla, G. K., Cook, C. & Carey, R. O. 2017 Aquatic nitrate retention at river network scales across flow conditions determined using nested in situ sensors. *Water Resour. Res.* **53** (11), 9740–9756.

Xu, J. & Shen, G. 2011 Growing duckweed in swine wastewater for nutrient recovery and biomass production. *Bioresour. Technol.* **102** (2), 848–853.

Xu, X. J., Lai, G. L., Chi, C. Q., Zhao, J. Y., Yan, Y. C., Nie, Y. & Wu, X. L. 2018 Purification of eutrophic water containing chlorpyrifos by aquatic plants and its effects on planktonic bacteria. *Chemosphere* **193**, 178–188.

First received 26 July 2019; accepted in revised form 29 October 2019. Available online 25 November 2019

Design and operation of submerged layer in bioretention for enhanced nitrate removal

Junyu Zhang IWA, Rajendra Prasad Singh IWA, Yunzhe Liu and Dafang Fu IWA

ABSTRACT

Bioretention, initially designed for treating discontinuous runoff pollution, faces considerable challenges in its trade-off between the hydraulic retention time (HRT) and its treatment capacity. In this study, six enhanced submerged media together with four HRTs were designed for bioretention cells to treat the highly nitrogenous river water in Tai lake basin in Yangtze River delta, China. Results revealed that bioretention with activated carbon has the highest removal of nitrate (NO_3^--N) (93–96%) compared with surfactant-modified activated carbon (SMAC), surfactant-modified zeolite (SMZ), zeolite, fly ash and ceramsite. Although the SMAC had the best absorption for NO_3^--N and could desorb NO_3^--N when its concentration was low in the submerged layer, the desorbed surfactant could inhibit the growth of denitrifying bacteria, which leads to low removal efficiency (49–66%). The dynamic balancing of NO_3^--N desorption and denitrifying system restrain in the SMAC device was observed and explained. The best activated carbon-gravel proportion in the submerged layer was 1:1 (150 mm). Such design could ensure the stable and efficient NO_3^--N removal rate (93–94%) under high inflow concentration (28.9 mg/L) and high hydraulic loading (8.2 cm/h).

Key words | bioretention, NO_3^--N, river treatment, submerged zone, surfactant modification, Tai lake basin

Junyu Zhang IWA
Rajendra Prasad Singh IWA
Dafang Fu IWA (corresponding author)
School of Civil Engineering,
Southeast University,
#2 Sipailou, Nanjing 210096,
China
E-mail: *fdf@seu.edu.cn*

Junyu Zhang
Rajendra Prasad Singh
Yunzhe Liu
Dafang Fu
Southeast University-Monash University Joint Research Centre for Future Cities,
Nanjing 210096,
China

INTRODUCTION

Due to the continuous deterioration of the water environment across China, the government carried out the action plan for 'Prevention and Control of Water Pollution' in 2015 to enhance the efforts for water quality improvement towards a long-term national strategy. This action plan aimed to restore all the contaminated water bodies in the Yangtze River delta by 2020. However, the inflowing rivers and tributaries of Tai lake (a major freshwater resource in the Yangtze River delta) are facing serious problems with pollution. Most of these rivers would not meet the National Standard for Surface Water Quality of China, especially on the total nitrogen (TN) (She *et al.* 2017; Tao *et al.* 2018). The annual average TN concentration for the 15 inflowing rivers ranges from 4.5 to 6.0 mg/L, which is 2–3 times the minimum standard, and 70–92% of the TN is composed of dissolved nitrogen (DN), in which 41–77% is NO_3^--N.

Bioretention systems have been studied and applied for removing pollutants, such as nitrogen, from runoff since the 1990s (Liu *et al.* 2014). The removal of ammonia nitrogen (NH_3-N) is mainly through adsorption or ion exchange in the soil and filler. Reports indicated that a regular bioretention system can remove 60–90% of NH_3^--N (Davis *et al.* 2001; Bratieres *et al.* 2008; Hunt *et al.* 2008; Zinger *et al.* 2013) and 45–75% of total kjeldahl nitrogen (Davis *et al.* 2001). The removal of NO_3^--N was generally poor, ranging from –4.7

doi: 10.2166/aqua.2019.164

to 24% due to its non-absorbability by the filler as well as the transformation from ammonia and organo-nitrogen between rainfall events (Davis *et al.* 2001; Kim *et al.* 2003; Hunt *et al.* 2008). Kim *et al.* (2003) conducted a study on the design of the submerged anaerobic zone (SAZ) and noted that bioretention systems with SAZ could remove 70–80% of NO_3^--N while the pre-added carbon source could ideally last 20 years. Further studies also proved that SAZ can provide a suitable environment for denitrifying bacteria and significantly improve the NO_3^--N removal rate (Ming & Chan 2012; Palmer *et al.* 2013). A recent study has also proved that most of the NO_3^--N from the inflow as well as that transformed from NH_3^+-N will be retained in SAZ.

However, SAZ also has negative impacts for bioretention systems. The anaerobic environment in the system will cause the leaching of phosphorus due to the pre-added carbon source (Zinger *et al.* 2013) or transformation of the undissolved phosphorus into dissolved ones (Correll 1998; Clark & Pitt 2012). The anaerobic zone also slows down the process of ammonification of organo-nitrogen as well as nitrification of ammonia (Zinger *et al.* 2007, 2013; Li *et al.* 2014; Ribas *et al.* 2015). Therefore, the removal rate of NH_3^+-N and organo-nitrogen is reduced to some extent. Nevertheless, the bioretention systems, especially those with SAZ, is originally designed for treating runoff, which takes advantage of the interval between two events to biodegrade or biologically uptake the nitrogen (N) (Davis *et al.* 2006, 2009; Palmer *et al.* 2013). Earlier studies suggested the capacity of removing NO_3^--N in the inflow water mainly depends on the size and hydraulic retention time (HRT) of SAZ, as the free-state NO_3^--N does not absorb on traditional bioretention media nor natural soil (Kim *et al.* 2003; Zinger *et al.* 2007; Blecken *et al.* 2009; Passeport *et al.* 2009; Lucas *et al.* 2010; Chen *et al.* 2013; Lefevre *et al.* 2015; Lynn *et al.* 2015). Therefore, to design a bioretention system for treating polluted rivers in Tai lake basin, the major challenge lies in the balance between HRT and treatment capacity. There are many well established evaluation methods available regarding bioretention design as the optimization of HRT or nitrogen removal capacity in the submerged layer has been investigated by many researchers.

This study is aimed at finding a practical long-term treatment method for the high nitrate river water in the Taihu basin other than traditional emergency methods (e.g. water diversion) or short-term methods (e.g. sediment dredging). The following work has been conducted: (1) selection and modification of six filter media in the submerged layer to provide NO_3^--N immobilization, enabling efficiency and continuous NO_3^--N removal; (2) monitor both the long-term and single-event performance to select the best modified filler; (3) monitor the vertical diversity of NO_3^--N removal performance to optimize the bioretention structure design; (4) monitor the performance under different hydraulic loadings to analyze the feasibility of treating river water using bioretention systems.

MATERIAL AND METHODS

Experimental device

In order to compare the performance of bioretention cells under different design features, a parallel experiment was carried out by designing and constructing a series of bioretention units, as shown in Figure 1. These units were created using polyvinyl chloride (PVC) tubes (250 mm in diameter, 1,200 mm in height, and 10 mm in thickness). The overflow outlet was set at 100 mm from the top (valve A) while the outlets were set at 0 and 450 mm from the bottom (valve E and valve F). Three sampling outlets were set at the height of 250, 500 and 700 mm from the bottom (valve D, valve C and valve B, respectively). Two transparent glass ports were placed at valve B and C for the piezometer.

Filter media

The detailed design of bioretention cells varied from case to case while the basic structure always stays the same and the size of each layer remains in a certain range. Thus, referring to current design guidelines could ensure the performance comparison with existing guidelines as there is no excessive difference in design. The experimental devices in this study were designed to be a typical bioretention system with a SAZ of 450 mm depth. The filter media was chosen from locally available materials and overlaid according to the design guidelines (Hatt *et al.* 2009). The setting of layers and composition of filler materials are presented in Tables 1 and 2. A mixture of 150 g wood shavings and 150 g recycled

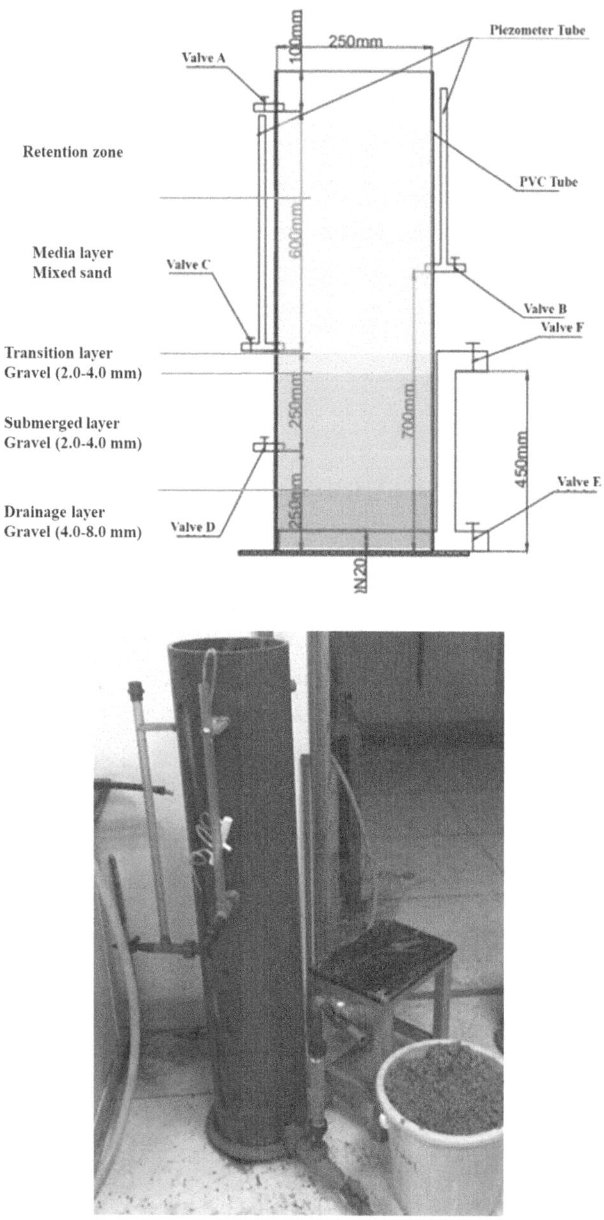

Figure 1 | Schematic design of bioretention unit.

newspaper was added in the submerged layer for extra carbon source.

Inflow water

As the concentration of pollutants in the river fluctuate dramatically across the year, it is very difficult to collect and store natural river water for the experiment. Therefore, the inflow water of the system was prepared by mixing pollutants with tap water according to a given concentration (Table 3). To ensure the reliability of the result, these standards were determined by investigating the highest annual average value of each indicator in all inflowing rivers in Tai lake basin.

Filler modification

According to existing studies, six NO_3^--N removal enhanced media were selected in this study due to their promising removal capacities in wastewater/runoff treatment: activated carbon (AC), surfactant-modified activated carbon (SMAC) (Choi *et al.* 2008, 2009; Hong *et al.* 2008), zeolites, surfactant-modified zeolites (SMZ) (Haggerty & Bowman 1994; Li 2003; Schick *et al.* 2010), fly ash, and ceramsite. The adoption of these media from literature could form a reliable base of modification options as well as offer comparability of results to existing studies.

A lab-scale application was primarily carried out to select feasible media for further experiments. The number of test sets was determined based on the minimum and maximum dosage as well as common intervals in the literature. Following the existing studies (Choi *et al.* 2008, 2009; Hong *et al.* 2008; Sarina *et al.* 2010), the SMAC used in this study was made up of particle activated carbon (2.0–4.0 mm) modified by cetylpyridinium chloride (CPC). Eight sets of SMAC: (1) for each set, 15 g activated carbon and 45 mL deionized water were added into a 250 mL conical flask; (2) 3.0, 6.0, 9.0, and 12.0 mmol technically pure CPC and analytically pure CPC were added into eight flasks, respectively (CPC concentration: 200, 400, 600, and 800 mmol/kg activated carbon); (3) the flasks were sealed and vibrated in a vapor-bathing vibrator at 150 r/min under 30 °C for 24 h; (4) the mixture in each flask was centrifuged at 4,000 r/min for 30 min to separate the solid and liquid; (5) the remaining solid was washed with deionized water repeatedly until no precipitate emerged when testing with $AgNO_3$; (6) the SMACs were dried at 60 °C for 12 h.

The SMZ used in this study was made up of zeolites (2.0–4.0 mm) modified by Hexadecyl trimethyl ammonium Bromide (HDTMA) according to existing studies (Haggerty & Bowman 1994; Li 2003; Schick *et al.* 2010). Twelve sets of SMZ were prepared: (1) for each set, 10 g zeolites and

Table 1 | Physical properties of filling materials

Layer	Depth (mm)	Filler	
		Traditional	Enhanced
Retention zone	200	–	–
Media layer	400	Mixed sand	
Transition layer	50	Gravel (1.0–2.0 mm)	
Submerged layer	300	Gravel (2.0–4.0 mm)	100 mm enhanced filler (2.0–4.0 mm) 200 mm gravel (2.0–4.0 mm)
Drainage layer	150	Gravel (4.0–8.0 mm)	

Table 2 | Composition of mixed sand in media layer

Particle size (mm)	< 0.15	0.15–0.30	0.30–0.55	0.55–1.00	1.00–2.00
Mass ratio	5%	10%	35%	35%	10%

45 mL deionized water were added into a 250 mL conical flask; (2) 1.0, 1.5, 2.0, 3.0, 4.0, and 5.0 mmol technically pure HDTMA and analytically pure HDTMA were added into the 12 flasks, respectively (HDTMA concentration: 100, 150, 200, 300, 400, and 500 mmol/kg zeolites); (3) the flasks were sealed and vibrated in a vapor-bathing vibrator at 150 r/min at 30 °C for 24 h; (4) the mixture in each flask was centrifuged at 4,000 r/min for 30 min to separate the solid and liquid; (5) the remaining solid with deionized water was washed 10 times; (6) the SMZs were dried at 60 °C for 12 h.

Filler selection

Enhanced filler selection

The performance of the modified fillers was tested and compared to determine the better modified pattern for different fillers. The experiment was carried out under the following procedures: (1) 1 g modified filler was added into a conical flask together with 20 mL standard potassium nitrate (KNO_3) solution (100 mg/L); (2) the flasks were sealed and vibrated in a vapor-bathing vibrator at 150 r/min at 30 °C for 24 h; (3) the supernatant was collected and the concentration of NO_3^--N was measured through ultraviolet (UV) spectrophotometry, the best performed SMAC and SMZ were selected as the candidate fillers. The performance of all fillers was tested and compared following the procedures as above. The better performed fillers were adopted in the practical experiments.

Performance of pollutant removal

After selecting the enhanced filler, the bioretention units were built according to Tables 1 and 2. The following preconditioning process was carried out to stabilize the structure of each layer as well as the properties of media, and to initiate the microbial community in the units. (1) The units were flushed with tap water four times per day for 7 days to simulate the saturation process during operation, as well as to precipitate fillers in each

Table 3 | Pollutant concentration in inflow water and monitoring method

Indicator	Target concentration (mg/L)	Source	Quantity (mg/L)	Monitoring method
COD	8	Glucose	8	Acidic potassium permanganate method
TN	8	–	–	UV spectrophotometry
ON (organic nitrogen)	1.6	$C_6H_5O_2N$	14.07	
NH_4^+-N	2.4	NH_4Cl	9.17	Nessler's reagent colorimetry
NO_3^--N	4	KNO_3	28.89	UV spectrophotometry
TP	0.4	KH_2PO_4	1.76	Phosphovanadomolybdate spectrophotometric method

layer to ensure stable structure in the following experiments and to wash out the salt and unsteady organics in the media. (2) The units were saturated with natural river water, which would be kept in the units for 48 h and drained out before the next saturation. The process was repeated for 2 weeks to initiate the relevant microbial community.

The long term performance of the bioretention units was monitored by the following procedure. (1) Units were saturated with inflow water (room temperature, 10–20 °C). The inflow rate (peristaltic pump) and the outflow rate (valve F) were adjusted to 4 L/h. (2) Systems were kept running while the outflow water quality was being monitored, the systems were stopped until the water quality remained stable for 4 h. (3) The experiments were repeated once or twice a week for 7 weeks.

Extra samples were taken during weeks 1, 3 and 7 to investigate the vertical diversity of NO_3^--N removal in the units. The samples were taken at the end of each experiment, from inflow, valve B, C, D and E (200, 400, 650, and 900 mm, respectively, from the surface of media layer).

Optimization of design and operation

After the selection of the best filler, the design of SAZ composition as well as the operation pattern were investigated by evaluating the system performance of bioretention units with various proportions of enhanced filler (Table 4) in SAZ under different hydraulic loading (4, 6, 8, and 12 L/h). After considering the similar preconditioning process mentioned in the section 'Performance of pollutant removal', the following procedures were carried out for the optimization of system operation: (1) the outflow valve F was fully opened to prevent saturation in media layer, providing the environment for the aerobic reaction; (2) the units were operated under a hydraulic loading of 4 L/h for 24 h and stopped for 48 h, and repeated five times to allow for system stabilization and adaptation of the microbial communities to such a loading environment; (3) the units were operated under 4 L/h loading for 24 h, and the outflow water quality was monitored during the experiment; and (4) the hydraulic loading was adjusted and steps 1–3 were repeated.

Table 4 | Composition of submerged layer

Layer	Media	Composition A (mm)	B (mm)	C (mm)	D (mm)
Submerged layer	Enhanced filler	50	100	150	200
	Gravel	250	200	150	100

RESULTS AND DISCUSSION

Enhanced filler selection

The NO_3^--N removal capacity of the eight sets of SMACs and 12 sets of SMZs was first investigated to determine the better modifying pattern for activated carbon and zeolites (Figure 2). For SMAC, the non-modified activated

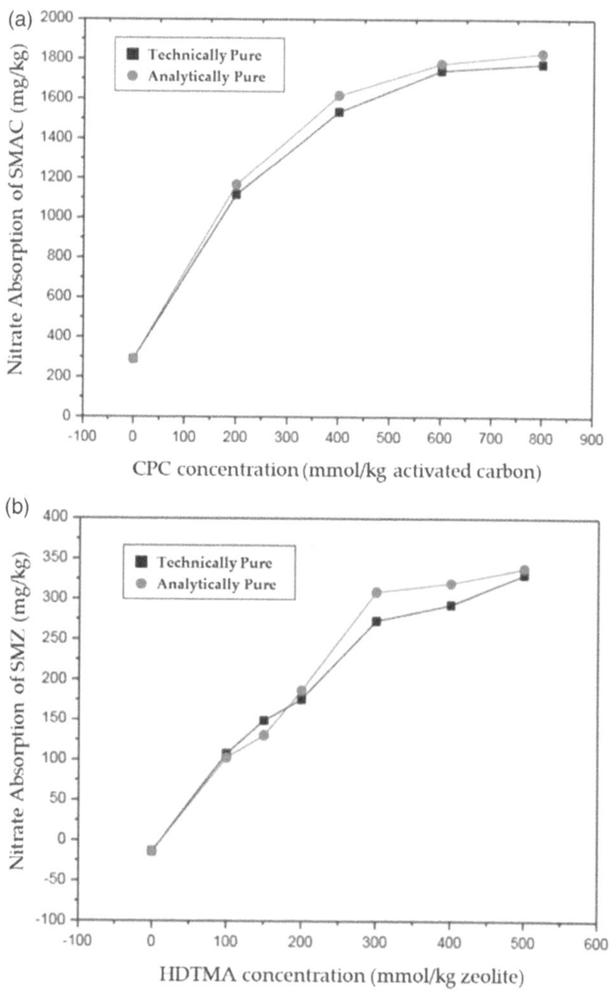

Figure 2 | NO_3^--N absorption capacity for different modifier concentrations: (a) activated carbon modified by CPC; (b) zeolite modified by HDTMA).

carbon has an NO_3^--N absorption capacity of 290 mg/kg. The capacity rises dramatically with the CPC concentration increasing from 0 to 400 mmol/kg and gradually declined after 400 mmol/kg. The improvement of NO_3^--N absorption from 600 to 800 mmol/kg was only 30 mg/kg (1.7%). The difference between modification by technically pure and analytically pure modifier was less than 85 mg/kg (5.6%). Thus, the efficient pattern of SMAC modification was 600 mmol technically pure CPC/kg activated carbon.

For SMZ, the non-modified zeolite has no NO_3^--N absorption capacity. The SMZs modified by analytically pure HDTMA perform better than those modified by technically pure HDTMA in most cases. For SMZs modified by analytically pure HDTMA, their capacity enhanced dramatically with the modifier concentration increasing from 0 to 300 mmol/kg and gradually slowed down after 300 mmol/kg. The improvement of NO_3^--N absorption from 300 to 500 mmol/kg was only 23 mg/kg (7.3%). Thus, the efficient pattern of SMZ modification was of 300 mmol analytically pure HDTMA/kg zeolite.

After determining the efficient modification pattern of SMAC and SMZ, the NO_3^--N absorption capacities of fillers were analyzed. As shown in Figure 3, SMAC has the highest absorption capacity (1,742 mg/kg) which was more than five times compared to the second ranked SMZ (311 mg/kg) and the third ranked activated carbon (290 mg/kg). The remaining three fillers have very limited capacity in NO_3^--N absorption. The result suggests higher absorption capacity of SMAC compared to SMZ, which is consistent with existing studies (Li 2003; Schick *et al.* 2010). Compared to previous studies, the SMAC and SMZ modified in the study have a lower capacity of NO_3^--N absorption capacity, which may be due to the larger particle size, smaller specific surface area, as well as higher pollutant concentration. Based on the absorption capacity, the SMAC, SMZ, activated carbon and zeolite (as compared to SMZ) were selected as the potential enhanced filler for further operation and experiments.

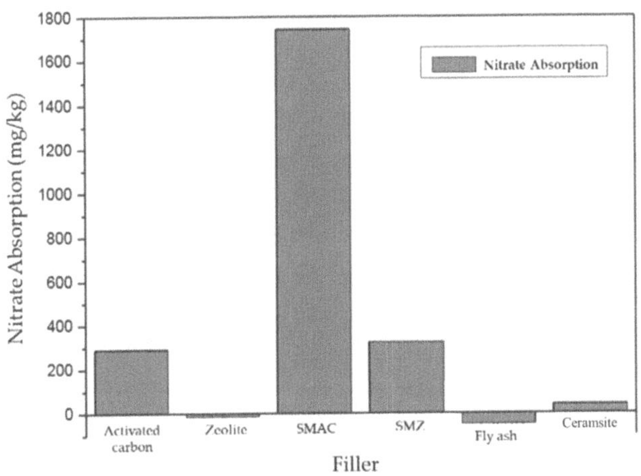

Figure 3 | NO_3^--N absorption capacity for different enhanced fillers.

Performance of bioretention units

Five bioretention units were built with the traditional sand filler and four enhanced fillers were selected after evaluation of their performance in primary stage experiments. Long-term performance as well as single-event performance of the units at the end of the experiment was investigated (Figure 4). At the initial stage, the SMAC device had the highest NO_3^--N removal efficiency (71%) while the other four devices had very poor performance (non-modified: 18%; zeolite: 16%; activated carbon: 32%; and SMZ: 12%). Such a difference could be due to the extraordinary absorption capacity of SMAC. The results show that after the 3-week preconditioning, SMAC units still had the capacity for effective removal of NO_3^--N.

With time, denitrification processes in the devices gradually occurred and stabilized at week 3. The NO_3^--N removal efficiency of SMAC device continuously decreased and reached 37% at day 21, while the other four devices had significant improvement on their performance (non-modified: 69%; zeolite: 52%; activated carbon: 73%; and SMZ: 29%). The decrease in performance of the SMAC device suggests that the high removal capacity at the initial stage was mainly due to the absorption of SMAC rather than denitrification. Notably, the modified fillers performed poorer than the non-modified fillers and the non-enhanced filler. The possible reason for such a phenomena may be the existence of cationic surfactant in the submerged area coming from the residual or desorb of the modifier which could restrain the denitrification process (Schick *et al.* 2010) in the submerged area. In this case, although the modified fillers can desorb the NO_3^--N during the operation interval, it could hardly be removed. Thus, the modified

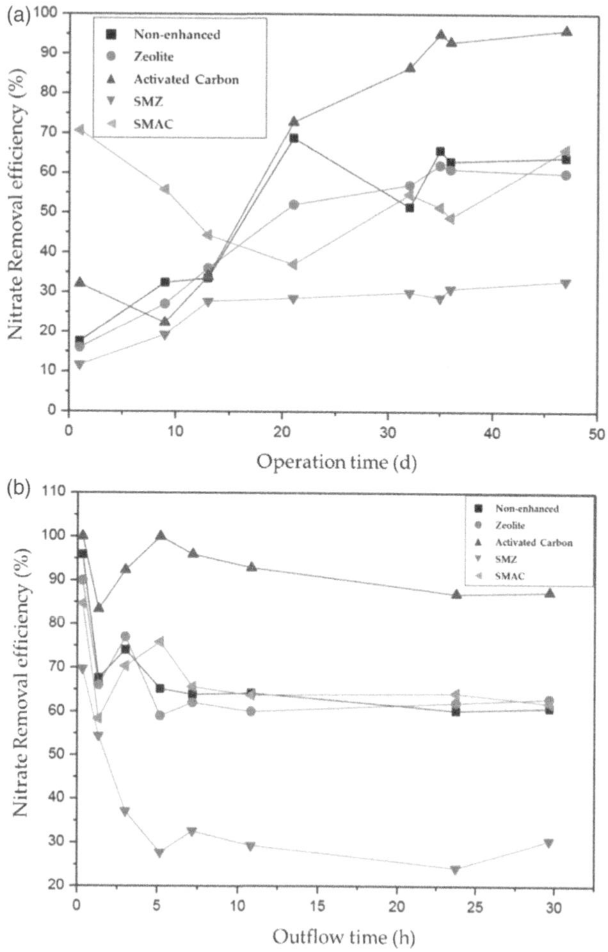

Figure 4 | NO_3^--N removal performance for candidate fillers: (a) long-term performance and (b) single event performance at week 7.

fillers could not provide satisfying performance due to the restrain of denitrification from modifiers.

At the later stage, the well-developed denitrification processes resulted in the stable performance of five devices. The activated carbon device had the highest NO_3^--N removal efficiency (93–96%) while the non-enhanced, SMAC and the zeolite device could remove 60–66% of the inflow NO_3^--N. The SMZ had the lowest performance since the beginning of the experiment and the final removal efficiency was only 29–33%. The SMAC device suffered a dramatic decrease on removal efficiency where its removal efficiency fluctuated and increased throughout the experiment period. According to the denitrification restrain phenomena (Schick *et al.* 2010), the reason for such fluctuations could be explained as follows. (1) Operation of the unit could flush out the dissolved surfactant which stopped restraining the denitrification process. (2) After the SMAC turned saturated, desorption may happen and the desorbed NO_3^--N could be denitrified in the submerged area. Desorption offers more absorption capacity in the next operation which made the NO_3^--N removal efficiency rise again. (3) With enhancement in the denitrification process, more NO_3^--N could desorb in one operation interval, which could enhance the surfactant desorption into the submerged area. This again restrained the denitrification system and reduced the NO_3^--N removal efficiency. Such processes will repeat until no more surfactant could be desorbed, and the final performance of the SMAC device will be the same as the activated carbon device (Schick *et al.* 2010).

The stabilized performance of the non-modified cell (63% removal) was consistent with existing studies (Randall & Bradford 2013), which indicates the NO_3^--N remove rate of traditional sand mix bioretention cells ranging at $62.7 \pm 18.8\%$. Considering the material of the stabilized performance of SMZ cells (33%), zeolite cells (60%) and SMZ cells (66%) were also consistent with the earlier findings ($-118.0 \pm 55.6\%$) for sorbtive cells (Randall & Bradford 2013). While considering the cell structure, these cells performed relatively weakly compared to traditional anoxic cells ($99.3 \pm 0.55\%$). The difference was caused by the running pattern where the performance system proposed in this section was monitored under continuous operation, which is not a strictly anoxic environment, while Randall & Bradford (2013) monitored their anoxic cells 24 hours after operation, which offered a sufficient anoxic environment for denitrification.

Notably, the activated carbon cells showed far better performance (96% removal efficiency) compared to the sorbtive cells in this study as well as in the literature. The extraordinary performance may be due to the high specific surface area of activated carbon which offered preferred conditions for simultaneous nitrification and denitrification (Mousavi *et al.* 2012). This may be due to the higher specific surface area of activated carbon which provided better NO_3^--N absorption capacity.

For a single event, the activated carbon device had the best NO_3^--N removal performance (stable at 87–96%) throughout the event followed by the non-enhanced, zeolite and SMAC devices having a similar performance (stable at

60–66%). The SMZ device had the poorest performance, stabilizing at 24–31%. The high removal efficiency at the beginning indicates that NO_3^--N could be efficiently denitrified within a 24 h operation interval and the dramatic performance drop could be due to: (1) NO_3^--N flushed out from the media layer, which was nitrified from ammonia during the operation interval; (2) inflow with high NO_3^--N had reached the outlet; and (3) the denitrifying bacteria in the SAZ were restrained due to lack of NO_3^--N at the end of the interval. With time, the NO_3^--N in the media layer being fully flushed out and the denitrifying bacteria being activated, the removal efficiency was gradually increased. With the continued inflow, the carbon source could not supply the consumption from denitrification and the removal rate was gradually decreased until stable.

Nitrogen removal efficiency distribution

The vertical diversity of NO_3^--N removal efficiency was also investigated to identify the effective part of the system.

The media layer offered relatively stable but limited capacity of NO_3^--N removal (13–20% at the bottom of the layer) throughout the experiment (Figure 5). In most cases, the removal capacity at the middle section of the layer was very similar to the bottom section. In some cases, such as the activated carbon device (week 7) and the SMAC device (week 7), the removal efficiency at the middle section was even higher than that at the bottom, which may be due to the flushing of ammonia-transformed NO_3^--N in the media layer. Such phenomena have also been monitored and verified in a runoff treatment case in the latest research.

At week 2, the non-enhanced, zeolite and SMZ devices had no significant change on removal efficiency in the submerged layer and the drainage layer, proving the denitrification systems were not formed yet and providing no contribution to NO_3^--N removal. This suggests that the removal of NO_3^--N in activated carbon and the SMAC device were mainly through absorption of the enhanced fillers. Notably, the well performed SMZ in the lab-scale experiments had a poor performance which indicates that the absorption power of SMZ was not strong enough under the hydraulic loading of 4 L/h and desorption occurred during the operation.

At week 4, the non-enhanced, zeolite and activated carbon devices had significant improvement on removal efficiency in the submerged layer. The increasing removal efficiency from the bottom of the media layer, through two-thirds of the submerged layer, to the bottom of the drainage layer is rather linear, indicating that the NO_3^--N in this area should be removed by the same route (denitrification). This suggests the formation of the denitrification system had been completed at week 4. The SMZ device had very limited improvement which indicates that the denitrification system had been restrained. Notably, the SMAC device had a dramatic decrease of removal efficiency compared to week 2. The upper two-thirds submerged layer had notably higher removal efficiency than the combination of the lower one-third submerged layer and the drainage layer. This suggests that the NO_3^--N removal in the SMAC device still depended more on the absorption of SMAC while the absorption capacity had a significant decrease over the past 2 weeks. The denitrification system was also restrained in the SMAC device, proving the cationic surfactant had a negative impact on denitrifying bacteria.

At week 7, the denitrification system had been stable for most devices besides the SMAC unit. The non-enhanced devices had a moderate decrease on the removal efficiency stabilizing at around 64% while the zeolite device slightly increased to 61%. The activated carbon cell had the highest removal capacity of 93% and the SMZ cell had the lowest at 28%. Remarkably, there was an increase of NO_3^--N removal in the SMAC device and the efficiency increment from the bottom of the media layer to the bottom of the drainage layer became more linear than in week 4. This proves that the denitrification system in the SMAC device had a dynamic balance between desorption of NO_3^--N (with cationic surfactant) and the activation of denitrifying bacteria, as discussed above.

Although the non-enhanced device had a similar performance compared to the zeolite and SMAC device at week 7, the overall trend of its performance was decreasing while the other two revealed increasing trends. This trend suggests a shorter life cycle and more careful maintenance of the non-enhanced units. Considering the long-term performance as well as the single-event performance, the activated carbon was selected as the enhanced filler for further experiments in the current study.

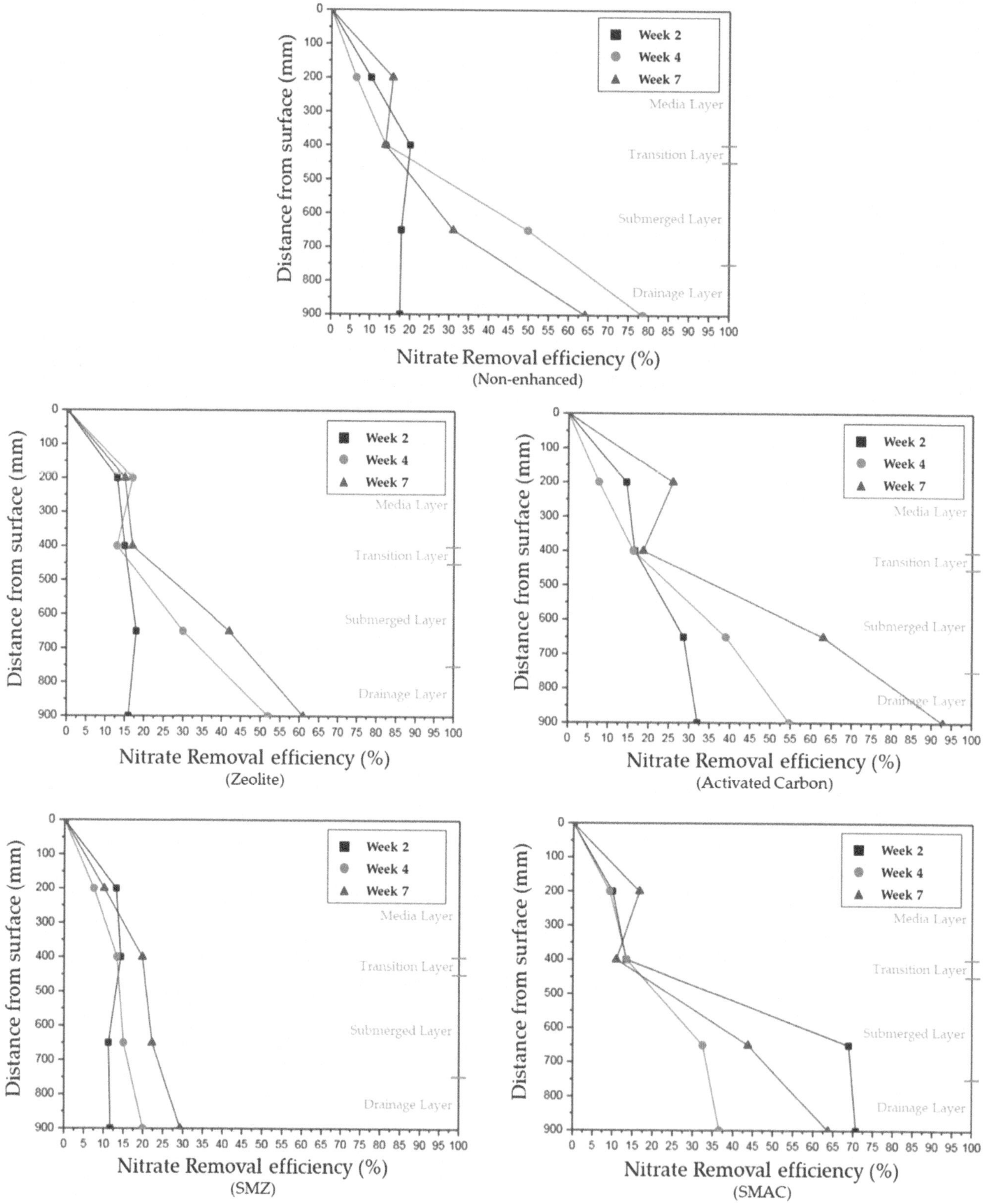

Figure 5 | Vertical diversity of NO_3^--N removal performance for candidate fillers.

Design and operation optimization

Hydraulic loading is the most influencing factor to adapt bioretention techniques to treat river water. Higher loading ensures the efficiency and feasibility of such treatment methods but can also disturb the performance of bioretention. Thus, different proportions of enhanced filler (50, 100, 150, and 200 mm) in the submerged layer were evaluated under different hydraulic loading rates (4, 6, 8, and 12 L/h) to determine a feasible design and operation pattern of application.

As shown in Figure 6, the NO_3^--N concentration in the outflow decreased with the proportion of activated carbon in the submerged layer increasing. The differences between 150 and 200 mm under different hydraulic loadings were –0.004 ± 0.109 mg/L ($p < 0.05$), which suggests limited improvement by replacing more gravel with activated carbon after the proportion reached 50%. Such a performance limit was due to the removal route of the NO_3^--N. The addition of activated carbon not only provided space for denitrifying bacteria but also absorbed NO_3^--N for more efficient denitrification. At the beginning of adding more activated carbon, the denitrification process worked together with the absorption. With more activated carbon added, more NO_3^--N was absorbed, and its concentration in the submerged layer became low enough to start

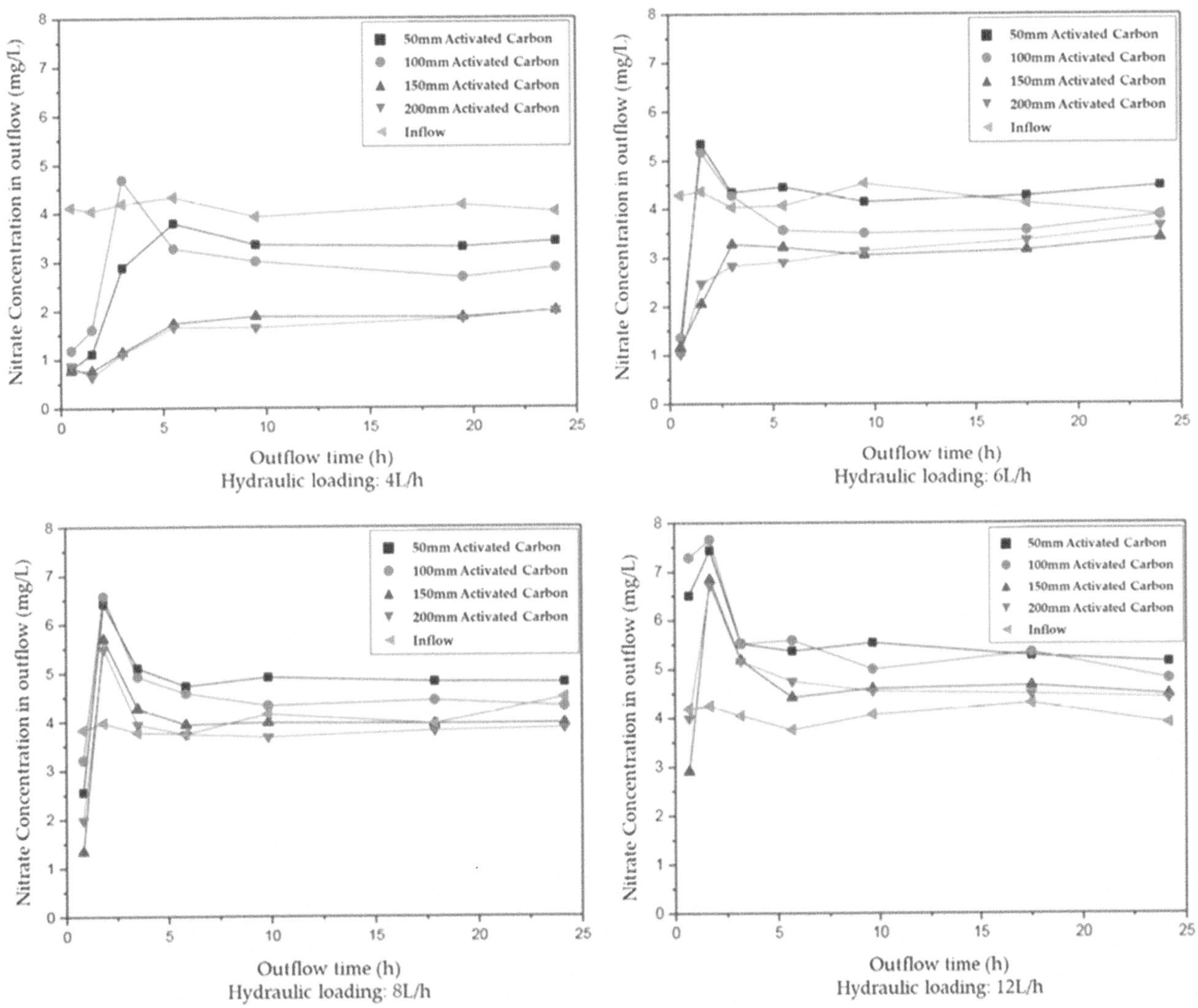

Figure 6 | NO_3^--N concentration in outflow under different filler layer designs and hydraulic loadings.

restraining the denitrification. Thus, the total removal performance would reach a certain limit with the increasing proportion of activated carbon and the most cost-efficient design would be 150 mm activated carbon +150 mm gravel.

The best hydraulic loading rate was suggested to be 4 L/h which equals to 81.5 mm/h (2.0 m/d), almost double the loading of a vertical constructed wetland (lower than 1.0 m/d) (Saeed 2012). As the NO_3^--N in the submerged layer could be fully denitrified within 24 h, the operation interval could be adjusted according to the need of the river water treatment or upcoming rainfall events. The overall hydraulic loading rate will be higher than a traditional vertical wetland and the location of the bioretention units could be decentralized along the river, providing more flexibility in design and construction.

Compared with existing studies (Table 5), the proposed design revealed a better and more stable performance than traditional enhanced bioretention designs. The results indicate that the proposed design could operate at a significantly higher hydraulic loading and NO_3^--N concentration while maintaining its performance. The dominant difference of the proposed design and an enhanced vertical wetland is that such a bioretention system can operate intermittently to ensure the proper removal of TN. Such a pattern enables the flexible operation between river treatment and runoff reduction, as well as a robust performance for both.

PRACTICAL APPLICATIONS AND FUTURE RESEARCH PERSPECTIVES

In application to river treatment, the proposed devices could be implemented on both sides of the river, integrated with the existing (distributed) landscapes or green areas along the river. As the bioretention cell systems are designed to be distributable, it could be easier and more flexible to allocate the systems than wetland systems. The two parts of the system could operate at intermittent conditions to provide enough operation time (to ensure treating volume) and sufficient interval time (to ensure nitrogen removal performance).

There are several limitations in this study which could be improved in future studies. First, although the SMAC showed a high capacity for absorbing NO_3^--N and feasibility of desorbing NO_3^--N during the operation interval for continuous denitrification, the conflict between the modifier and the denitrifying bacteria is the major problem in application. As found in the current study, the performance of the SMAC unit fluctuates but is improving, and future studies can be carried out on the life cycle of different enhanced bioretention units (total volume of river water that can be treated between maintenances). This study also shows the possibility of using the bioretention structure to treat highly nitrogenous rivers. Notably, the proposed design in this study was only focused on treating polluted water in the Taihu lake basin, therefore, future studies could be conducted on the improvement in design criteria

Table 5 | Performance of enhanced bioretention systems/wetlands for NO_3^--N removal

Reference	System	Hydraulic loading	NO_3^--N concentration (mg/L)	Removal efficiency
Kim *et al.* (2003)	Bioretention	6.0 h 4.0 cm/h	2.00	80%
Palmer *et al.* (2013)	Bioretention	3.2 h 13.3 cm/h	0.40–0.74	45.2% (mean)
Sun *et al.* (2018)	Vertical wetland	20 d 4.16 cm/h	10.04 ± 1.29	93.1–99.0%
Tian *et al.* (2019)	Bioretention	21.0 h 5.5 cm/h	10.7–12.4	66.2–92.9%
This study	Bioretention	24.0 h 8.15 cm/h	28.89	93.2–94.1%

or the optimization of bioretention units for river treatment in a wider sense.

Nevertheless, it was a limitation of the study that the data from the experiment could not carry out a reliable sensitivity analysis. This was due to the following. (1) The experiments were designed as a variable-controlled experiment instead of a comprehensive experiment. The parameters above were modified step by step and the poor performing parameters would not be adopted in the following steps. Thus, the data could not support a sensitivity analysis process. (2) The performance of bioretention cells is also strongly dependent on the construction quality (e.g. compactness uniformity), which is quite difficult to consider in the pilot-scale experiments. Further study could be carried out on the design parameters such as composition of the submerged zone, the enhancement pattern of the filler media, and the operation time and intervals to support more applicable and effective designs.

CONCLUSIONS

The current study was conducted to optimize the performance of a bioretention system in treating river water: (1) modification of filter media; (2) identify the best modified filler by monitoring the long-term and single-event performance; (3) optimize the bioretention structure design; and (4) identify the best operation pattern. The results reveal that the SMAC (modified with 600 mmol technically pure CPC/kg activated carbon) has the highest absorption capacity (1,742 mg/kg) which was five times higher than second ranked SMZ (311 mg/kg). For long-term performance, the activated carbon, non-modified, zeolite and SMZ gradually improved their NO_3^--N removal efficiency due to the formation and stabilization of the denitrification system throughout the experiment. The SMAC device started with a high removal efficiency of 71% due to its strong absorption capacity. The most cost-efficient design would be 150 mm activated carbon +150 mm gravel operating under hydraulic loading of 4 L/h. Such loading equals to 81.5 mm/h (2.0 m/d) which is far higher than a traditional vertical wetland. The operation interval could be adjusted according to the need for river treatment or upcoming rainfall events.

ACKNOWLEDGEMENTS

This work was financially supported by the National Key Research & Development Program of China (Grant No. 2018YFC0809900).

REFERENCES

Blecken, G. T., Zinger, Y., Fletcher, T. D. & Viklander, M. 2009 Impact of a submerged zone and a carbon source on heavy metal removal in stormwater biofilters. *Ecol. Eng.* **35** (5), 769–778.

Bratieres, K., Fletcher, T. D., Deletic, A., Alcazar, L., Coustumer, S. L. & Mccarthy, D. T. 2008 Removal of nutrients, heavy metals and pathogens by stormwater biofilters. In: *Proceedings of the 11th International Conference on Urban Drainage*, Edinburgh, Scotland, UK.

Chen, X., Peltier, E., Sturm, B. S. & Young, C. B. 2013 Nitrogen removal and nitrifying and denitrifying bacteria quantification in a stormwater bioretention system. *Water Res.* **47** (4), 1691–1700.

Choi, H. D., Shin, M. C., Kim, D. H., Jeon, C. S. & Baek, K. 2008 Removal characteristics of reactive black 5 using surfactant-modified activated carbon. *Desalination* **223** (1), 290–298.

Choi, H. D., Jung, W. S., Cho, J. M., Ryu, B. G., Yang, J. S. & Baek, K. 2009 Adsorption of Cr(VI) onto cationic surfactant-modified activated carbon. *J. Hazard. Mater.* **166** (2), 642–646.

Clark, S. E. & Pitt, R. 2012 Targeting treatment technologies to address specific stormwater pollutants and numeric discharge limits. *Water Res.* **46** (20), 6715–6730.

Correll, D. L. 1998 Role of phosphorus in the eutrophication of receiving waters: a review. *J. Environ. Qual.* **27** (2), 261–266.

Davis, A. P., Shokouhian, M., Sharma, H. & Minami, C. 2001 Laboratory study of biological retention for urban stormwater management. *Water Environ. Res.* **73** (1), 5–14.

Davis, A. P., Shokouhian, M., Sharma, H. & Minami, C. 2006 Water quality improvement through bioretention media: nitrogen and phosphorus removal. *Water Environ. Res.* **78** (3), 284–293.

Davis, A. P., Hunt, W. F., Traver, R. G. & Clar, M. 2009 Bioretention technology: overview of current practice and future needs. *J. Environ. Eng.* **135** (3), 109–117.

Haggerty, G. M. & Bowman, R. S. 1994 Sorption of chromate and other inorganic anions by organo-zeolite. *Environ. Sci. Technol.* **28** (3), 452–458.

Hatt, B. E., Morison, P., Fletcher, T. D. & Deletic, A. 2009 *Adoption Guidelines for Stormwater Biofiltration Systems*. Summary report: Cities as Water Supply Catchments – Sustainable Technologies. CRC for Water Sensitive Cities, Australia.

Hong, H. J., Kim, H., Baek, K. & Yang, J. W. 2008 Removal of arsenate, chromate and ferricyanide by cationic surfactant modified powdered activated carbon. *Desalination* **223** (1), 221–228.

Hunt, W. F., Smith, J. T., Jadlocki, S. J., Hathaway, J. M. & Eubanks, P. R. 2008 Pollutant removal and peak flow mitigation by a bioretention cell in urban Charlotte. *N.C. J. Environ. Eng.* **134** (5), 403–408.

Kim, H., Seagren, E. A. & Davis, A. P. 2003 Engineered bioretention for removal of nitrate from stormwater runoff. *Water Environ. Res.* **75** (4), 355–367.

LeFevre, G. H., Paus, K. H., Natarajan, P., Gulliver, J. S., Novak, P. J. & Hozalski, R. M. 2015 Review of dissolved pollutants in urban storm water and their removal and fate in bioretention cells. *J. Environ. Eng.* **141** (1), 04014050.

Li, Z. 2003 Use of surfactant-modified zeolite as fertilizer carriers to control nitrate release. *Microporous Mesoporous Mater.* **61** (1), 181–188.

Li, M. H., Swapp, M., Kim, M. H., Chu, K. H. & Sung, C. Y. 2014 Comparing bioretention designs with and without an internal water storage layer for treating highway runoff. *Water Environ. Res.* **86** (5), 387–397.

Liu, J., Sample, D. J., Bell, C. & Guan, Y. 2014 Review and research needs of bioretention used for the treatment of urban stormwater. *Water* **6** (4), 1069–1099.

Lucas, W. C., Greenway, M. & Palmer, R. N. 2010 Nitrogen retention in bioretention mesocosms with outlet controls. In: *Proceedings of the World Environmental and Water Resources Congress, Providence*, 16–20 May, 2010, Rhode Island, USA, pp. 3038–3047.

Lynn, T. J., Yeh, D. H. & Ergas, S. J. 2015 Performance of denitrifying stormwater biofilters under intermittent conditions. *Environ. Eng. Sci.* **32** (9), 1–10.

Ming-Han, L. I. & Chan, Y. S. 2012 Performance of bioretention system in treating urban highway runoff: a comparison study of designs with and without an internal water storage layer. *Landsc. Archit.* **102**, 140–147.

Mousavi, S., Ibrahim, S. & Aroua, M. K. 2012 Sequential nitrification and denitrification in a novel palm shell granular activated carbon twin-chamber upflow bio-electrochemical reactor for treating ammonium-rich wastewater. *Bioresour. Technol.* **125**, 256–266.

Palmer, E. T., Poor, C. J., Hinman, C. & Stark, J. D. 2013 Nitrate and phosphate removal through enhanced bioretention media: mesocosm study. *Water Environ. Res.* **85** (9), 823–832.

Passeport, E., Hunt, W. F., Line, D. E., Smith, R. A. & Brown, R. A. 2009 Field study of the ability of two grassed bioretention cells to reduce storm-water runoff pollution. *J. Irrig. Drain. Eng.* **135** (135), 505–510.

Randall, M. T. & Bradford, A. 2013 Bioretention gardens for improved nutrient removal. *Water Qual. Res. J.* **48** (4), 372–386.

Ribas, D., Calderer, M., Martí, V. & Rovira, M. 2015 Effect of different seasonal conditions on the potential of wetland soils for groundwater denitrification. *Desal. Water Treat.* **53** (4), 994–1000.

Saeed, T. F. 2012 A comparative study of pollutant removal stability in hybrid wetland columns. *Wetland Sci.* **10** (2), 142–148.

Sarina, J. E., Sengupta, S., Siegel, S. R., Pandit, A., Yao, Y. & Yuan, X. 2010 Performance of nitrogen-removing bioretention systems for control of agricultural runoff. *J. Environ. Eng.* **136** (10), 1105–1112.

Schick, J., Caullet, P., Paillaud, J. L., Patarin, J. & Mangold-Callarec, C. 2010 Batch-wise nitrate removal from water on a surfactant-modified zeolite. *Microporous Mesoporous Mater.* **132** (3), 395–400.

She, D., Zhang, L., Gao, X., Yan, X., Zhao, X., Xie, W., Cheng, Y. & Xia, Y. 2017 Limited N removal by denitrification in agricultural drainage ditches in the Tai lake region of China. *J. Soils Sediments* **18** (1), 1–10.

Sun, H., Yang, Z., Wei, C. & Wu, W. 2018 Nitrogen removal performance and functional genes distribution patterns in solid-phase denitrification sub-surface constructed wetland with micro aeration. *Bioresour. Technol.* **263**, 223–231.

Tao, Y., Dan, D., Kun, L., Chengda, H., Haibing, C., Guo, F., Quijin, X., Fuhong, S. & Fengchang, W. 2018 Δ15n and nutrient stoichiometry of water, aquatic organisms and environmental implications in Tai lake, China. *Environ. Pollut.* **237**, 166–173.

Tian, J., Jin, J., Chiu, P. C., Cha, D. K., Guo, M. & Imhoff, P. T. 2019 A pilot-scale, bi-layer bioretention system with biochar and zero-valent iron for enhanced nitrate removal from stormwater. *Water Res.* **148**, 378–387.

Zinger, Y., Fletcher, T. D., Deletic, A., Blecken, G. T. & Viklander, M. 2007 Optimisation of the nitrogen retention capacity of stormwater biofiltration systems. *Surg. Endosc. Intervent. Techniq.* **17** (6), 891–895.

Zinger, Y., Blecken, G. T., Fletcher, T. D., Viklander, M. & Deletić, A. 2013 Optimising nitrogen removal in existing stormwater biofilters: benefits and tradeoffs of a retrofitted saturated zone. *Ecol. Eng.* **51** (2), 75–82.

First received 21 April 2019; accepted in revised form 29 October 2019. Available online 2 December 2019

Experimental study on filter media using locally available materials in bioretention

Feikai Yang, Rajendra Prasad Singh and Dangfang Fu

ABSTRACT

Bioretention systems and selection of effective filter media are very important in implementation of sponge cities. The current study was carried out to find proper composition of filter media using locally available materials, which acclimate to the special/local climate, environmental and geographical conditions in Yangtze River Delta region. Results revealed that sand with discontinuous gradation and containing a certain amount of clay led to unsatisfactory hydraulic performance (hydraulic conductivity ranged from 423 mm/h to 1,054 mm/h, and 1,500 mm/h to 29 mm/h). In contrast, a mixture of locally available sand, which consisted of continuous gradation of coarse sand (40–70%, by mass), fine sand (0–40%, by mass), very fine sand (10–60%, by mass) and nutrient soil (0–3%, by mass), had a hydraulic conductivity ranging from 200 to 400 mm/h and relatively stable structure. During the 70 days' flooding test, the hydraulic conductivity changed in the first 20 days due to the migration of particles (mainly <0.6 mm) and then became stable; the stable value was close to the initial. Moreover, easy access and simple production processes made it easier to promote. Findings could be used as a guideline for implementation of bioretention systems and selection of locally available and effective filter material.

Key words | bioretention, filter layer gradation, hydraulic conductivity, sponge city, stormwater control

Feikai Yang
Rajendra Prasad Singh
Dangfang Fu (corresponding author)
School of Civil Engineering,
Southeast University,
Nanjing 210096,
China
E-mail: *fdf@seu.edu.cn*

INTRODUCTION

Stormwater management concepts have been proposed and developed around the world to deal with water-related issues caused by rapid urbanization (Liu *et al.* 2017). Dominant concepts with respect to the performance include low-impact development (LID – USA and Canada), water-sensitive urban design (WSUD – Australia) and sustainable urban drainage systems (SUDS – UK) (Zhang *et al.* 2016; Ariza *et al.* 2019). Low-impact development or best management practices (BMPs) have been widely adopted and proven successful in addressing hydrologic and water quality issues (Eckart *et al.* 2017). In 2013, the government of China proposed and called for the comprehensive promotion of the construction of sponge cities, which emphasized the importance of integrated management of water-related issues. Moreover, bioretention with good adaptive mechanisms for runoff reduction (Charlesworth *et al.* 2013; Haile *et al.* 2016; Gao *et al.* 2018; de Macedo *et al.* 2019) and water quality improvement (Li *et al.* 2014; de Macedo *et al.* 2019; Hermawan *et al.* 2019; Singh *et al.* 2019) is a common stormwater control measure used to mitigate the impacts of urban runoff, and an important technology for the construction of sponge cities.

A bioretention system is a low energy treatment technology with the potential to provide both water quality and quantity benefits. The typical structure of bioretention is composed of a detention layer, filter layer, transition layer

doi: 10.2166/aqua.2019.210

and drainage layer, from top to bottom. Stormwater is diverted from a kerb or pipe into the bioretention system, where it flows through dense vegetation temporarily ponding on the surface, before slowly filtering down through the filter media, followed by purification through physical and biological processes (Archer *et al.* 2002; Hatt *et al.* 2009). Filter media is an important factor affecting the operation and efficiency of bioretention systems (Hunt *et al.* 2006; Bratieres *et al.* 2008). Permeability of the filter material affects total runoff reduction rate, peak flow reduction rate, peak time of the entire system (Tahvonen 2018) and, as well, as the pollutant removal efficiency. However, the surface soil type varies from region to region, and choosing the mixing ratio also differs accordingly.

The fillers in the current study were initially selected with respect to the hydraulic conductivity of the filter layer and its removal impact on the contaminants. In existing designs, it is recommended to use a mixture of filler with good permeability, soil-based and certain organic matter. The recommended mixing ratio of fillers in North Carolina (85–88% sand, 8–12% clay and silt, and 3–5% organic matter) seems economically viable. However, the recommended filler ratio in Maryland (50% sand, 30% topsoil and 20% organic matter: wood chips, leaves compost (MDE 2000)) is relatively expensive. The Facility for Advancing Water Biofiltration (FAWB), Australia has recommended the use of sandy loam as filler and suggested that it is better to have all gradation particles in the filler to prevent system collapse caused by loose particles. At the same time, it is also suggested that the hydraulic conductivity of filter layer fillers should be between 100 and 300 mm/h. The optimal value ranged between 100 and 200 mm/h after the system has stabilized (Hatt *et al.* 2009).

Based on the FAWB guidelines, the current study attempts to develop a composition of the filter layer suitable for climate and soil conditions in the Yangtze River Delta region of China by using locally available materials. Considering the difference in climate between Australia and local conditions, Australia has relatively clean stormwater runoff and low mean annual rainfall, while in the Yangtze River Delta region, the rainfall is abundant and stormwater runoff is highly contaminated. Therefore, permeability of the filter layer should not be very low for effective flood protection. Severe runoff pollution in China and other countries also leads to a serious clogging problem in the filter (Xu *et al.* 2016; Zhao *et al.* 2017; Yi *et al.* 2019; Yang *et al.* 2019; Zhou *et al.* 2019). Due to clogging, it is very hard for many pollutants to be fully removed from the treatment systems and advanced technologies or new materials for further treatment are needed (Ali *et al.* 2015, 2016; Basheer 2017, 2018; Burakova *et al.* 2018).

Research studies have mainly focused on the composition of the filter layer in recent years. Winston *et al.* (2018) found that higher underlying soil K_{sat} directly affected the volume reduction and peak flow mitigation in bioretention. Wang *et al.* (2011) found that the hydraulic conductivity of silty sand decreases with the increase of the fines' content. It is also noteworthy to mention that the climatic, economic and geographical conditions in countries like Australia or America are different from China. Therefore, design guidelines developed in these countries cannot be implemented directly in China or other developing countries. Moreover, locally available filter media are also different from the materials used in developed countries which have already been investigated for their quality and pollutant removal capacity. Hence, the current study aimed to find out the proper composition of filter media using locally available materials, which are suitable for the specific/local climate and environmental conditions in the Yangtze River Delta region. Another goal was to meet the requirements of sponge city construction in China, and not only focusing on effective performance but also considering the cost issues.

MATERIALS AND METHODS

Experimental design

Information regarding bioretention filter layers was highly available from previous studies, but they all had their own drawbacks such as inadaptability and inaccessibility of materials in different areas, which makes it hard to replicate a certain design in practice. Thus, three typical sand samples from the Yangtze River were chosen to find the optimal composition for filter layers in bioretention. The experiment was divided into four parts (Figure 1).

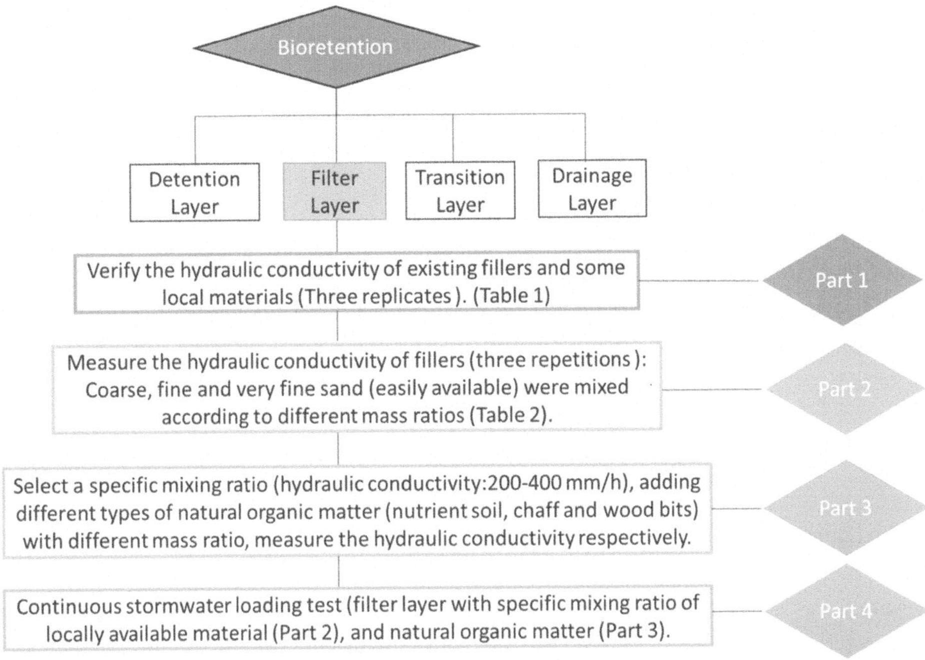

Figure 1 | Flow chart of the experiment.

Previous studies have suggested that some plant species are very useful in maintaining the infiltration capacity in bioretention due to their thick roots, which can be extended into the drainage layer (Hatt *et al.* 2009). This study further stated that a strong root system can increase the hydraulic conductivity while fine roots do not necessarily increase. Plants with thick roots (e.g., *Melaleuca*) demonstrated an ability to maintain hydraulic conductivity over time. Other species studied, with finer roots, had no such beneficial effects (Le Coustumer *et al.* 2012). Mainly, vegetation with fine roots is often used in bioretention in the Yangtze River Delta. Although plants may have some constraints on the bioretention performance they are not considered in this experiment because the main goal was to find an optimal mixing ratio of locally available materials to get a suitable initial hydraulic conductivity. Schematic diagrams of lab-scale experimental units designed and implemented in the current study and TST-70 analyser (Nanjing Soil Instrument Co., Ltd, Nanjing, China) are presented in Figure 2. TST-70 analyser was used to determine the permeability coefficient of sand and gravel non-cohesive soil under the unchangeable water pressure during the experiment.

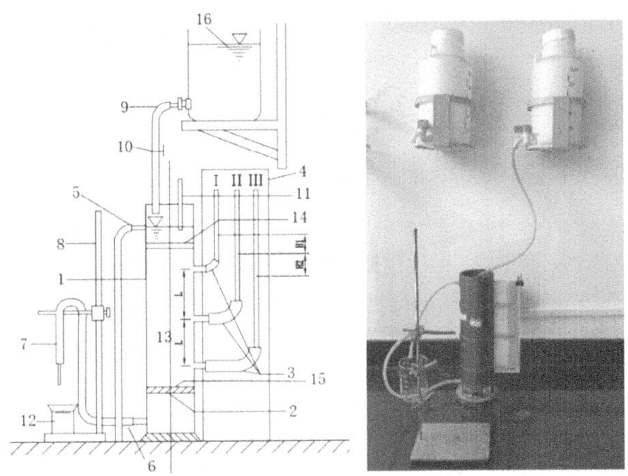

Figure 2 | Schematic design and lab-scale set-up of the TST-70 soil hydraulic conductivity tester (1. metal cylinder; 2. metal orifice; 3. pressure tap; 4. piezometer pipe; 5. spillway hole; 6. infiltrating hole; 7. regulating tube; 8. sliding support; 9. water supply pipe; 10. flatjaw pinchcock; 11. thermometer; 12. measuring cup; 13. sample; 14. gravel layer; 15. copper wire gauze; 16. water bottle).

Water quality

The inflow for the experiment was stormwater runoff collected from the roof of the laboratory. The average pollutant concentrations were as follows: total suspended solids (TSS) 3 mg/L; NH_3-N 0.8 mg/L; total nitrogen (TN)

1.5 mg/L and total phosphorus (TP) 0.34 mg/L. Nutritive soil (1 g) was added into 500 mL of water and soaked for 24 hours. The mean pollutant concentrations were as follows: NH_3-N, 0.58 mg/L; TN, 3.01 mg/L and TP, 0.22 mg/L. Therefore, the nutritive soil had little effect on the effluent quality, and only occurred in the early operation of the bioretention system (Hatt *et al.* 2009).

Hydraulic conductivity was obtained by Darcy seepage formula (Equation (1)), calculating the hydraulic gradient of the filter layer between each two piezometer pipes and the flow rate of the whole infiltration system.

$$v = ki \quad (1)$$

where v is the infiltration velocity; k is the hydraulic conductivity; $i = \frac{h}{L}$ hydraulic gradient, which is the head loss per unit distance along the seepage direction; h is the water heads at both ends of the filter layer; and L is the length of the seepage path.

The flow was measured by a measuring cylinder and a stopwatch. After the flow was determined, the flow rate was calculated according to the cross-section area of the infiltration column. The filler experiments were carried out with the location of each piezometer pipe as the boundary, and the thickness of the filler L was calculated according to the type of fillers. The head difference was determined according to the reading data of the piezometric pipe in each layer, and then the hydraulic gradient was calculated according to Darcy seepage formula (Equation (1)).

RESULTS AND DISCUSSION

Analysis of different existing materials

Considering that the permeability of surface soil in Kunshan is extremely poor, a filter layer with low hydraulic conductivity has very little impact on alleviating flooding. The fillers are also prone to clogging due to the highly polluted storm runoff. Details on the locally available materials, previous designs in Kunshan and FAWB design requirements are presented in Table 1. However, the original filter media used in Kunshan included the surface soil, coarse sand (particle size between 1 and 2 mm) and

Table 1 | Different types of filter layer

No.	Type of filter layer
#1	Composition of the filter layer in FAWB
#2, 3	Composition of filter layer used in Kunshan, China
#4	Coarse sand
#5	Fine sand
#6	Very fine sand

medium sand (particle size between 0.25 and 1 mm), and the proportion was 2:3:15 (by volume). The average hydraulic conductivity of different fillers is shown in Figure 3. Findings presented in Figure 3 and Table S1 (Supplementary material) reveal that material of the filter layer recommended by FAWB is not continuous graded sand, although the hydraulic conductivity in engineering application meets the requirements. However, secondary screening of raw materials prolong the production process and escalate the production cost. Similarly, the original fillers in Kunshan were mixed with 10% surface soil. The initial hydraulic conductivity was estimated as >1,500 mm/h, and after a 20-day operation, the hydraulic conductivity decreased to 29 mm/h. This may be due to the low permeability of the surface soil, which is easy to agglomerate, resulting in a sharp drop of hydraulic conductivity which may affect the life cycle of the whole bioretention system.

A continuous gradation of fine sand was chosen as the basic filler of the filter layer, and a certain ratio of coarse

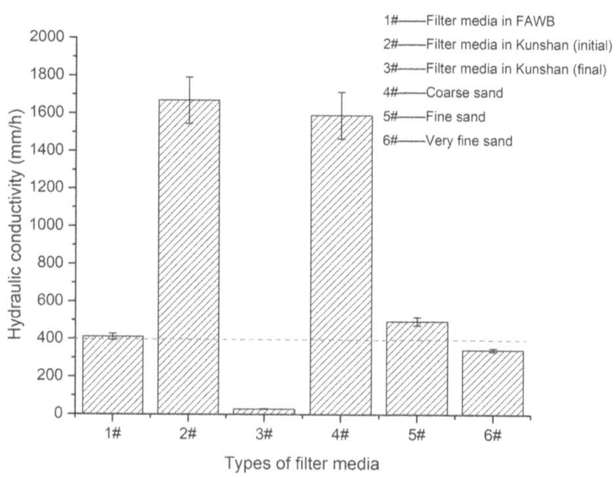

Figure 3 | Mean hydraulic conductivity of local and previously used filter layer.

sand (1.18–2.36 mm) and medium sand (0.3–1.18 mm) which are not continuous graded were added. The test results revealed that the hydraulic conductivity was 423 mm/h after the addition of 5% coarse sand and 25% medium sand, which is close to the required composition recommended by FAWB. The hydraulic conductivity increased sharply to 1,054 mm/h at day 20 of the continuous stormwater loading test. Gradation curves and the hydraulic conductivity of different compositions are shown in Figure 4(a) and Table S2 (Supplementary material). Results revealed that the filler volume (#10) greatly reduced after the continuous flooding, due to the gap gradation of sand, which affected the structural stability of natural sand. Because fine sand was the main part of the filler, the structural stability of the filter layer was consequently insufficient. The change of the filler structure is the key reason for the rapid increase of hydraulic conductivity. It is evident from Figures 3 and 4(b) that it was difficult for locally available materials to meet the requirements of FAWB, except fine sand which has a similar hydraulic conductivity and gradation curves. However, for local fine sand, particle size in the range of 0.15–0.6 mm accounted for >85%. The proportions of big particles were too low to stabilize the filler, and the long-lasting stormwater loading may lead to the collapse of the system.

Earlier, Water by Design (2014) recommended a 100% sand and loam mix while Australian guidelines recommended 100% sandy loam. However, Davis *et al.* (2009) highlighted the fact that when the use of loam was more than 30% (by volume), it may lead to collapse of the system due to the high clay content. Because of the lower average annual rainfall in Australia, the loam-based filter media may be applicable. As China has higher average annual rainfall, this filter media might not be suitable for application. Therefore, the previous design (local or international) may not adapt to the special conditions in the Yangtze River Delta region, China.

Analysis of fillers with different mixing ratio

As recommended by Australian guidelines (Hatt *et al.* 2009), initial hydraulic conductivity of filter media should be between 100 and 300 mm/h. Considering the difference of climate and environment conditions between China and Australia, this study selected the filter media with hydraulic conductivity that ranged between 200 and 400 mm/h. The experiment result shows that local materials cannot be directly used as filter layers in bioretention. Moreover, the filler, which is mixed with soil, will significantly affect the service life of the bioretention. Payne *et al.* (2014) suggested that sand with different particle size (between 0.05 and 3.4 mm) would provide sufficient infiltration rate and hold enough water to support the growth of plants. Therefore, local coarse, fine and very fine sand were used and mixed with a certain ratio in this section (Table 2). The mean hydraulic conductivity of these compositions is shown in Figure 5. It revealed that the hydraulic conductivity was

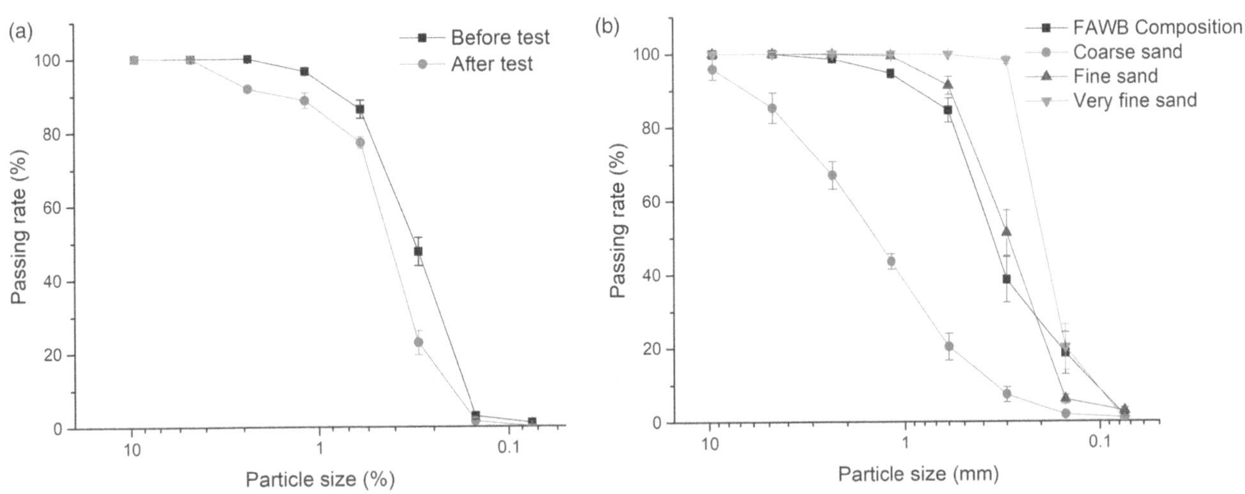

Figure 4 | (a) Gradation curves of #10 filler and (b) #3 locally available materials and filter layer in FAWB.

Table 2 | Mixing ratio of the filter layer using locally available materials

No.	Types (%) Coarse sand	Fine sand	Very fine sand
#1	90	0	10
#2	90	10	0
#3	80	0	20
#4	80	10	10
#5	70	0	30
#6	70	10	20
#7	60	0	40
#8	60	10	30
#9	60	20	20
#10	60	30	10
#11	60	40	0
#12	50	50	0
#13	50	40	10
#14	50	30	20
#15	50	20	30
#16	50	0	50
#17	40	0	60
#18	30	0	70

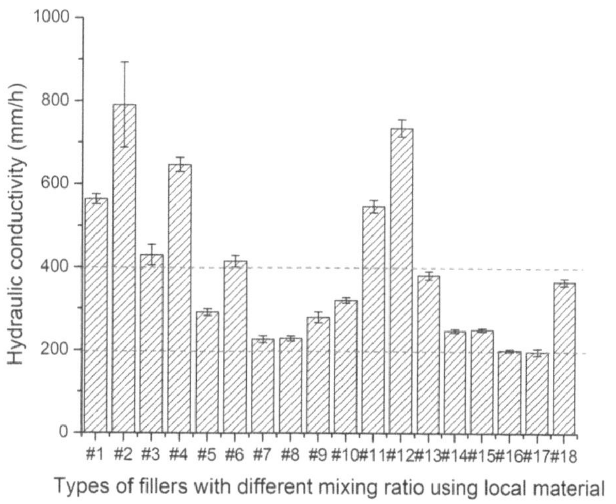

Figure 5 | Mean hydraulic conductivity of fillers with different mixing ratio using locally available materials.

appropriate in fillers #5 #7, #8, #9, #10, #13, #14, #15, #16, #17 and #18. The mixing ratios of the coarse, fine and very fine sand were estimated at 70:0:30, 60:0:40, 60:10:30, 60:20:20, 60:30:10, 50:40:10, 50:30:20, 50:20:30, 50:0:50, 40:0:60 and 30:0:70, respectively. Out of these, coarse sand ranged from 30 to 70%, fine sand ranged from 0 to 40% and very fine sand ranged from 10 to 70%. The grading range is presented in Table S3 (Supplementary material).

It can be seen from Figure 6(a) that when there are only two local materials involved in the filter layer and the coarse sand is fixed, if another local material is very fine sand, the hydraulic conductivity of the mixed filler is far lower than that of fine sand. This indicated that very fine sand is the key content to control the hydraulic conductivity of the whole layer. The reason is that very fine particles continuously fill the pores of the coarse sand while the fine particles will create their own pores beyond a certain ratio. In unison with Figure 6(b), it can be seen that, with the increase of the mixing ratio of very fine sand, the hydraulic conductivity of the filler tends to decline. As the very fine sand mixing ratio reached 60%, the hydraulic conductivity reached its lowest point, then it started to increase and reached close to the values of the original very fine sand. This result showed that coarse sand does not contribute much to the hydraulic conductivity of the mixed composition, and that the very fine sand plays the leading role, which leads to a poor structure stability. Therefore, the coarse, fine and very fine sand should be controlled between 40 and 70%, 0 and 40% and 10 and 60%, respectively, to meet the requirements of hydraulic conductivity from 200 to 400 mm/h and a relatively stable structure. This composition can meet the local requirements.

Analysis of different organic matter and mixing ratios

Fillers are usually filled with nutrients to keep the plant growing. In this section, a certain mixing ratio of coarse, fine and very fine sand (50:20:30) was chosen to start the experiment for screening the optimal mixing ratio of natural organic matter (Figure 7). Findings revealed that different types of organic matter have specific impacts on the hydraulic conductivity of the filler. In respect to nutrient soil and wood chips, the hydraulic conductivity at first decreased and then increased with addition of their respective content. It reached the lowest point when the content of nutrient soil was 4% while that of wood chips was 1%. The hydraulic conductivity of the lowest point was 189.8 mm/h and

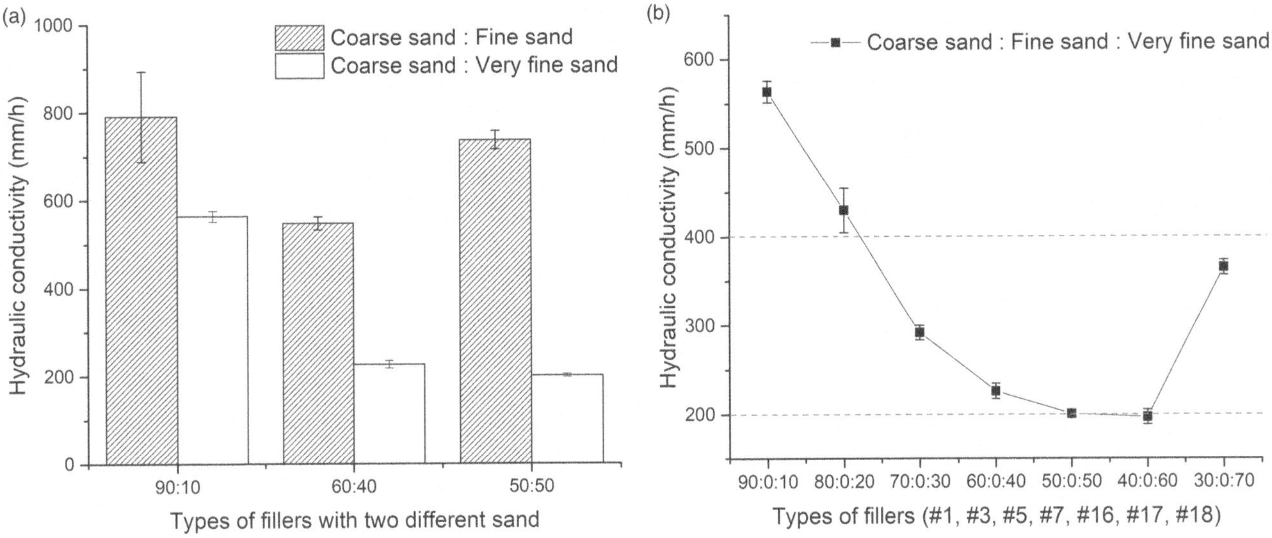

Figure 6 | Comparison of hydraulic conductivity of fillers.

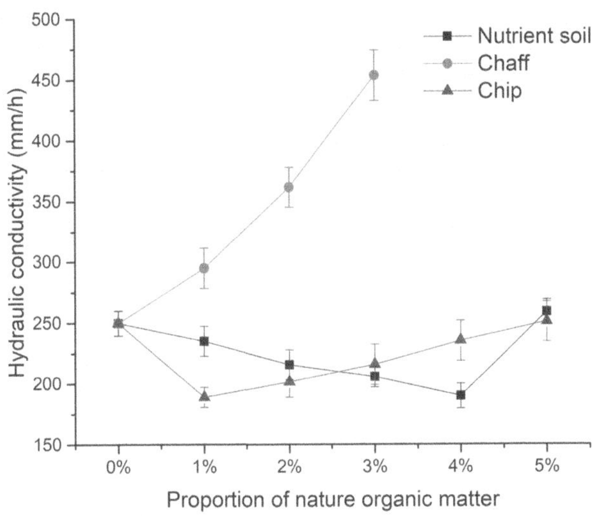

Figure 7 | Mean hydraulic conductivity of filler with the increase of natural organic matter.

188.8 mm/h, respectively. It showed a tendency towards rapid increase with the increase of chaff content. This was highly attributed to the nature of the organic matter added into the filler.

The nutritive soil used in the experiment was a fermentation product of animal excrement and litter compost, and its actual organic matter content was >85%. At the same mass ratio, the volume of the nutritive soil was low because of the grinding process before addition. Therefore, the nutritive soil continuously fills the pores of the original media at the beginning and the hydraulic conductivity declines. When the content of the additive exceeds 4%, the porosity of the nutritive soil itself begins to contribute to the porosity of the entire filler, resulting from its fleecy structure, which leads to the rise of the hydraulic conductivity of the fillers. Wood chips have small density, and their volume under the same mass percentage is much larger than that of nutrient soil. Therefore, when the content of wood chips was 1%, the hydraulic conductivity of the composition reaches its lowest point. The original, ending and lowest points of the mixing ratio are basically the same as that of the filler mixed with nutrient soil, except that the lowest point comes earlier. Different from nutrient soil and wood chips, chaff particles are larger and cannot fill the pores of the original filler. Therefore, the hydraulic conductivity increases at the beginning. In addition, due to the large size and low density of chaff particles, volume ratio of the filler was also increased, resulting in a sharp increase in the hydraulic conductivity.

In comparison to the filler mixed with nutrient soil, the actual organic matter content of wood chips and chaff varied from 70 to 80%. Hydraulic conductivity of filler mixed with chaff also exceeded the control range, because the chaff particle could not enter into the pores of coarse sand. Therefore, nutrient soil is more suitable as a natural organic matter added to the composition of the filter layer in bioretention. The hydraulic conductivity was lower than

200 mm/h when nutrient soil addition was 4%, and it does not meet the requirements. Therefore, the range of the nutrient soil content should be controlled within 0–3%. Results presented in Figure 7 revealed that when the addition ratio of nutrient soil exceeded 4%, the hydraulic conductivity started to increase, because the filler pores were filled up and new pores came into being. While this mixing ratio cannot meet the requirements (>3%) of FAWB (Hatt *et al.* 2009), considering that road runoff in China is rich in organic matter (Xu *et al.* 2016; Zhao *et al.* 2017) and the mean annual rainfall is high, the organic matter content of the filter layer would fulfil the requirements of plant growth in the respective periods.

After a prolonged operation, pollutants in runoff mainly maintained the plant growth and development. The moisture content of the nutrient soil in the experiment was around 43% without dry process. Average particle size distribution is presented in Table S4 (Supplementary material). According to FAWB (Hatt *et al.* 2009), the filter layer should contain some organic matter which has low nutrient content for increasing water holding capacity. Otherwise, leaching would occur if the nutrient contents (TN and PO_4^{3-}) were high, which would lead to deterioration of the water quality. In this study, organic matter content in the filter layer was <3%, and has a slight effect on effluent quality; de Macedo *et al.'s* (2019) research findings revealed that the soil organic matter did not affect contaminant concentrations, and also suggested that filter media with higher organic matter reduced the mass flux of contaminants more than that with lower organic matter.

Analysis of filter stability with long-term operation

The stability of the filter layer directly affects the service life of the bioretention system. In this experiment, the selected mixing ratio of coarse, fine and very fine sand was 50:20:30. Nutrient soil (2%, by mass) was chosen as natural organic matter. After mixing, it was loaded into a TST-70 analyser for continuous flooding test. Following that, the change in hydraulic conductivity was monitored and results are presented in Figure 8(a).

Results revealed that during the first 4 days, the hydraulic conductivity of the filler showed a sharp increase (Figure 8(a)). The possible reason could be that in the early stage, part of sand in the filler migrates with the flow, and in the process, the originally dense pores may have temporary capillary channels which can increase hydraulic conductivity. Later, it began to fall, by process of recomposing and stabilizing of some of the fillers. Under continuous flooding condition, it takes nearly one month for the filter layer to reach a stable structure. After one month, the hydraulic conductivity of the filler remained

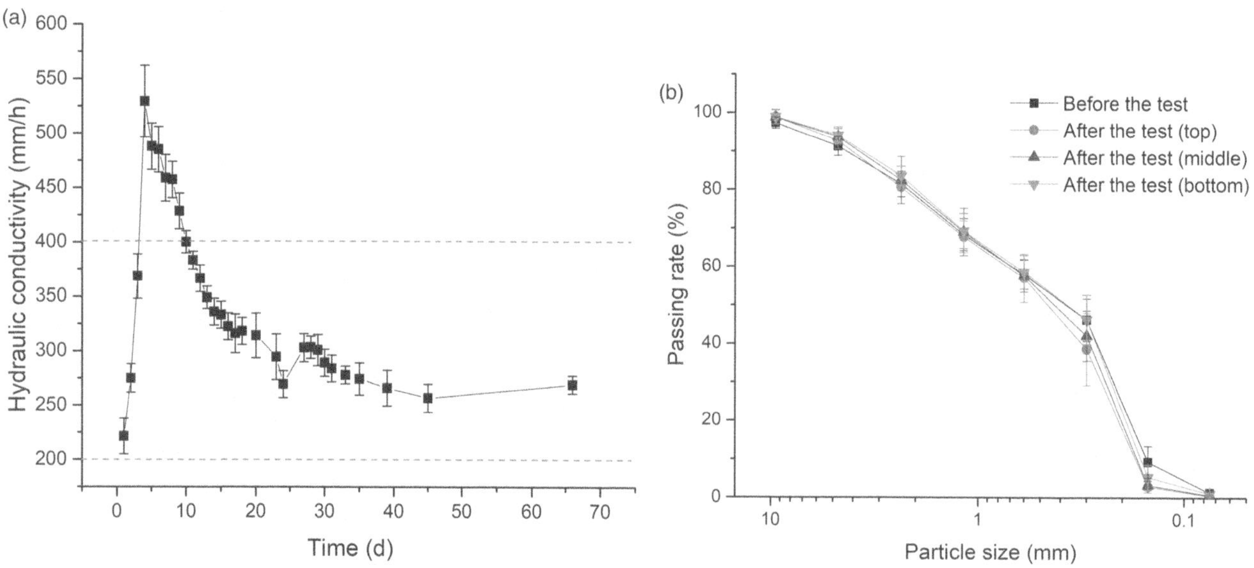

Figure 8 | (a) Hydraulic conductivity of the filter layer with time and (b) gradation curve before and after the experiment.

almost stable, slightly larger than the initial value, indicating that the initial hydraulic conductivity of the filter layer is an important factor. It can be concluded from the results that the stable hydraulic conductivity was close to the initial value.

At the final stage of the experiment, the sand column in the hydraulic conductivity tester was completely removed, and the original state was maintained. The top, middle and lower parts of the filler were taken for screening, and the grading curve before the test was compared, as shown in Figure 8(b). Results revealed that when the particle size was >0.6 mm, the four curves practically coincided, indicating that the sand in this particle size rarely moved with the flow. However, fillers with particle sizes <0.6 mm have different degrees of loss. The grading curve of the bottom filler was closest to the hydraulic conductivity of the initial one, followed by the middle and top filler. Also indicated was that the bottom filler controlled the hydraulic conductivity of the whole filter layer.

The excellent stability of this composition enables a longer service life, which also improves its economic efficiency. Compared with recent studies on filter layers in bioretention, this study provides an adaptable composition by using locally available materials in the Yangtze River Delta, which have suitable hydraulic conductivity, stable structure and a simple mass production process. Jiang *et al.* (2018a, 2018b) used four types of modifiers, which they mixed with traditional bioretention soil (65% sand +30% soil +5% sawdust, by mass), and ended up with a hydraulic conductivity of around 120–2,300 mm/h. However, that is too large a range and economically unviable. Liu & Fassman-Beck (2018) measured the saturated hydraulic conductivity of 14 engineered media with different compositions, mixed with fine pumice, coarse pumice, pumice sand, marine sand, compost, zeolite and topsoil. Their hydraulic conductivities varied between 468 and 9,612 mm/h, and although the mixing materials were simple, the hydraulic conductivity was too high to maintain plant growth and development. However, the effect of filter layer on plants was not investigated in this study. Tirpak *et al.* (2018) noted that, when bioretention was coupled with suspended pavement systems, which provide tree roots with an uncompact soil matrix that enhances root access to oxygen and water, the transpiration of plants was enhanced. This can lead to a better hydrological effect of the whole bioretention system. However, it is not clear whether plant transpiration has an effect on the hydraulic conductivity of the filter layer in a bioretention implemented in fields under long-term operation and thus needs further research.

PRACTICAL APPLICATIONS AND FUTURE RESEARCH PERSPECTIVES

The research results have been applied as a standard guideline in the construction of a sponge city in Kunshan, China. Current findings have also been used as reference by some other cities in China (Nanjing, Zhenjiang, etc.). Some bioretention units constructed in Kunshan city under the current project (bioretention units in Kunshan Lab, Xiaolin Road, Boshi Road and Fortune Plaza) are presented in Figure 9. The results of the current study are very significant and applicable to the Yangtze River Delta region and other regions with similar geophysical, climatic and soil conditions. The current study provides more accurate information on suitable particle size and hydraulic conductivity of filter media materials. It also helps to improve the construction quality to a certain extent and provides a basis for the detection unit. Finally, it contributes towards reduction of production difficulties and improving cost efficiency to make it economically viable, which is a key problem in the promotion of sponge city construction in China. Considering that the studied filler has not yet been used in a specific project in Kunshan for a long time, the actual service life needs to be further studied. In future research, the removal effect of pollutants will be taken into consideration to optimize the composition.

CONCLUSIONS

The performance of established filter materials in bioretention (used in developed countries) showed poor hydraulic performance in local conditions. Therefore, the current study aimed to find proper filter materials adapted to local climatic, environmental and soil conditions. Results revealed that optimal hydraulic conductivity ranged

Figure 9 | Bioretention implemented in Kunshan, China.

between 200 and 400 mm/h and had stable structure under continuous flooding when the mixing ratios of coarse, fine, very fine sand and nutrient soil were 40–70%, 0–40%, 10–60% and 0–3% (by mass), respectively. Easy access and simple production processes make these materials more economic and easier to promote their utilization in the construction of sponge cities in China.

ACKNOWLEDGEMENTS

This work was financially supported by the National Key Research & Development Program of China (Grant No. 2018YFC0809900).

SUPPLEMENTARY MATERIAL

The Supplementary Material for this paper is available online at https://dx.doi.org/10.2166/aqua.2019.210.

REFERENCES

Ali, I., Al-Othman, Z. A. & Al-Warthan, A. 2015 Removal of secbumeton herbicide from water on composite nanoadsorbent. *Desalination and Water Treatment* **57** (22), 1–13.

Ali, I., Alothman, Z. & Al-Warthan, A. 2016 Sorption, kinetics and thermodynamics studies of atrazine herbicide removal from water using iron nano-composite material. *International Journal of Environmental Science & Technology* **13**, 733–742.

Archer, N. A. L., Quinton, J. N. & Hess, T. M. 2002 Below-ground relationships of soil texture, roots and hydraulic conductivity in two-phase mosaic vegetation in south-east Spain. *Journal of Arid Environments* **52** (4), 535–553.

Ariza, S. L. J., Martinez, J. A., Munoz, A. F. & Baron, J. P. Q. 2019 A multicriteria planning framework to locate and select Sustainable Urban Drainage Systems (SUDS) in consolidated urban areas. *Sustainability* **11** (8), 2312.

Basheer, A. A. 2017 Chemical chiral pollution: impact on the society and science and need of the regulations in the 21st century. *Chirality* **30**, 402–406.

Basheer, A. A. 2018 New generation nano-adsorbents for the removal of emerging contaminants in water. *Journal of Molecular Liquids* **261**, 583–593.

Bratieres, K., Fletcher, T. D., Deletic, A. & Zinger, Y. 2008 Nutrient and sediment removal by stormwater biofilters: a large-scale design optimization study. *Water Research* **42** (14), 3930–3940.

Burakova, E. A., Dyachkova, T. P., Rukhov, A. V., Tugolukov, E. N., Galunin, E. V., Tkachev, A. G., Basheer, A. A. & Ali, I. 2018 Novel and economic method of carbon nanotubes synthesis on a nickel magnesium oxide catalyst using microwave radiation. *Journal of Molecular Liquids* **253**, 340–346.

Charlesworth, S. M., Perales-Momparler, S., Lashford, C. & Warwick, F. 2013 The sustainable management of surface water at the building scale: preliminary results of case studies in the UK and Spain. *Journal of Water Supply Research and Technology-AQUA* **62** (8), 534–544.

Davis, A. P., Hunt, W. F., Traver, R. G. & Clar, M. 2009 Bioretention technology: overview of current practice and future needs. *Journal of Environmental Engineering* **135** (3), 109–117.

de Macedo, M. B., do Lago, C. A. F. & Mendiondo, E. M. 2019 Stormwater volume reduction and water quality improvement by bioretention: potentials and challenges for water security in a subtropical catchment. *Science of the Total Environment* **647**, 923–931.

Eckart, K., Mcphee, Z. & Bolisetti, T. 2017 Performance and implementation of low impact development – a review. *Science of the Total Environment* **607–608**, 413–432.

Gao, J., Pan, J., Hu, N. & Xie, C. 2018 Hydrologic performance of bioretention in an expressway service area. *Water Science and Technology* **77** (7), 1829–1837.

Haile, T. M., Hobiger, G., Kammerer, G., Allabashi, R., Schaerfinger, B. & Fuerhacker, M. 2016 Hydraulic performance and pollutant concentration profile in a stormwater runoff filtration systems. *Water Air and Soil Pollution* **227**, 34.

Hatt, B. E., Fletcher, T. D. & Deletic, A. 2009 Hydrologic and pollutant removal performance of stormwater biofiltration systems at the field scale. *Journal of Hydrology* **365** (3–4), 310–321.

Hermawan, A. A., Talei, A., Leong, J. Y. C., Jayatharan, M., Goh, H. W. & Alaghmand, S. 2019 Performance assessment of a laboratory scale prototype biofiltration system in tropical region. *Sustainability* **11** (7), 1–16.

Hunt, W. F., Jarrett, A. R., Smith, J. T. & Sharkey, L. 2006 Evaluating bioretention hydrology and nutrient removal at three field sites in North Carolina. *Journal of Irrigation and Drainage Engineering* **132** (6), 600–608.

Jiang, C. B., Li, J. K., Zhang, B. H. & Li, H. E. 2018a Phosphorus removal of retrofit bioretention systems on urban surface runoff. *Desalination and Water Treatment* **126** (2), 39–47.

Jiang, C. B., Li, J. K., Zhang, B. H., Ruan, T. S., Li, H. E. & Dong, W. 2018b Design parameters and treatment efficiency of a retrofit bioretention system on runoff nitrogen removal. *Environmental Science and Pollution Research* **25** (33), 33298–33308.

Le Coustumer, S., Fletcher, T. D., Deletic, A., Barraud, S. & Poelsma, P. 2012 The influence of design parameters on clogging of stormwater biofilters: a large-scale column study. *Water Research* **46** (20), 6743–6752.

Li, M. H., Swapp, M., Kim, M. H., Chu, K. H. & Sung, C. Y. 2014 Comparing bioretention designs with and without an internal water storage layer for treating highway runoff. *Water Environment Research* **86** (5), 387–397.

Liu, R. F. & Fassman-Beck, E. 2018 Pore structure and unsaturated hydraulic conductivity of engineered media for living roofs and bioretention based on water retention data. *Journal of Hydrologic Engineering* **23** (3), 04017065.

Liu, Y., Engel, B. A., Flanagan, D. C., Gitau, M. W., McMillan, S. K. & Chaubey, I. 2017 A review on effectiveness of best management practices in improving hydrology and water quality: needs and opportunities. *Science of the Total Environment* **601–602**, 580–593.

Maryland Department of the Environment 2000 *Maryland Stormwater Design Manual*, Vols. I and II. Center for Watershed Protection and the Maryland Department of the Environment, Water Management Administration, Baltimore, MD, USA.

Payne, E. G., Pham, T., Cook, P. L., Fletcher, T. D., Hatt, B. E. & Deletic, A. 2014 Biofilter design for effective nitrogen removal from stormwater – influence of plant species, inflow hydrology and use of a saturated zone. *Water Science and Technology* **69** (6), 1312–1319.

Singh, R. P., Zhao, F., Ji, Q., Saravanan, J. & Fu, D. F. 2019 Design and performance characterization of roadside bioretention systems. *Sustainability* **11** (7), 2040.

Tahvonen, O. 2018 Adapting bioretention construction details to local practices in Finland. *Sustainability* **10** (2), 1–17.

Tirpak, R. A., Hathaway, J. M. & Franklin, J. A. 2018 Evaluating the influence of design strategies and meteorological factors on tree transpiration in bioretention suspended pavement practices. *Ecohydrology* **11** (8), 1–8.

Wang, L., Li, X. & Tian, H. 2011 Effect of fines content on permeability coefficient of sand and its correlation with state parameters. *Rock and Soil Mechanics* **32**, 21–26.

Water by Design 2014 *Bioretention Technical Design Guidelines (Version 1.1)*. Health Waterways Ltd, Brisbane, Australia.

Winston, R. J., Dorsey, J. D., Smolek, A. P. & Hunt, W. F. 2018 Hydrologic performance of four permeable pavement systems constructed over low-permeability soils in Northeast Ohio. *Journal of Hydrologic Engineering* **23** (4), 04018007.

Xu, P., He, J. C., Zhang, Y. J., Zhang, J. Q. & Sun, K. P. 2016 Runoff pollutant characteristics and first flush analysis in different urban functional areas: a case study in China. *Fresenius Environmental Bulletin* **25** (7), 2444–2453.

Yang, Y., Zeng, Z. T., Zeng, G. M., Huang, D., Xiao, R., Chen, Z., Zhou, C., Xiong, W., Wang, W., Cheng, M., Xue, W., Guo, H., Tang, X. & He, D. 2019 Ti_3C_2 Mxene/porous g-C_3N_4 interfacial Schottky junction for boosting spatial charge separation in photocatalytic H_2O_2 production. *Applied Catalysis B: Environmental* **258**, 117956.

Yi, H., Yan, M., Huang, D. L., Zeng, G., Lai, C., Li, M., Huo, X., Qin, L., Liu, S., Liu, X., Li, B., Wang, H., Shen, M., Fu, Y. & Guo, X. 2019 Synergistic effect of artificial enzyme and 2D nano-structured Bi_2WO_6 for eco-friendly and efficient biomimetic photocatalysis. *Applied Catalysis B: Environmental* **250**, 52–62.

Zhang, J., Wang, Y., Hu, Q. & He, R. 2016 Discussion and views on some issues of the sponge city construction in China. *Advances in Water Science* **27** (6), 793–799.

Zhao, F., Cao, M., Fan, K. E., Liu, Z., Feng, M. & Li, W. 2017 Analysis of pollution characteristics and first flush effect in rainfall runoff in separate storm sewer system in upstream of Tangxi River, Chaohu catchment. *Journal of Lake Sciences* **29** (2), 285–296.

Zhou, C. Y., Xu, P., Lai, C., Zhang, C., Zeng, G., Huang, D., Chen, M., Hu, L., Xiong, W., Wen, X., Qin, L., Yuan, J. & Wang, W. 2019 Rational design of graphic carbon nitride copolymers by molecular doping for visible-light-driven degradation of aqueous sulfamethazine and hydrogen evolution. *Chemical Engineering Journal* **359**, 186–196.

First received 27 July 2019; accepted in revised form 17 November 2019. Available online 2 December 2019

Nitrogen removal performance of microbial fuel cell enhanced bioretention system

Yajun Wang, Rajendra Prasad Singh IWA, Junyu Zhang IWA, Yan Xu and Dafang Fu IWA

ABSTRACT

Bioretention cell (BRC) and an enhanced system combining bioretention cell with microbial fuel cell (BRC-MFC) were used to treat domestic wastewater. Nitrogen removal characteristics and permeation characteristics of two systems were investigated by adjusting influent carbon/nitrogen ratio ($C/N = 2$–20). Results showed that nitrification and denitrification performances were mainly influenced by organic matter and system combination, which further effected the nitrogen removal. When optimal operating parameters were: electrode space of 30 cm, hydraulic load of 1.0 $m^3/(m^2 \cdot d)$ and inlet/reaction time of 1/8 in BRC-MFC system, chemical oxygen demand (COD), total nitrogen (TN) and NH_4^+ removal efficiencies still reached 97.63, 64, and 42.26%, respectively and achieved high removal efficiency of organic matter and nitrogen simultaneously compared to the BRC system. Efficient supply of electron and phylogenetic diversity of bacterial communities in BRC-MFC process was the main reason to achieve deep denitrification removal. After the V3-V4 variable region of 16S rRNA gene was sequenced by the Miseq high-throughput sequencing method, introduction of MFC enhancement technology affected the microbial community structure in the system. The presence of MFC contributed to an increase in community diversity (from 14 to 19 phyla). The results provide a simple method without kinetic energy for simultaneous denitrification and steady infiltration of bioretention.

Key words | bioelectricity enhancement technology, bioretention cell, nitrogen removal, permeation characteristics, wastewater

Yajun Wang
School of Civil Engineering,
Lanzhou University of Technology,
Lanzhou 730050,
China

Yajun Wang
Rajendra Prasad Singh IWA
Junyu Zhang IWA
Yan Xu
Dafang Fu IWA (corresponding author)
School of Civil Engineering,
Southeast University,
Nanjing 210096, runon
China
and
Southeast University-Monash University Joint Research Centre for Future Cities,
Nanjing 210096,
China
E-mail: *fdf@seu.edu.cn*

INTRODUCTION

Media filtration is one of the oldest forms of treatment technology used in the production of potable water and it is still widely used due to its reliability and low cost. Bioretention (often referred to as biofiltration systems, biofilters, or raingardens) has demonstrated high removal of metals and suspended sediment from stormwater, while the reduction in nutrient concentrations displays greater variability (Payne *et al.* 2014a). However, the removal of stormwater pollutants is a relatively recent practice in situations that protect waterways. This includes nitrogen and pathogen treatment of stormwater for water reuse in water stressed areas (Payne *et al.* 2014a), or to meet the standards required for discharge into sensitive water courses and drinking water protected areas (Zhou *et al.* 2017). Treating wastewater by bioretention is much more challenging than for stormwater due to the higher resources and energy required, much of which is organic in nature. To illustrate, the average influent nitrogen for a stormwater bioretention system is typically around 2.1 mg/L with occasional spikes up to 3.2 mg/L (Payne *et al.* 2014b). However, domestic effluents typically have total nitrogen between 40 and 50 mg/L which causes formidable loading in most conventional bioretention.

In China, the pollutant concentrations in municipal wastewater are lower in the south than in the north, especially the ratio of chemical oxygen demand to total nitrogen in influent ($C/N \leq 4$–6). Therefore, it is extremely

doi: 10.2166/aqua.2019.069

difficult to achieve the efficient and stable removal of nitrogen in conventional biological treatment processes (Cao *et al.* 2013). A lack of electron donor is always the key limiting factor of nitrogen removal efficiency during denitrification process. New electrons or their donor are significant in nitrogen removal. The introduction of microbial fuel cell (MFC or 'biogeobattery') and catalytic iron has been recommended because it has demonstrated the capacity to enhance denitrification and thereby increase overall nitrogen removal. The biogeobattery phenomenon is caused by microbes driving the electron flow that is coupled to spatially separate biogeochemical processes. It is worth noting that some innovative hybrid systems were also reported recently based on the similarities in the design of constructed wetland integrated with biogeobattery for ensuring the efficiency of wastewater treatment and power generation. At the same time, a catalytic iron system has the advantages of high reduction ability, convenient in use, persistent in transporting electron and wide obtaining paths. Therefore, it has received significant attention in recent years.

In the classical biogeobattery model (Figure 1(a)), the conductor is a bacterial biofilm and/or conductive biominerals (e.g. magnetite, Fe_3O_4) straddling a redox gradient (e.g. the water table or the plume fringe), with the return current pathway through the surrounding pore fluid. The suggested mechanism is a microbial fuel cell (BRC-MFC) (Figure 1(b)), where organic matter at the anodic region was oxidized, electrons (e^-) and protons (H^+) were produced and then moved to the cathode, with the return current pathway via insulated copper wires connecting the cathode and anode.

The study on bioretention and biogeobattery individually has been investigated on nitrogen removal extensively (Ge *et al.* 2019). However, the integration of bioretention and biogeobattery is still in its infancy, and it is highly likely to provide a promising technical path for improving the low nitrogen removal efficiency of bioretention. Hence, the study of biogeobattery integrated into bioretention is highly significant as it a promising technology to treat domestic wastewater with bioelectricity generation. In addition, to improve the understanding of the role of C/N ratio in nitrogen removal of bioretention coupling with biogeobattery treating sewage, all processes under natural conditions need to be studied. Therefore, in the present study, bioretention systems were operated to investigate the effect of influent different carbon nitrogen ratios and different combinations on nitrogen removal performance.

MATERIALS AND METHODS

Bioretention cell and microbial fuel cell assemble

Three bioretention (BRC and BRC-MFC) columns were constructed from nonopaque perspex. These columns were

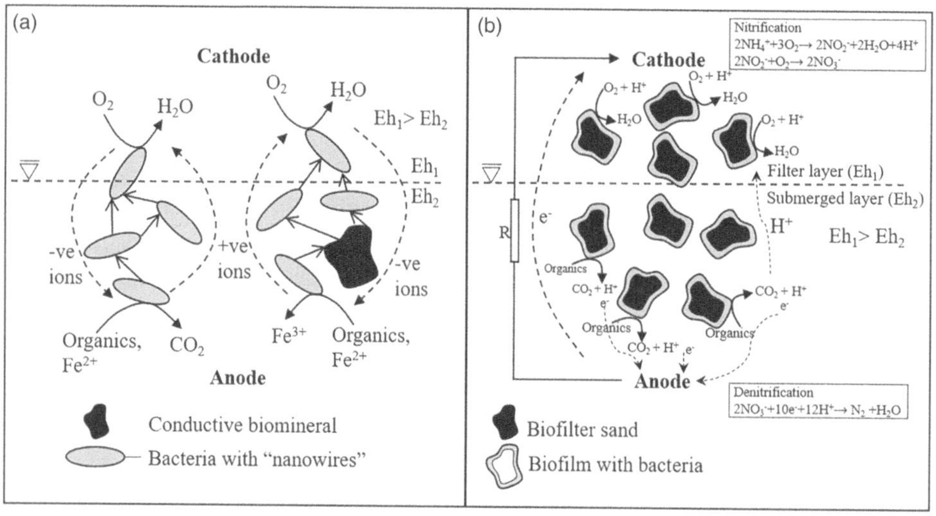

Figure 1 | Conceptual models of redox-associated mechanisms: (a) classical biogeobattery model*; (b) BRC-MFC model (*Revil *et al.* 2015).

fabricated using an acrylic column (8 cm depth × 70 cm height). Figure 2 shows the schematic diagram of BRC and BRC-MFC. These columns were located indoors at an average temperature of 30 ± 4°. The composition of bioretention media layers for saturated designs follow the Australian FAWB adoption guidelines (Payne *et al.* 2014b): a filter layer of washed sand (300 mm, diameter 0.15–1.00 mm), a transition layer of coarse sand (100 mm, diameter 0.50–1.00 mm), a sand submerged layer (200 mm, diameter 0.25–0.50 mm), and the bottom gravel drainage layer (100 mm, diameter 3–5 mm). Three columns with an elevated outlet provided a saturated zone in the lower half of the column ('saturated' design). These columns included a supplementary carbon source of pine chips comprising approximately 5% by volume of saturated zone.

The electrodes (anode and cathodes) employed in the current study were made of activated carbon. Activated carbon is a common carbon material used in the water treatment process. The high specific surface makes it a good medium for the attachment of microorganisms and characteristics of activated carbon such as good biocompatibility as well as moderate electrical conductivity make it a promising biocathode material (Yu *et al.* 2017). Activated carbon is one of the common materials in sewage treatment that has low resistivity $3.2 \times 10^6\ \Omega$cm, high specific area 796 m^2/g and a high C/O mass ratio of 15.2. Activated carbon was immersed in the mixed cultivated sludge and fed with synthetic wastewater. The activated carbon reached saturation during the inoculation period, and acts as a medium for attachment of microbes for biological activity during operation. Thus, the virgin activated carbon transformed into biological activated carbon. Meanwhile, the activated carbon was only placed in the cathode and anode area, and its dosage was very little relative to the whole packing layer. Therefore, although the activated carbon may have an effect on the removal of nitrogen and phosphorus, its interference was negligible in the current experiment. The volume for each layer of anode and cathode electrode was 251.20 cm^3. Insulated copper wire was used to connect both the anode and cathode with an external resistance of 1,000 Ω to complete the circuit.

In order to create an anaerobic condition for the anodic region, the anode was positioned at 15 cm from the bottom of the column. Besides that, an elevated outlet was located for providing a saturated zone in the lower half of the column. The cathode was situated at 60 cm from the bottom of the column, complying with the aerobic condition. Three columns were kept unplanted in order to only assess the influence of different carbon nitrogen ratios and different combinations on nitrogen removal performance.

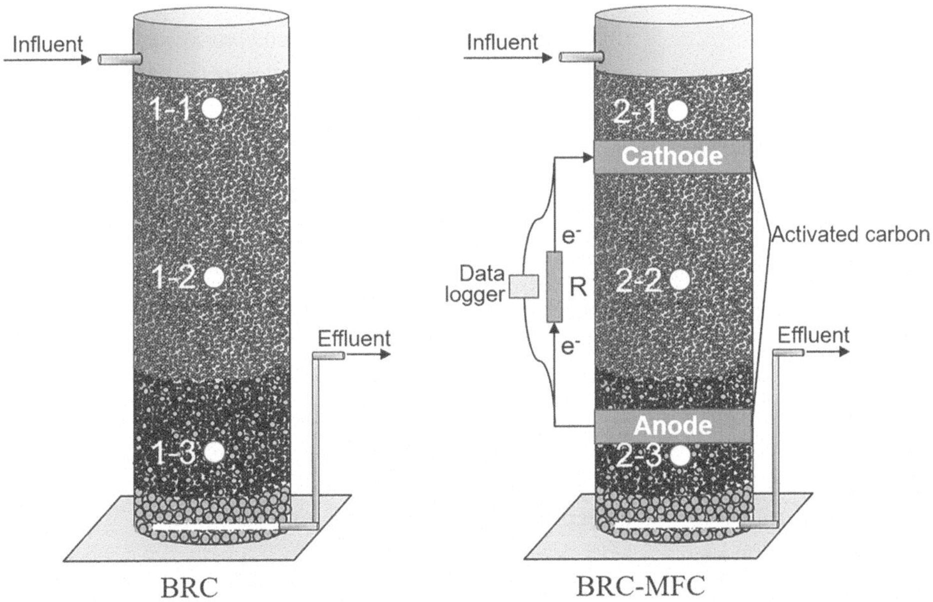

Figure 2 | Schematic diagram of bioretention cell.

Synthetic wastewater

A synthetic wastewater containing the desirable amount of glucose as a carbon source was used throughout the study. The concentrations were targeted to typical domestic wastewater nutrient levels (concentration in g/L): varying amounts of $C_6H_{12}O_6$ (0.15 ± 0.003, 0.24 ± 0.02, 0.48 ± 0.01, 0.6 ± 0.005, and 1.2 ± 0.01); NH_4Cl (22.24 ± 0.03); K_2HPO_4 (2.18 ± 0.05); $NaHCO_3$ (8.73 ± 0.04); $FeCl_2{\cdot}4H_2O$ (4.82 ± 0.05); $CoCl_2{\cdot}6H_2O$ (0.39 ± 0.01); $NiCl_2{\cdot}6H_2O$ (0.19 ± 0.007); $MnCl_2{\cdot}4H_2O$ (0.01 ± 0.001); $CuSO_4$ (0.01 ± 0.001); $ZnCl_2$ (0.01 ± 0.001); $CaCl_2$ (0.04 ± 0.002); Humic acid (0.12 ± 0.005). All the chemical reagents used in this study were of analytical reagent (AR) grade (Wang *et al.* 2013).

Inoculation

Inoculation medium in the submerged layer and activated carbon were immersed in the mixed cultivated sludge for inoculation for 45 days, which was obtained from an anaerobic reactor in the local wastewater treatment plant (Nanjing Banqiao Water Group, Nanjing, China). During the inoculation period, synthetic wastewater was fed to the anaerobic inoculation tank to support the growth of the microbial community.

Experimental procedure

Synthetic wastewater was supplied to the top of the bioretention using a peristaltic pump (Longer BT100-2 J, China) at a flow rate of 3.5 mL/min (1.0 m^3/m^2 s hydraulic load (HLR)), operated in intermittent mode, which was maintained at 12 hours hydraulic retention time (HRT). Treated wastewater was discharged and collected from an elevated outlet.

The experiment was divided into five stages (over a total of 112 days): (1) Stage I, $C/N = 2{:}1$; (2) Stage II, $C/N = 4{:}1$; (3) Stage III, $C/N = 8{:}1$; (4) Stage IV, $C/N = 10{:}1$; (5) Stage V, $C/N = 20{:}1$. Experiments were conducted in triplicate with three treatment cells. Nitrogen and carbon levels under various C/N ratios were analyzed and compared.

Sample collection and determination

Influent and effluent of the experimental cells were sampled to evaluate transformation and treatment performance three times during each stage. The samples were immediately transported and analyzed at the Laboratory of Southeast University-Monash University Joint Research Centre. Chemical oxygen demand (COD) was determined with a HACH DR2800 (HACH Co., USA). Total nitrogen (TN), NH_4^+-N, NO_3^--N, and NO_2^--N were measured using a spectrophotometer UV-1800 (SHIMADZU Co., Kyoto, Japan), and total phosphorus (TP) was determined by the ammonium molybdate spectrophotometry method. All the parameters mentioned above were determined using the methods and procedures described in the Standard Method for Examination of Water and Wastewater (MoEE China 2002).

Microorganism samples were collected from submerged layers at the end of 150 days. For this sampling, three micro organism samples were taken from each column. After on-site collection, the samples were transported immediately to the laboratory for the purpose of total DNA extraction. The genomic DNA was extracted and purified using a MOBIO PowerSoil® DNA Isolation Kit (12888-50 Soil DNA Kit MOBIO, USA). The extracted genomic DNA was kept at $-20°$ and detected using 1% agarose gel electrophoresis.

Data analysis

Influent and effluent concentrations were used to calculate the removal efficiencies and transformation or accumulation rates of COD, TN, NH_4^+-N, NO_2^--N, and NO_3^--N. Stepwise regression models were built to determine the multiple linear regression equation between nitrogen transformation rates and functional genes using SPSS Statistics 20 (IBM, USA). All data were analyzed using analysis of variance (ANOVA). Data were checked to ensure ANOVA assumptions were not violated and were transformed appropriately when needed prior to analysis.

RESULTS AND DISCUSSION

Effect of influent C/N ratio on COD removal

The trends of COD concentration and removal efficiency under different C/N operating conditions are presented in

Figure 3. The results revealed that when the influent C/N ratio fluctuates between 2.54 and 4.16, the average COD concentration of effluent from BRC group and BRC-MFC group was 24–42 and 17–27 mg/L, respectively. The average COD removal rates of BRC group and BRC-MFC group were 69.86 and 83.04% ($C/N = 2.54$); 79.82 and 92.04% ($C/N = 4.16$); 80.89 and 95.27% ($C/N = 7.90$); 80.26 and 97.63% ($C/N = 10.42$); 82.94 and 95.55% ($C/N = 19.36$), respectively.

It was noted in an earlier study that that the effective utilization rate of COD was 74% ($C/N < 5.0$) in the intermittent system (Cao *et al.* 2013). In contrast, the BRC-MFC enhanced group has obvious advantages in enhancing the utilization of carbon sources when treating wastewater with low C/N ratio under the same conditions. This is probably due to the fact that in the anode (anaerobic submerged layer), organic compounds are almost completely oxidized by microorganisms (Oon *et al.* 2016). Liu *et al.* (2019) determined that the COD removal rate of vertical flow constructed wetland coupled with MFC system was only 79.65% (about 505 mg/L of influent COD). In the current study, the COD removal rate was 95.27% (about 500 mg/L of influent COD) higher than previous research results. At the same time, activated carbon functional groups will enhance the attachment of microorganisms on the surface of activated carbon (Wong *et al.* 2018). In addition, a closed loop was formed in the BRC-MFC strengthening system, which enhanced the electron transfer from the anode to the cathode and further affected the biodegradability of organic pollutants. Zhang *et al.* (2016) also found that the performance of constructed wetland coupled MFC system was 12–20% better than that of an open-circuit system, and 27–49% better than that of an ordinary constructed wetland without MFC. Therefore, this study indicates that the BRC-MFC system has a significant impact in organic degradation.

Effect of influent C/N ratio on nitrogen removal

As shown in Figure 4, NH_4^+-N and TN removal rates are closely related, mainly because the nitrification process provides an electron acceptor for denitrification. Complete oxidation of NH_4^+-N is the basis for effective TN removal. The average removal efficiencies of NH_4^+-N were 32.19% (BRC) and 42.26% (BRC-MFC) under $C/N = 2.54$. In terms of average removal efficiency of TN, the BRC-MFC

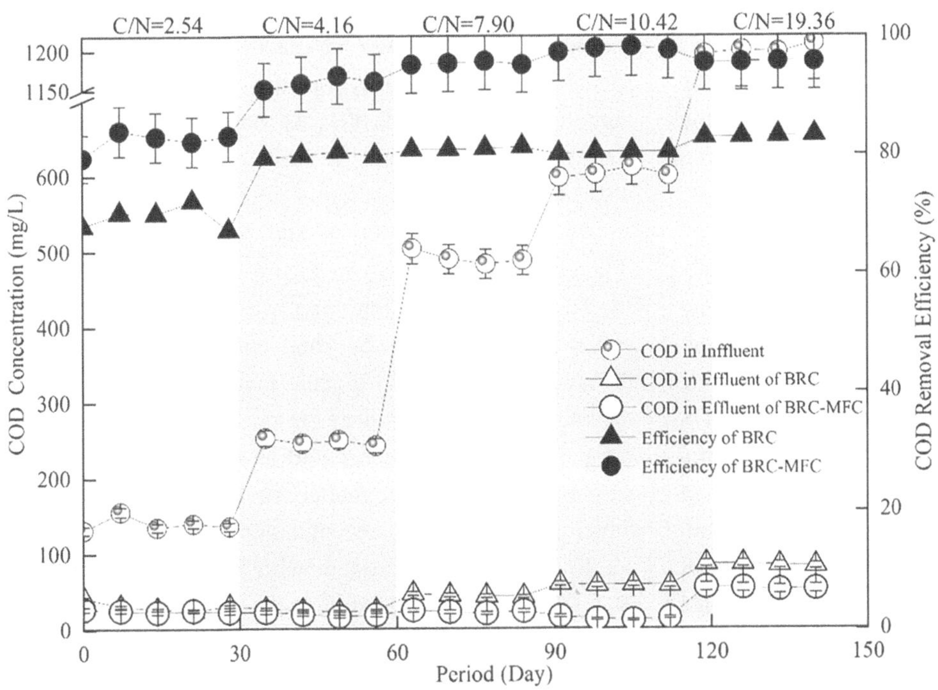

Figure 3 | Removal performance of COD at different C/N ratios.

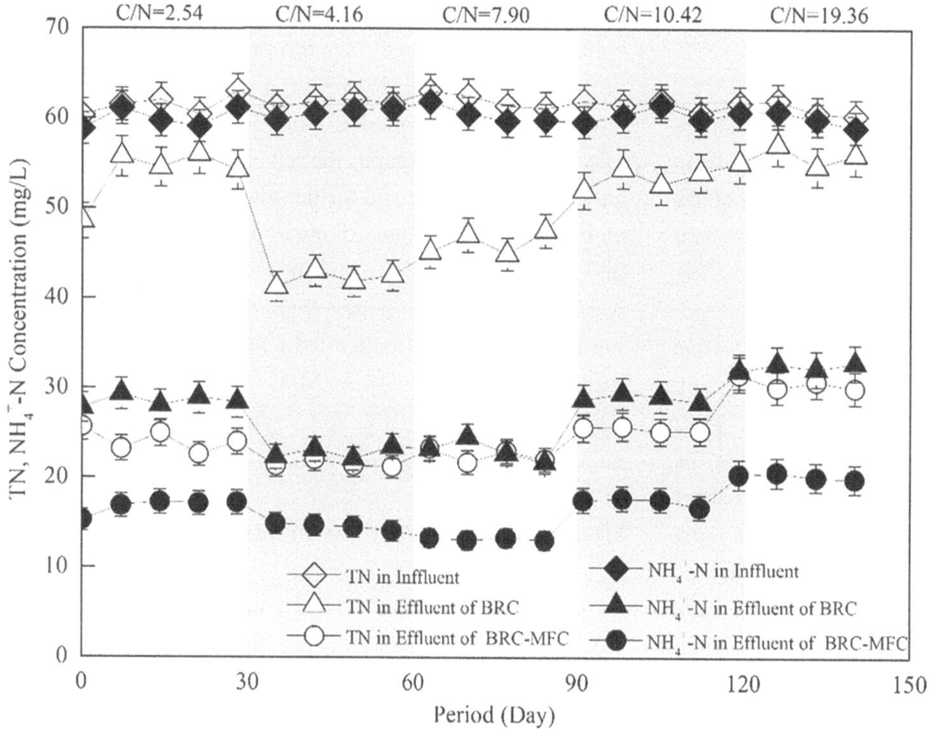

Figure 4 | Removal performance of TN and NH_4^+-N at different C/N ratios.

system (60.75%) was 75% higher than that of the BRC system. The average removal efficiency of TN and NH_4^+-N was significantly higher than that of other C/N ratios except for 4.16 and 7.90. The removal rates of TN and NH_4^+-N were hardly improved further, mainly because the cathode was located 10 cm below the top of the reaction column, which limits the electron acceptor (oxygen) and slows down the metabolism of nitrifying bacteria (Shen *et al.* 2018).

The results reveal that when the C/N ratio was increased to 10.42, the nitrification was intensified, and NH_4^+-N concentrations in the effluent of the BRC system and the BRC-MFC system gradually accumulated, reaching 46.22 and 28.86 mg/L, respectively. The average TN removal rates were 24.41 and 52.80%, respectively. Combined with the utilization of COD, the decrease in nitrogen removal efficiency was mainly due to the incomplete nitrification caused by the excessive loading of influent organic matter, resulting in the decrease of TN removal rate. The removal rates of TN and NH_4^+-N were difficult to further improve, mainly because the cathode was located 10 cm below the top of the column, which limits the electron acceptor (oxygen) and then slows down the metabolism of nitrifying bacteria.

The above results reveal that, in a certain range, improving the influent C/N was an effective measure to improve the TN removal efficiency, but this must be based on the complete oxidation of NH_4^+-N, otherwise the excessive C/N will not only cause the waste of carbon sources, but also lead to the deterioration of nitrogen removal. Figure 5 shows the trend of NO_3^--N and NO_2^--N changing with different C/N. As shown in Figure 5, the high plateau of the NO_3^--N concentration curve was observed in the BRC system under $C/N = 2.54$, and because of insufficient demand for denitrifying organic matter, it is more advantageous to produce nitrite from ammonia than from nitrate. A possible reason could be that low C/N influent cannot meet the organic matter required for denitrification, which also leads to the accumulation of NO_3^--N.

In the BRC-MFC system with influent C/N of 4.16, NO_2^--N accumulated significantly, but NO_3^--N did not accumulate. The possible reason was that COD decreased and the supply of electron donors was insufficient. This trend is consistent with the findings of Wu *et al.* (2009),

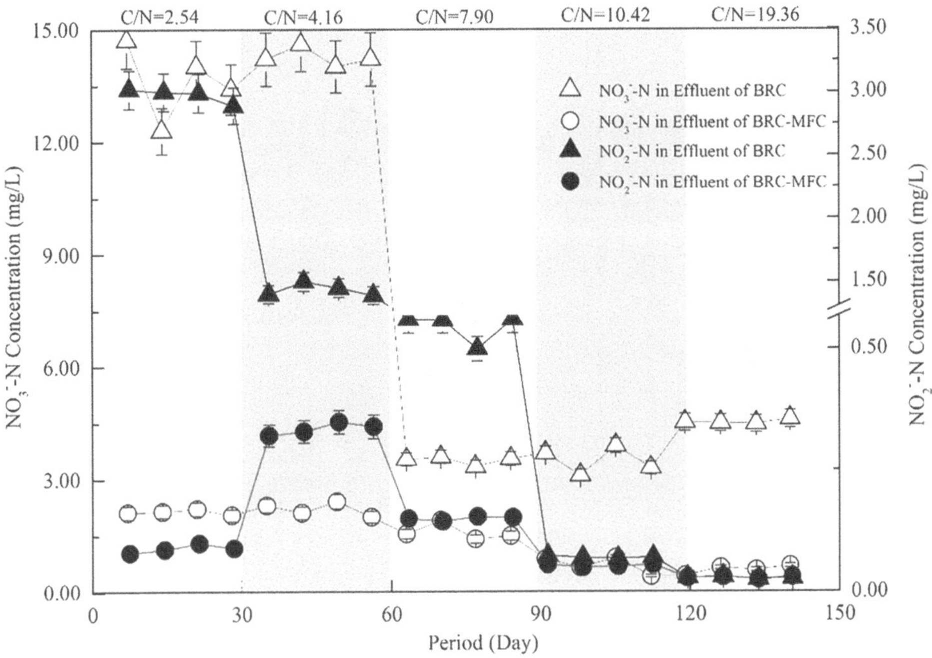

Figure 5 | Removal performance of NO_3^--N and NO_2^--N at different C/N ratios.

where they noted that NO_2^--N increases during denitrification when the supply of electron donors was insufficient. When influent C/N was 10.42 and 19.36, the concentration of NO_3^--N decreases (Figure 5), and the concentration of NO_3^--N increases gradually (Figure 4). The possible reason could be that the nitrate was changed by dissimilatory nitrate reduction to ammonium (DRNA) process, and DRNA dominates the reduction of NO_3^- and NO_2^- in anoxic environments. This process is in competition with denitrification, especially under high influent C/N conditions (Xu *et al.* 2017). The average voltage of BRC-MFC system reaches 35.5 mV, and the maximum power density was 103.83 mW/m^3. Hence, it could be concluded that under the electric enhancement condition, the electron migration in the system is beneficial to denitrification and improves the overall nitrogen removal performance.

Effect of different C/N ratio and C/P ratio on simultaneous removal of nitrogen and phosphorus

The results showed that the change of influent C/N had little effect on the removal of COD and TP, but the simultaneous removal of nitrogen and phosphorus was based on sufficient electron acceptor, BRC-MFC system had obvious advantages in enhancing the utilization of carbon sources and the ability of electron transfer under low influent C/N. Therefore, the correlations between the value of C/N and C/P and the removal rate of TN and TP were compared.

As shown in Figure 6, when C/N < 3 or C/P < 40, the TN and TP removal rates of BRC were less than 20 and 70%, respectively. The TN removal rate of BRC-MFC was lower than 60%, and the TP removal rate was about 97%. In C/N > 4 or C/P > 60 conditions, the TN removal rates of both systems revealed a downward trend, and TP removal rates were basically stable. This trend is consistent with the earlier findings (Xu *et al.* 2019). Therefore, under $C/N = 3–4$ or $C/P = 30–60$ conditions, the removal rates of TN and TP in BRC system were 33 and 75%, respectively, and the treatment efficiency was limited. The removal rates of TN and TP in BRC-MFC system reached 65 and 98%, respectively, which basically realized the simultaneous removal of nitrogen and phosphorus. Meanwhile, the removal efficiency of the BRC system was improved by 30%.

Although the average voltage and maximum power density were not as good compared to the findings of Xu *et al.* (2018) in the CW-MFC (the average voltage was

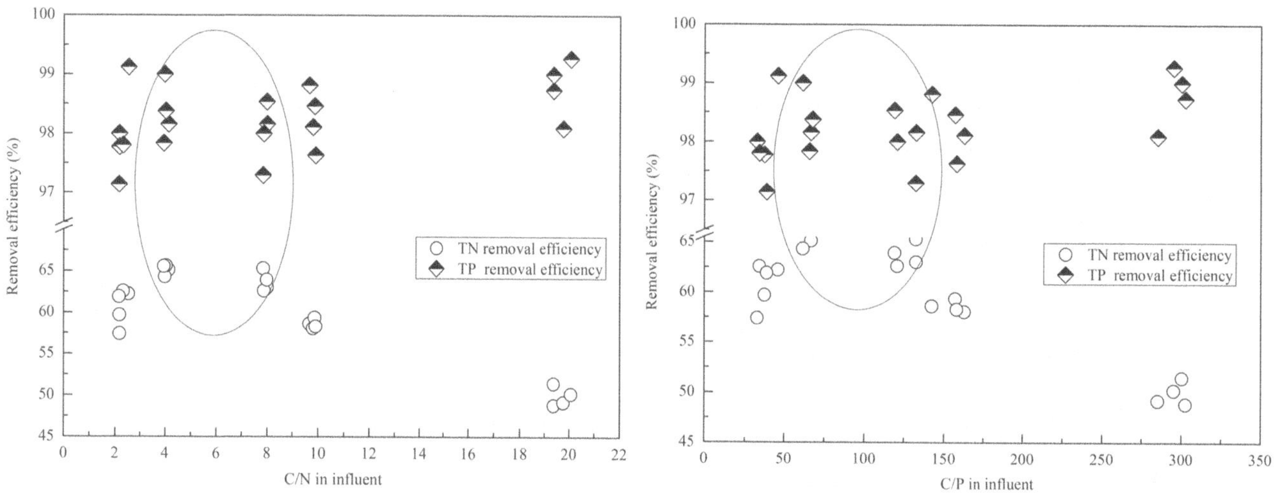

Figure 6 | Correlations of TN and TP removals with influent C/N and C/P ratios.

265.77 ± 12.66 mV and the highest power density was 3,714.08 mW/m^2), the removal rate of all pollutants by BRC-MFC in the current study was higher than that by CW-MFC. Under different electric enhancement conditions, the electron migration in the system was beneficial to denitrification and improves the overall nitrogen removal performance.

Comparative study on permeation characteristics of MFC strengthened group and BRC control group

The BRC-MFC system and the BRC system were tested for sewage infiltration with 1.0 m^3/(m^2·d) hydraulic load (porosity = 41%, initial permeability coefficient $K_{ini} = 3.893 \times 10^{-4}$ m/s, influent $C/N = 4$, inlet/reaction time = 1/8, intermittent operation cycle), the results are presented in Figure 7. As shown in Figure 7, the standard permeability coefficient K_{20} of BRC-MFC and BRC systems decreases with the running time, but the changing process was different. The BRC system has experienced the three following stages: the permeability coefficient of the first stage (0–57 d) dropped sharply (slope $K_{2\text{-}1} = -0.0416$); decreased slowly in the second stage (59–121 d) (slope $K_{2\text{-}2} = -0.0055$); but reached a more stable state in the third stage on the 123rd day, where K_{20} (BRC) reduced to 0.89×10^{-4} m/s (accounting for 23% of K_{ini}).

In contrast, the BRC-MFC system experienced four stages, the permeability coefficient of the first stage (0–25 d) system dropped sharply (slope $K_{1\text{-}1} = -0.0332$). The slope for the BRC system is slightly lower than the BRC-MFC system and revealed low permeability coefficient values in the BRC system. The reason could be that the structure and diversity of microbial community in the filler layer were changed by the increase of electron transport. *Proteobacteria* (31%), *Bacteroidetes* (21%) and *Chloroflexi* (14%) were the dominant bacteria in the BRC-MFC system, while the dominant bacteria in the BRC system were mainly *Chlorobi* (45%) and *Proteobacteria* (21%). The second stage (27–49 d) appeared to be a plateau stage, mainly because MFC enhanced the ability of microorganisms to degrade pollutants, and the microbial biomass could be evenly distributed throughout the reaction column region, thus making the permeability more stable. In the third stage (51–111 d), the permeability coefficient of the system continued to decline again. A possible reason could be the coupling effect of multiple influencing factors. The accumulation of inorganic/organic compounds and the growth of biofilm would inevitably lead to the gradual decrease of the pore space in the reaction column, and consequently the permeability coefficient gradually decreased.

Although the decrease was greater than that of the second stage of BRC (slope $K_{1\text{-}3} = 0.0102 > K_{2\text{-}2} = 0.0055$), compared to the BRC system, BRC-MFC reached the stable platform 10 days ahead of schedule, the $K_{20\ (BRC\text{-}MFC)}$ was reduced to 1.86×10^{-4} m/s (48% of K_{ini}), which was

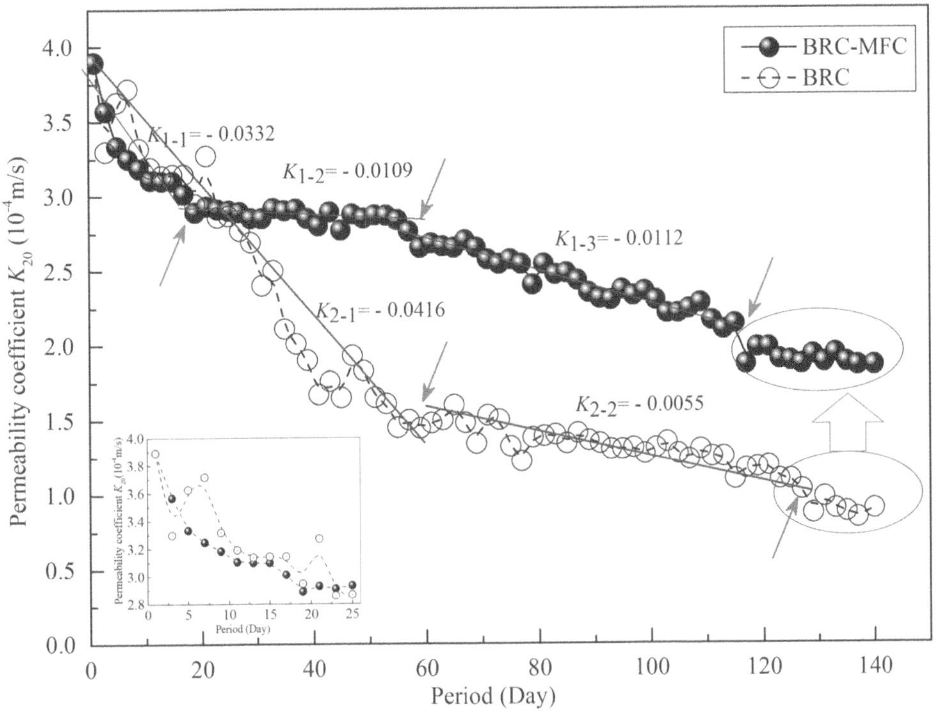

Figure 7 | Trend of the permeability coefficient of BRC and BRC-MFC.

twice that of the BRC system in the stable stage. Therefore, the BRC-MFC system can delay the 50% reduction of permeability coefficient, and can enter the stable stage of permeability coefficient earlier.

In a previous study about the BRC treatment (Coustumer *et al.* 2009), the removal rate of TN decreases with the decrease of permeability coefficient. The highest removal rate of TN (51%) occurs in the middle region of the permeability coefficient. Similar changes were also found in the current study. The difference is that the highest TN removal rate occurred in the low permeability area. This trend is different from the previous study (Coustumer *et al.* 2009). BRC-MFC can not only delay the decrease of permeability coefficient, but also enhance the TN removal efficiency. Therefore, BRC-MFC has the effect of simultaneous denitrification and stable permeability.

Biomass distribution on the sand of differential reactors

The biofilm samples were analyzed on the 90th and 150th day of the experiment by opening a BRC column as a control group and a BRC-MFC column, and then different locations were collected from the top (filter layer), the middle (submerged layer) and the bottom (drainage layer) and directly observed by Confocal Laser Scanning Microscope (CLSM), as shown in Figure 8.

As shown in Figure 8 and Table 1, the distribution of biofilm on the sand surface of the two systems shows an increasing trend with time. However, the growth/decrease in biofilm thickness and its spatial distribution in each system were different. Compared to BRC system, the biofilm thickness of BRC-MFC system was 2-3 > 2-2 > 2-1, but the values were basically close. As microorganisms catalyze substrate denitrification by respiration to produce electrons, electrodes act as temporary electron acceptors to absorb electrons and transfer electrons to the cathode surface through a loop for pollutant degradation in the cathode region. Thus nutrients can be effectively degraded and removed when they firstly contact the 2-1 region (upper filter layer). At the same time, the microbial community in the 3-1 sampling site was in the unsaturated filter layer, without sufficient HRT and long-term mass transfer of nutrients. On the contrary, the 2-2 and 2-3 regions were saturated (with submerged area) which could ensure the contact

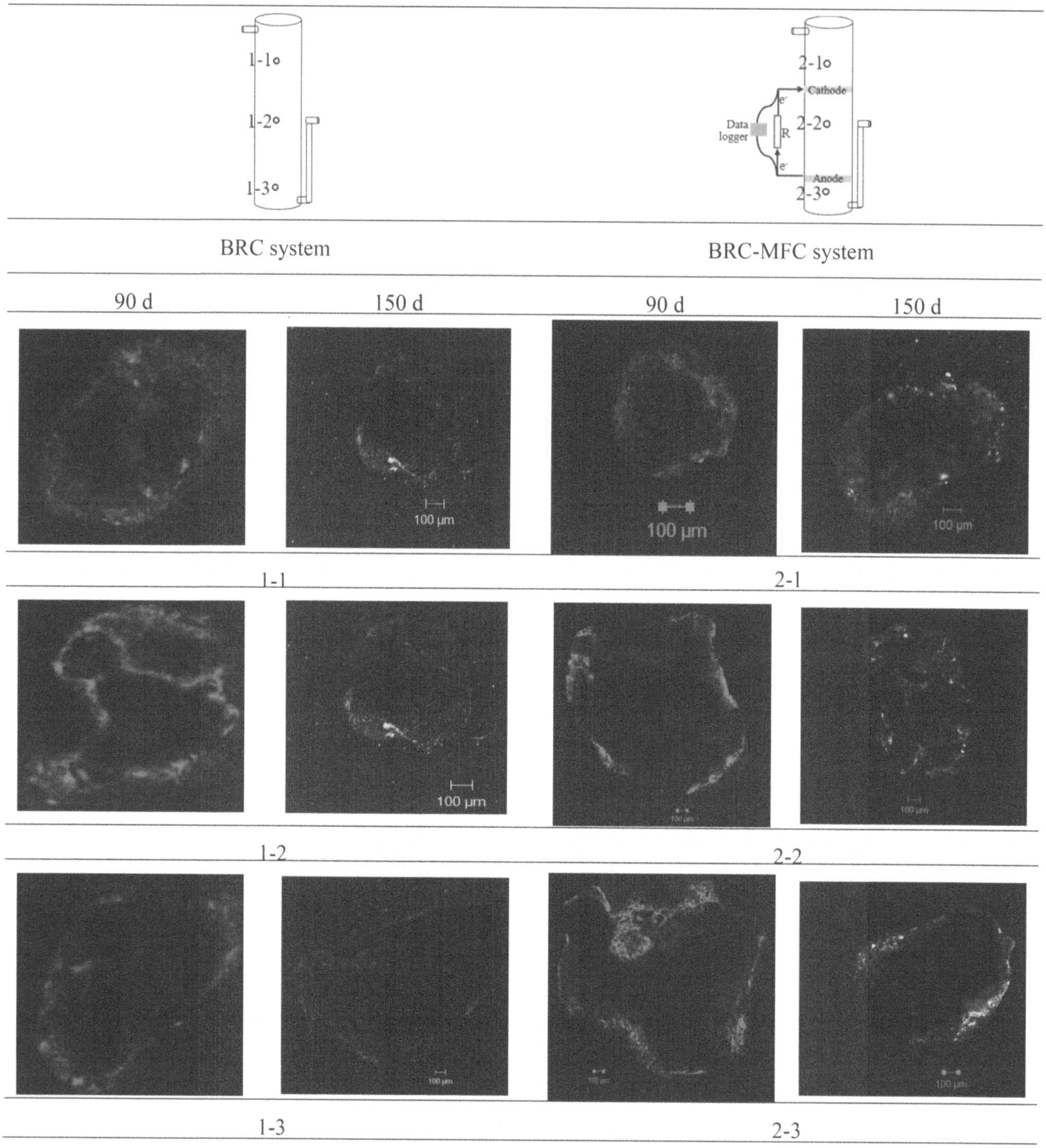

Figure 8 | Biomass distribution on the sand of differential reactors.

adsorption and degradation of microorganisms and nutrients, and promote microbial growth. Compared with the BRC system, the BRC-MFC system had a considerable increase in the thickness of biofilm and the proportion of sand surface area, and formed a uniform 'upper thin and lower thick' spatial distribution, which enhanced the ability to resist the accumulation of pollutants on the surface of the reaction system.

Table 1 | Thickness of biofilm on sand after 90 and 150 day inoculations

Sampling location	Time (days)	Biofilm thickness range (μm)	Biofilm thickness (μm)	Biofilm specific surface ratio (%)
1–1	90	0–75	36.5	80
	150	0–80	38.0	85
1–2	90	0–85	55.1	87
	150	15–120	57.2	95
1–3	90	0–60	25.5	68
	150	5–65	31.0	75
2–1	90	0–100	55.0	90
	150	0–95	75.3	95
2–2	90	0–200	102.5	95
	150	31–190	90.3	100
2–3	90	23–150	82.0	100
	150	41–186	94.0	100

Phylogenetic diversity of bacterial communities

Many scholars have proved that *Proteobacteria*, *Firmicutes*, *Acidobacteria* and *Bacteroidetes* have electrochemical activity (Lee *et al.* 2003), *Citrobacter* (Chen *et al.* 2017) and *Geobacter* of *Proteobacteria* and *Clostridium* were the dominant bacteria producing electricity. All of these were found to exist in the BRC-MFC system. As shown in Tables 2 and 3, the relative abundance of *Proteobacteria* and *Bacteroidets* in the BRC-MFC system increased gradually with time, and by the 200th day, the relative abundance was 41 and 13%, respectively. The relative abundance of *Dok59* and *Geobacter* was high in all bacteria (9 and 6%, respectively). It indicates that the electricity producing bacteria group is not limited to the *Geobacter* bacteria. Therefore, the discovery of electro bacterium *Dok59* has expanded the scope of electro-generated microorganisms. The difference in the microbial community structure may be due to the composition of the fuel cell packed with sediment and its degradation products (Dong *et al.* 2017). In the BRC system, the *Rhodanobacter* content was relatively stable over time, with a content of 3%. It can be concluded that the composition and structure of the microbial community in the system did not change significantly without the intervention of bioelectric intensification, and anaerobic denitrifying bacteria were the dominant species.

Practical applications and future research perspectives

Current research findings provide a technical guideline to enhance the long term operation reliability and denitrification performance of BRC systems for an intermittent domestic sewage treatment in China and other developing countries.

Table 2 | Relative abundance distribution of bacterial species in experimental group (unit: %)

	BRC system		BRC-MFC system	
	(90 d)	(200 d)	(90 d)	(200 d)
Top ten	*Chlorobi*, 45	*Proteobacteria*, 40	*Proteobacteria*, 31	*Proteobacteria*, 41
	Proteobacteria, 21	*Bacteroidetes*, 21	*Bacteroidetes*, 21	*Chloroflexi*, 18
	Chloroflexi, 10	*Chloroflexi*, 9	*Chloroflexi*, 14	*Bacteroidetes*, 13
	Bacteroidetes, 8	*Chlorobi*, 4	*Fibrobacteres*, 11	OD1, 7
	Spirochaetes, 4	OD1, 4	*Chlorobi*, 10	*Chlorobi*, 5
	Firmicutes, 3	*Acidobacteria*, 3	*Spirochaetes*, 4	*Fibrobacteres*, 2
	Acidobacteria, 2	*Firmicutes*, 3	*Acidobacteria*, 1	*Nitrospirae*, 2
	WS6, 1	*Verrucomicrobia*, 3	*Firmicutes*, 1	OP11, 2
	Verrucomicrobia, 1	*Chlamydiae*, 2	*Verrucomicrobia*, 1	*Acidobacteria*, 1
	OD1, 1	*Cyanobacteria*, 2	OD1, 1	*Actinobacteria*, 1
Other	*Nitrospirae*	*Spirochaetes*, 2	*Candidate division*, WS6, TM7	*Cyanobacteria*
	Cyanobacteria	*Actinobacteria*	*Cyanobacteria*	*Firmicutes*
	TM7	*Planctomycetes*	*Nitrospirae*	*Planctomycetes*
		Fibrobacteres		*Spirochaetes*
		Gemmatimonadetes		*Verrucomicrobia*
		Nitrospirae		*Chlamydiae*
		OP11, TM6		*Gemmatimonadetes*
				TM6

Table 3 | Relative abundance distribution of bacterial genus in experimental group (unit: %)

	BRC system (200 d)	BRC-MFC system (200 d)
Top ten	Dok 59, 4	Dok 59, 9
	Rhodanobacter, 3	*Geobacter*, 6
	Chlorobaculum, 2	*Chlorobaculum*, 3
	Dechloromonas, 2	*Dechloromonas*, 2
	Simplicispira, 2	*Devosia*, 2
	Syntrophus, 2	*Anaerolinea*, 1
	T78, 2	*Thiobacillus*, 1
	Clostridium, 1	*Treponema*, 1
	Flavihumibacter, 1	*Uliginosibacterium*, 0.2
	Sediminibacterium, 1	*Syntrophus*, 0.2

In future research, the long-term running permeability coefficient change test under the clean water condition should be added, so as to make a comparative study with the results of the permeability coefficient change test under sewage conditions, so as to better reveal the change rule of BRC internal permeability coefficient. The future research should also carry out the operation of the BRC system under the real sewage conditions so that the findings could be beneficial in the actual field application environment.

CONCLUSIONS

The results showed that removal rates of TN and TP were 33 and 75%, respectively when the C/N ratio was 3–4 or the C/P ratio was 30–60 in BRC system, but the treatment efficiency was limited. The removal rates of TN and TP in BRC-MFC system reached 65 and 98%, respectively, with simultaneous removal of nitrogen and phosphorus. Permeability characteristics results showed that BRC-MFC system can delay the decrease of permeability coefficient by 50%, and can enter in the stable stage earlier. The BRC-MFC system has better stabilizing percolation performance and denitrification effects than the BRC system under long-term intermittent operation.

ACKNOWLEDGEMENTS

The authors are grateful to Jinxi Wang for her immense help and suggestions on preparing this paper.

FUNDING

This research was co-funded by the National Natural Science Foundation of China (No. 41967043), National Key Technologies R&D Program, China (No. 2015BAL02B05), Priority Academic Program Development of Jiangsu Higher Education Institutions, National Natural Science Foundation of China (No. 51550110231), and the Key Technology Program of Gansu province, China (No. JK2018-16).

REFERENCES

Cao, G., Wang, S., Peng, Y. & Miao, Z. 2013 Biological nutrient removal by applying modified four step-feed technology to treat weak wastewater. *Bioresour. Technol.* **128** (1), 604–611.

Chen, C. Y., Tsai, T. H., Wu, P. S., Tsao, S. E., Huang, Y. S. & Chung, Y. C. 2017 Selection of electrogenic bacteria for microbial fuel cell in removing Victoria blue r from wastewater. *J. Environ. Sci. Health A* **53**, 1–8.

Coustumer, S. L., Fletcher, T. D., Deletic, A., Barraud, S. & Lewis, J. F. 2009 Hydraulic performance of biofilter systems for stormwater management: influences of design and operation. *J. Hydrol.* **376** (1), 16–23.

Dong, G. R., Kariminia, H. R., Chen, Z. W., Parker, W., Pritzker, M. D. & Legge, R. L. 2017 Enhancement of electricity generation by a microbial fuel cell using a highly active non-precious-metal nitrogen-doped carbon composite catalyst cathode. *Energy Fuels* **31** (1), 959–967.

Ge, Z., Wei, D., Zhang, J., Hu, J., Liu, Z. & Li, R. 2019 Natural pyrite to enhance simultaneous long-term nitrogen and phosphorus removal in constructed wetland: three years of pilot study. *Water Res.* **148**, 153–161.

Lee, J., Phung, N. T., Chang, I. S., Kim, B. H. & Sung, H. C. 2003 Use of acetate for enrichment of electrochemically active microorganisms and their 16 s rdna analyses. *FEMS Microbiol. Lett.* **223** (2), 185–191.

Liu, F., Sun, L., Wan, J., Tang, A., Deng, M. & Wu, R. 2019 Organic matter and ammonia removal by a novel integrated process of constructed wetland and microbial fuel cells. *RSC Adv.* **9** (10), 5384–5393.

Ministry of Ecology and Environment of the People's Republic of China 2002 *Standard Method for the Examination of Water and Wastewater.* Environmental Science Press of China, Beijing.

Oon, Y. L., Ong, S. A., Ho, L. N., Wong, Y. S., Dahalan, F. A., Oon, Y. S. & Thung, W. E. 2016 Synergistic effect of up-flow constructed wetland and microbial fuel cell for simultaneous wastewater treatment and energy recovery. *Bioresour. Technol.* **203**, 190–197.

Payne, E. G., Pham, T., Cook, P. L., Fletcher, T. D., Hatt, B. E. & Deletic, A. 2014a Biofilter design for effective nitrogen removal from stormwater – influence of plant species, inflow hydrology and use of a saturated zone. *Water Sci. Technol.* **69** (6), 1312–1319.

Payne, E. G. I., Fletcher, T. D., Cook, P. L. M., Deletic, A. & Hatt, B. E. 2014b Processes and drivers of nitrogen removal in stormwater biofiltration. *Crit. Rev. Environ. Sci. Technol.* **44** (7), 796–846.

Revil, A., Mendonca, C. A., Atekwana, E. A., Kulessa, B., Hubbard, S. S. & Bohlen, K. J. 2015 Understanding biogeobatteries: where geophysics meets microbiology. *J. Geophys. Res. Biogeosci.* **115** (G1), 379–384.

Shen, X., Jian, Z., Liu, D., Zhen, H. & Liu, H. 2018 Enhance performance of microbial fuel cell coupled surface flow constructed wetland by using submerged plants and enclosed anodes. *Chem. Eng. J.* **351**, 312–318.

Wang, Q., Ye, L., Jiang, G. & Yuan, Z. 2013 A free nitrous acid (fna)-based technology for reducing sludge production. *Water Res.* **47** (11), 3663–3672.

Wong, S., Ngadi, N., Inuwa, I. M. & Hassan, O. 2018 Recent advances in applications of activated carbon from biowaste for wastewater treatment: a short review. *J. Clean. Prod.* **175**, 361–375.

Wu, J., Jian, Z., Jia, W., Xie, H., Gu, R. R., Cong, L. & Gao, B. 2009 Impact of COD/N ratio on nitrous oxide emission from microcosm wetlands and their performance in removing nitrogen from wastewater. *Bioresour. Technol.* **100** (12), 2910–2917.

Xu, D., Xiao, E., Xu, P., Lin, L., Zhou, Q., Xu, D. & Wu, Z. 2017 Bacterial community and nitrate removal by simultaneous heterotrophic and autotrophic denitrification in a bioelectrochemically-assisted constructed wetland. *Bioresour. Technol.* **245** (Pt A), 993–999.

Xu, F., Cao, F. Q., Kong, Q., Zhou, L. L., Yuan, Q., Zhu, Y. J., Wang, Q., Du, Y.-D. & Wang, Z.-D. 2018 Electricity production and evolution of microbial community in the constructed wetland-microbial fuel cell. *Chem. Eng. J.* **339**, 479–486.

Xu, R., Zhang, Y., Liu, R., Cao, Y., Wang, G., Ji, L. & Xu, J. 2019 Effects of different substrates on nitrogen and phosphorus removal in horizontal subsurface flow constructed wetlands. *Environ. Sci. Pollut. Res.* **2**, 1–10.

Yu, Y. Y., Zhai, D. D., Si, R. W., Sun, J. Z., Liu, X. & Yong, Y. C. 2017 Three-dimensional electrodes for high-performance bioelectrochemical systems. *Int. J. Mol. Sci.* **18** (1), 90–98.

Zhang, M., Peng, Y.-Z., Zhang, J.-H., Wang, C., Wang, S.-Y. & Zeng, W. 2016 Effect of influent C/N ratios on denitrifying phosphorus removal characteristics in the A2/O-BCO process. *China Environ. Sci.* **36** (5), 1366–1375.

Zhou, F., Cheng, S., Hui, W., Xian, C. & Li, X. 2017 Feasibility study of simultaneous azo dye decolorization and bioelectricity generation by microbial fuel cell-coupled constructed wetland: substrate effects. *RSC Adv.* **7** (27), 16542–16552.

First received 20 April 2019; accepted in revised form 30 October 2019. Available online 29 November 2019

Removal of low-concentration phosphorus by efficient phosphorus removal composite-based ecological floating beds

Yan Liu, Juanjuan Lv and Rajendra Prasad Singh

ABSTRACT

In order to strengthen the effect of ecological floating beds on the removal of low-concentration phosphorus (P) in water, the efficient P removal composite (EPRC), which is a high-efficiency P removal material developed in the current work as the substrate for the *Ipomoea aquatica* floating beds, was introduced into the solar chamber. The EPRC dosage (2 g/L) and the optimal number of *Ipomoea aquatica* plants (6 plants/10 L) suitable for the floating beds were determined experimentally. Results revealed that EPRC and *Ipomoea aquatica* composite floating beds had the best P removal effect among the three floating beds. Moreover, the root growth ratio of *Ipomoea aquatica* in the composite floating bed and the plant-only floating bed was 1.90 and 1.25, respectively. The stem growth ratio of *Ipomoea aquatica* in the composite floating bed and plant-only floating beds was 1.54 and 1.21, respectively. The leaching experiments showed that the leaching of heavy metals from the EPRC was negligible.

Key words | efficient P removal composite (EPRC), floating bed, *Ipomoea aquatica*, matrix, P removal

Yan Liu
Juanjuan Lv (corresponding author)
Rajendra Prasad Singh
School of Civil Engineering,
Southeast University,
Nanjing 21000,
China
E-mail: *1036649868@qq.com*

INTRODUCTION

Due to the rapid economic development, the problem of eutrophication of water bodies has become critical. The excess nutrients discharged into the water bodies seriously affected the self-purification ability of the lake, resulting in algal blooms (Mehner & Benndorf 1995; Yang *et al.* 2008). Eutrophication leads to the sharp decline of dissolved oxygen in water and the deterioration of water quality, which destroy the freshwater ecosystems. Prominent methods widely used in practical engineering to control the eutrophication mainly include: physical methods, chemical methods, biological methods, and adsorption methods (Li *et al.* 2014; Santiago Sampaio *et al.* 2015; Liu *et al.* 2016a, 2016b; Hamzah *et al.* 2018; Tian *et al.* 2018; Wang *et al.* 2019).

As one of the key biological measures, ecological floating beds have been widely used in the pollution control and ecological restoration of surface water bodies due to easy to construct, simple, and cost-effective characteristics (Yeh *et al.* 2015; Chang *et al.* 2017). Performance, however, is unavoidably restricted by the growth rate and limited stand biomass of the plant. Therefore, improving the ability of artificial floating beds to remove pollutants is one of the research focus in studies related to the control of eutrophication (Lu *et al.* 2015). Liu *et al.* (2016a, 2016b) reported a novel artificial floating island which was equipped with an aerator device powered by a solar power. Their findings revealed that the artificial floating island can effectively break down the stratified water into homogenized water and inhibit the growth of algae. Li *et al.* (2010) introduced a novel approach to enhance the performance of traditional planted floating bed by the introduction of filter-feeding bivalve and biofilm

doi: 10.2166/aqua.2019.056

carrier. Therefore, ecological floating beds are a good development direction in the future.

However, most of the previous studies focused on the removal effect of ecological floating beds on the water bodies with higher pollutant concentration (Chang *et al.* 2017), and there were few studies on the lower pollutant concentration (P concentration is less than 0.8 mg/L) (Chang *et al.* 2014; Wu *et al.* 2016). The previous study showed that for mildly eutrophic water bodies, due to their inability to provide sufficient nitrogen (N) and phosphorus (P) for plant growth, the biomass was relatively low and the purification capacity was correspondingly decreased (Ge *et al.* 1999). In addition, other studies revealed that the focus must be on reducing the inputs of P to control the eutrophication (Schindler *et al.* 2008; Schindler 2012).

Therefore, in order to solve the problem of poor P removal efficiency of traditional floating beds under low P concentration (Zhang *et al.* 2014), this study used the efficient P removal composite (EPRC) as the growth substrate of the ecological floating island for the first time. Through the enrichment of P by the EPRC, the uptake rate of P by plants in the ecological floating islands at low P concentration was improved. The main components of the EPRC were fly ash and steel slag, which were developed previously as a part of this research work (Liu *et al.* 2015). Industrial waste slags were recovered which also contribute to reducing environmental pollution. The roots of *Ipomoea aquatica* were buried in the EPRC particles. When the dissolved P in the water was low, the P absorbed by the EPRC was released through chemical equilibrium and the acidic substance secreted by the roots caused the hydroxyapatite precipitate to be converted into soluble P which was easily absorbed by the plant, thereby continuously transporting P to the plant. Therefore, the influence of the EPRC particle dosage on P removal was studied to determine the amount of EPRC particles used in the floating bed. The influence of different numbers of *Ipomoea aquatica* plants on the removal of low-concentration P (0.13 mg/L) was compared to select the optimal number of plants for floating beds. Based on these findings, the removal efficiency of low-concentration P in actual water bodies by the EPRC-based ecological floating bed was also conducted. The heavy metal leaching test was conducted after an absorption efficiency test.

MATERIAL AND METHODS

Water quality measuring methods and experimental equipment

The concentration of the PO_4^{3-}-P and total phosphorus (TP) was determined through a colorimetric ammonium molybdate method (State Environmental Protection Administration of the People's Republic of China 2002). The amount of heavy metals leaching in beds was determined by inductively coupled plasma mass spectrometry. The main instruments used in the experiment are shown in Table 1.

The floating bed systems consist of a plastic pot (15 L in volume), and a foam board and a plastic mesh were designed and implemented in the current work as shown in Figure 1. Figure 2(a) shows the smallest unit of the

Table 1 | Model and manufacturer of main instruments

Name of instruments	Model no.	Manufacturer
Ultraviolet–Visible spectrophotometer	752	Shanghai Jinghua Technology Instrument Co., Ltd (China)
ICP-MS	Optima 5300DV	PerkinElmer (USA)
Electronic balance	AR124CN	Ohaus Instruments Co., Ltd (China)
Portable stainless steel pressure steam sterilizer	YX280A	Shanghai Sanshen Medical Devices Co., Ltd (China)
High-speed multi-function pulverizer	JP-400A-8	Yongkang Jiubin Industry and Trade Co., Ltd (China)
pH meter	101-B	Jiangyan Guorui Analysis Instrument Co. (China)

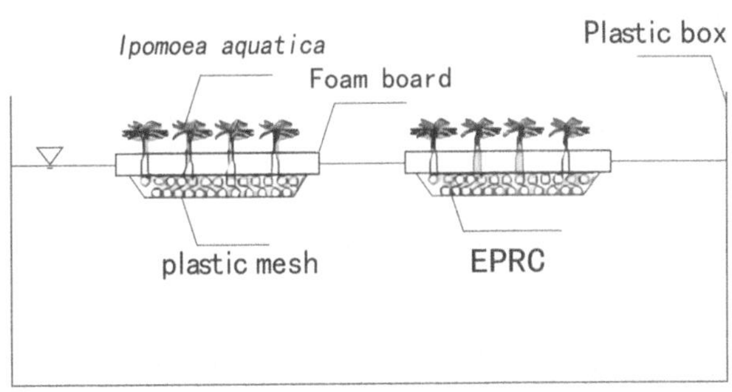

Figure 1 | Schematic diagram of the floating bed.

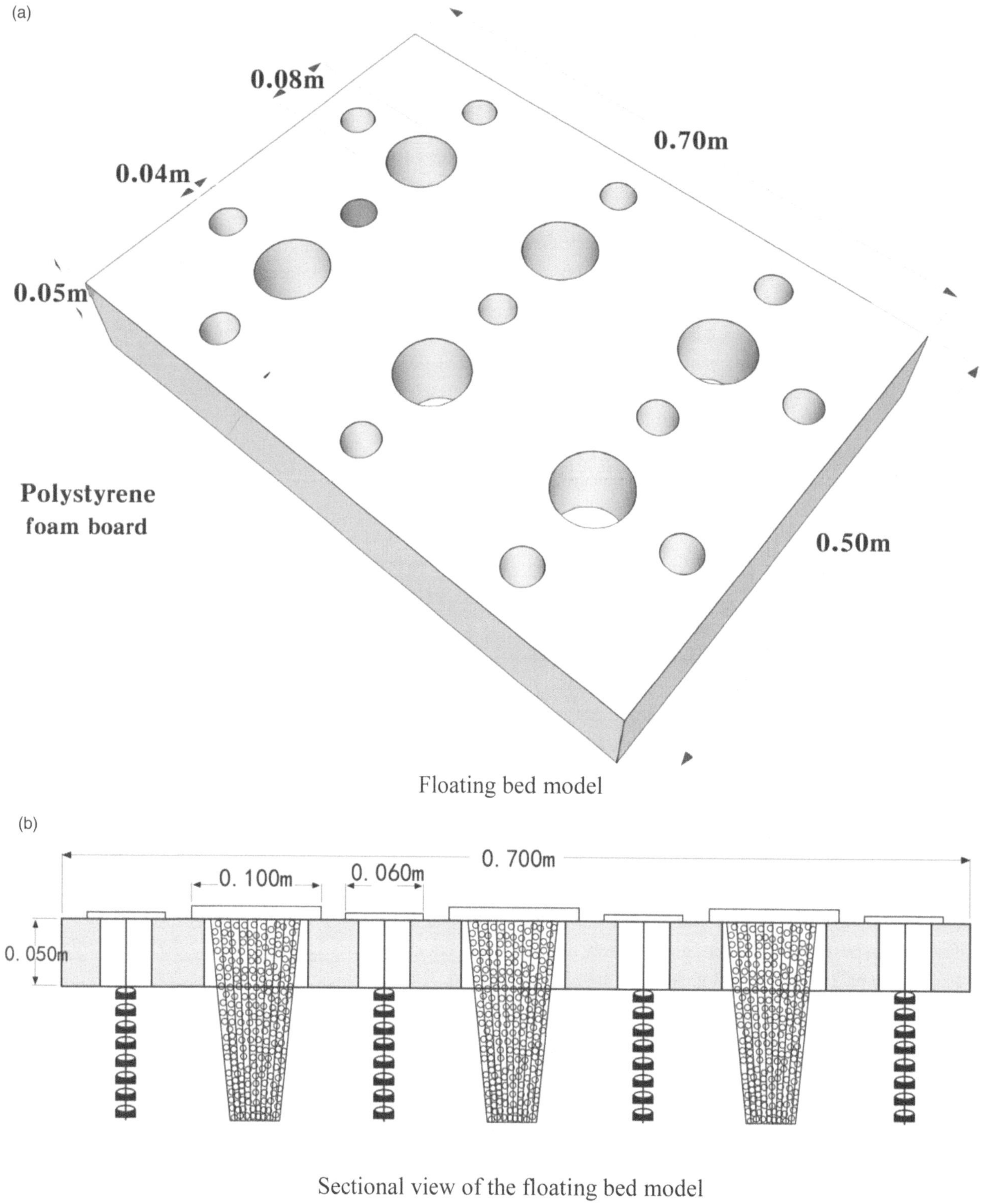

Figure 2 | Floating bed model: (a) single experimental unit and (b) sectional view.

designed floating bed. The bed material used in the current work was polystyrene foam with the size of 0.7 m × 0.5 m × 0.05 m (length × width × height). It was perforated according to needs, and the diameter of the large hole was 8 cm and the diameter of the small hole was 4 cm. The EPRC particles and plants were placed in the plastic basket, and plant roots were grown in the EPRC. The plastic basket has an upper diameter of 10 cm, a lower diameter of 4 cm, and a total height of 15 cm, which was embedded in the large hole in the bed. Figure 2(b) shows a sectional view of the floating bed that has been placed in a plastic basket. The filler used in the current study was a bioelastic three-dimensional material. As a growth carrier of the microorganisms, the filler also facilitates the removal of nitrogen from the water. The upper end diameter of the filler and the diameter of the filling ring were 6 and 2 cm, respectively. The filler was suspended under the ring, and the length of the packing section was 15 cm. The fillers were embedded in the small holes of the floating bed.

Preparation of matrix materials

The principal component of the EPRC was fly ash, which is an industrial by-product from salt-refining by the chlorine-alkali process. According to the previous research, EPRC particles with a size of 5 mm were selected, and the optimum ratio of fly ash:steel slag:cement (mass ratio) was 12:2:1 (Liu *et al.* 2015). Foaming agent and water were added appropriately to the homogenator and mixed until the solution attains the appearance of a milky foam. Particles of specific diameter were sieved after the raw materials grounded mechanically and then added a certain amount of cement and an appropriate amount of foam, granulating in the drum granulator. Water was sprayed onto the sphere after it had gelled for 6–8 h and conserved it for 1 week at standard temperature (25 °C) and atmospheric pressure. Spraying was conducted regularly with water to maintain the certain humidity.

Statistical analysis

The removal rate (R) was calculated using the following equation:

$$R(\%) = (C_0 - C_t)/C_0 \times 100\% \quad (1)$$

where C_0 (mg/L) is the initial concentration of a parameter in each group and C_t (mg/L) is the concentration at time t (days). All experiments were set up with three sets of parallel experiments. All figures were plotted by using the Origin 8.5 software. The data were analyzed by SPSS, version 12.0. Comparisons in pH, PO_4^{3-}, and TP concentrations, root growth ratio, and stem growth ratio between different groups of floating beds were evaluated using a one-way analysis of variance (ANOVA). *Post hoc* multiple comparisons were performed using Fisher's least significant difference (LSD) test. Differences between floating beds were deemed significant if $p < 0.05$.

Effect of the EPRC particle dosage on P removal

EPRC particles with a size of 5 mm were selected, and EPRC particles of 2, 3, 4, 5, and 6 g quantity were added, respectively, to five beakers with a specification of 1 L capacity. Subsequently, the simulated solution with a concentration of 0.5 mg/L of P in deionized water was added to five beakers with the additional amount of 1 L capacity, oscillated in an oscillator at constant temperature, and sampled at regular intervals. The P concentration of the supernatant was measured by a 752 UV spectrophotometer (Shanghai Jinghua Technology Instrument Co., Ltd, China). Through this, the influence of different EPRC particle dosages on the P removal effect was investigated.

Influence of the number of plants on the P removal effect

In the current study, four plastic pots numbered with 1, 2, 3, and 4 were utilized. All experimental pots were filled with 10 L of raw water collected from Jiulong Lake, Nanjing, China. The concentration P, total nitrogen (TN), and NH_4^+-N in the raw water was 0.13, 2.60, and 1.37 mg/L, respectively, and the pH value was 7.5. A foam board was placed in each of the plastic pots, and 2, 4, 6, and 8 plants of *Ipomoea aquatica* with similar length and growth conditions were sequentially planted. The TP concentration in the water was measured in 3-day intervals, and a certain amount of distilled water was added to the previously marked scale before each sampling to make up the evaporation amount. Following this process, the removal effect of different numbers of *Ipomoea aquatica* on low concentration of P was investigated.

P removal effect of the composite plant floating bed with EPRC and *Ipomoea aquatica*

In the current study, 15 plastic pots were selected and divided into three groups, numbered A, B, and C, and were utilized to investigate the P removal effect of the composite plant floating bed with EPRC and *Ipomoea aquatica*. The five pots of each group were numbered sequentially from 1 to 5. Raw water (10 L) was added (which was taken from Jiulong Lake) to each pot. The P concentration was 0.10 mg/L, the TN concentration was 3.22 mg/L, the ammonia nitrogen concentration was 1.53 mg/L, and the pH value was 7.8 in the raw water of Jiulong Lake. Since the time of the water sampling in the current work differs from the time of the sampling of the above water samples by nearly 1 month, there was a certain difference in the TP concentration of the two water samples. Considering the difference of P concentration under different water systems, the current experiment was designed for a certain gradient of P concentration. As the current study mainly focused on water with low P concentration, the designed concentration gradient was 0.1, 0.2, 0.4, 0.6, and 0.8 mg/L. A certain amount of KH_2PO_4 particles were added to the pots numbered 1 to 5 of groups A, B, and C, respectively, to ensure the P concentrations in the No. 1 to No. 5 experimental pots were 0.1, 0.2, 0.4, 0.6, and 0.8 mg/L, respectively. Then, a floating bed was placed in each of the plastic pots. Group A was a planted floating bed, and the number of *Ipomoea aquatica* plants selected were 6 based on the previous study. Group B consists of the plant + EPRC particles floating bed, and the number of plants was 6 and the amount of EPRC particles was 20 g. Group C consists of a floating bed with only EPRC particles, and the amount of EPRC particles was 20 g. The concentration of TP and PO_4^{3-}-P in the water was monitored every 3 days. A certain amount of distilled water was added to the previously marked scale before sampling to compensate for the evaporation, and the test period was 18 days. Through this process, the P removal effect of the composite floating bed by EPRC and plants was investigated.

Heavy metal leaching test

The purpose of the composite floating island was to control P in water source, so the safety of materials was required. Considering that the heavy metals contained in fly ash and steel slag of matrix materials were mainly Zn and Cu, the test mainly measured these two indexes. The optimal EPRC dosage (2 g/L) taken as an example to determine the EPRC with a particle size of 0.5 cm about 2 g was taken and placed into 1 L of KH_2PO_4 solution with a concentration of 0.5 mg/L. The heavy metal leaching test was conducted after absorption equilibrium and analyzed for heavy metals using inductively coupled plasma mass spectrometry (ICP-MS, PerkinElmer, USA).

RESULTS AND DISCUSSION

Effect of the EPRC particle dosage and *Ipomoea aquatica* floating bed on P removal

As shown in Figure 3(a), results revealed that with the increase of the dosage of EPRC particles, the P removal rate increased sequentially, while the P removal amount per unit mass of particles decreased in a sequential manner. When the EPRC dosage was increased, the usable surface area of the EPRC was increased. More adsorption sites for P exist, and the P was more easily combined with the active sites and hence removed. When the dose was low, the removal rate of P increased rapidly. The removal rate continued to increase with gradually decreased dosage until P was completely removed. For example, the P removal rate was the highest when the dose was 6 g, which was 69.8%, but the P removal amount per unit mass of particles was only 0.058 mg/g, which was the lowest value of the five groups. While at the 2 g dose, the P removal amount per unit mass of particles was 0.157 mg/g and the P removal rate was 62.6%. Therefore, considering the P removal rate and the P removal amount per unit mass of particles, from an economic point of view, the 2 g/L dosage was selected in the subsequent experiments as the most economical.

The P removal rate of different quantities of *Ipomoea aquatica* is shown in Figure 3(b). Under the condition of different numbers of plants ranging between 2 and 8, the P

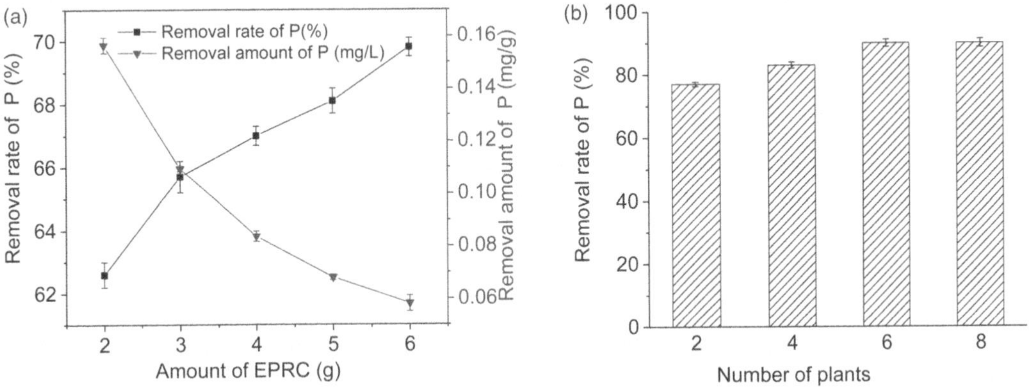

Figure 3 | Influence of (a) the EPRC dosage and (b) the number of *Ipomoea aquatica* on P removal.

concentration of the four devices decreased from the initial 0.13 to 0.030, 0.021, 0.013, and 0.012 mg/L, respectively. The P removal rates were 76.6%, 84.1%, 90.0%, and 90.8%, respectively. The P removal rate did not show further increase with increase of plant numbers due to the limited total amount of P in the solution. Therefore, 6 plants/10 L were selected as the optimal number of plants in this experiment when influent P concentration was 0.13 mg/L.

The comparison of the removal rates of PO_4^{3-} and TP of three floating beds under different P concentrations is shown in Figure 4. Results revealed that the composite floating bed with EPRC particles and plants had the best P removal effect, followed by the EPRC-only floating bed and finally plant-only floating bed. There was no significant difference in TP and PO_4^{3-} concentration between three groups of floating beds when initial P concentration was 0.1 mg/L ($p > 0.05$). There was a significant difference in TP and PO_4^{3-} concentration between the plant-only floating bed and other floating beds when initial P concentration was 0.2 mg/L ($p < 0.05$). The TP and PO_4^{3-} concentration in the EPRC + plant floating bed was significantly lower than that of other floating beds ($p < 0.05$) when initial P concentration was 0.4, 0.6, and 0.8 mg/L. It is indicated that the EPRC can significantly improve the P removal efficiency of plant-only floating beds. The removal pathways of P in the plant-only floating bed mainly include plant absorption, precipitation, adsorption, and the fixation of microorganisms in roots of plants (Wu *et al.* 2016; Li *et al.* 2009). The mechanism of the EPRC-only floating bed for the removal of P is mainly manifested in the joint process of physical adsorption and chemical adsorption (Oguz 2005). The large specific surface area and many active sites which are easy to react with ions in water (such as silicon and aluminum) determine the strong adsorption capacity of steel slag and fly ash. The

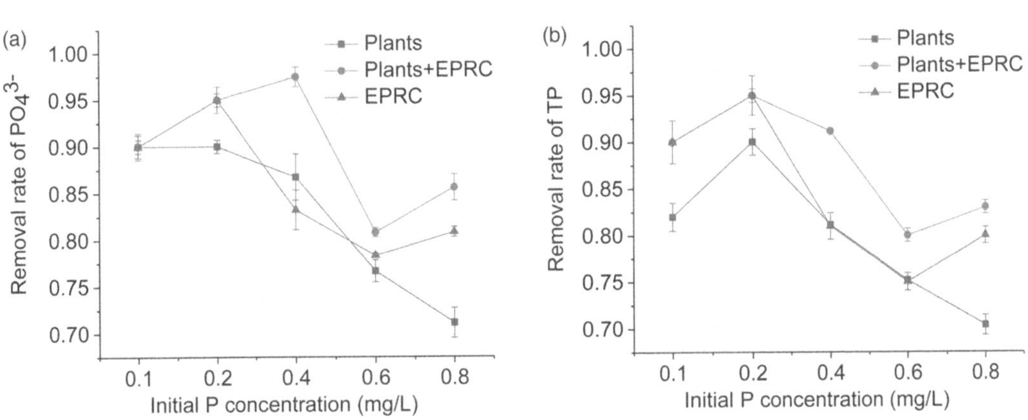

Figure 4 | Comparison of (a) PO_4^{3-} and (b) TP removal among three floating beds at different initial P concentrations.

physical adsorption effect mainly depends on the porosity and specific surface area of fly ash, which is positively correlated with the specific surface area (Zeng *et al.* 2015). The adsorption of P by the EPRC is mainly chemical adsorption, which is an exothermic process. It is mainly achieved by chemical precipitation and ligand exchange (Blanco *et al.* 2016). Studies have shown that P immobilization by fly ash and steel slag was governed by Ca ingredient (especially CaO) and Fe ingredient (especially Fe_2O_3) (Chen *et al.* 2007; Bowden *et al.* 2009). When the EPRC was mixed with the P solution, calcium ions and phosphate ions form calcium phosphate precipitation (Blanco *et al.* 2016) which can be expressed as shown in Equation (2):

$$3Ca^{2+} + 2OH^- + 2HPO_4^{2-} = Ca_3(PO_4)_2 + 2H_2O \quad (2)$$

The generated calcium phosphate precipitation can be further converted into stable hydroxyl calcium phosphate precipitation which can be expressed as shown in Equation (3):

$$5Ca^{2+} + 4OH^- + 3HPO_4^{2-} = Ca_5(OH)(PO_4)_3 + 3H_2O \quad (3)$$

Iron oxides in the EPRC enter the solution and combine with coordinating water to form hydroxides or hydroxides of hydroalloys, resulted in a large number of -OH groups on the surface of the EPRC. An earlier study revealed that phosphate ions can be adsorbed by exchanging with the coordination groups of iron ions (Li *et al.* 2016). The specific reaction process is shown in Equations (4) and (5):

$$Fe_2O_3 + 3H_2O = 2Fe(OH)_3 \quad (4)$$

$$Fe(OH)_3 + H_2PO_4^- = FePO_4 + 2H_2O + OH^- \quad (5)$$

In the composite floating bed, plants did not easily absorb the crystalline P formed by the chemisorption of the EPRC, and it was necessary to complete the absorption of P by plants through two pathways. Earlier studies (Lei *et al.* 2018; Cichy *et al.* 2019) considered that crystalline P was unstable, because it may form unstable colloids. The plant firstly absorbed the dissolved P in the water. When the dissolved P dropped to a certain concentration, the colloidal P was released to maintain the chemical balance between the solution and the solute in the water and the released P can be absorbed by the plant. The second pathway is that plant roots secreted acidic substances (such as citric acid) under the condition of P deficiency to promote the release of insoluble P (Lyu *et al.* 2016; Xing *et al.* 2016). Previous studies have shown that the adsorbed phosphate could be successfully recovered by 2% citric acid solution, which implied that recovered phosphate could be reused as a fertilizer (Jiang *et al.* 2017). The roots of *Ipomoea aquatica* were buried in the EPRC particles, with a small amount of soluble P. Under the P deficiency, the roots secreted acidic substances and dissolved the crystal P into dissolved P which was easily absorbed by plants. This was equivalent to increasing the concentration of dissolved P around the roots, promoting the absorption of P, and thus improving the P removal rate.

Influences of three groups of floating beds on the pH of raw water

The experiment also examined the influences of three groups of floating beds on the pH of the water. Figure 5 shows the changes in the pH of raw water of the three floating beds under different initial P concentrations (where the pH of the raw water at 24 °C was 7.8). It can be seen from Figure 5 that at the end of experiment, the pH of the floating bed of C was the highest, reaching to about 8.5. The floating bed of A had little change compared with the pH of raw water. The floating bed of B had a slight change compared with the original water, but it was also stable at around

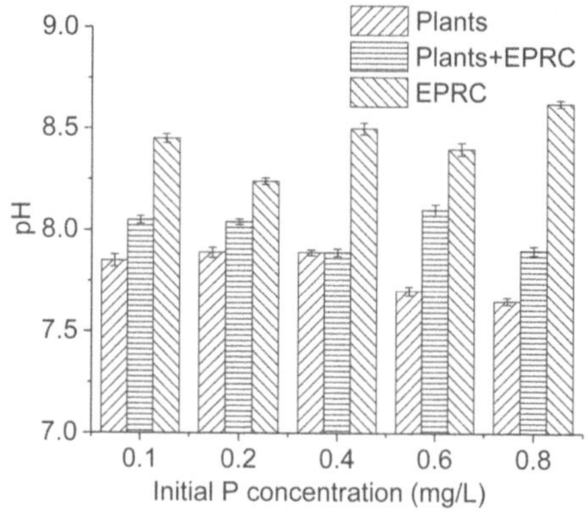

Figure 5 | Influence of the floating bed on the pH.

8.0. There were significant differences in the pH values in different groups of the floating bed ($p < 0.005$). At the initial stage of the reaction in the floating bed of C, alkaline substances such as CaO in the EPRC precipitated into the solution and hydrolyzed:

$$CaO + H_2O = Ca^{2+} + 2OH^- \tag{6}$$

This reaction caused the pH of the solution to rise rapidly. As the reaction proceeded further, the pH of the solution dropped to some extent. The reason could be that Ca^{2+} combined with phosphate ions to form stable precipitate, such as hydroxyl calcium phosphate, after the pH reached a certain value, which consumed a part of OH^-. When the reaction was completed in the later stage of adsorption, the pH began to stabilize. The specific reaction process is shown in Equations (7) and (8) (Khadhraoui *et al.* 2002):

$$5Ca^{2+} + 7OH^- + 3H_2PO_4^- = Ca_5(OH)(PO_4)_3 + 6H_2O \tag{7}$$

$$3Ca^{2+} + 4OH^- + 2H_2PO_4^- = Ca_3(PO_4)_2 + 4H_2O \tag{8}$$

From the experimental results, the pH of the floating bed of group C was finally stabilized at around 8.5. Whereas in group B, the roots of the plants released acidic substances, neutralizing a part of the alkalinity (Wang *et al.* 2018), so that the pH had a certain degree of decline compared with the EPRC-only floating bed.

Growth status of plants in groups A and B under different initial P concentrations

In the experiment, the growth of the roots and stems of the *Ipomoea aquatica* in the floating beds of groups A and B were also measured to verify that the EPRC can enhance the P removal of the *Ipomoea aquatica* floating beds. Figure 6 shows the comparison of roots and stems growth ratio of plants in the floating beds of groups A and B under different initial P concentrations. It can be seen from the figure that under the conditions of five different initial P concentrations, the root length and the stem length of group B floating bed (plants + EPRC) plants were larger than that of group A floating bed (plants-only). The root growth ratio and stem growth ratio in the EPRC + plant floating bed was significantly higher than that in the plant-only floating bed ($p < 0.05$).

This further demonstrated that EPRC particles promote the growth of plants. Deficiency in plant-available P is considered to be a major limiting factor to the growth of many plants (Arcand & Schneider 2006). A study conducted by Vidal-Beaudet *et al.* (2018) also indicated that available P content was sufficient for plant development. In this study, the enrichment of P by EPRC particles allowed the roots in contact with the particles to absorb P more effectively; therefore, the plants of group B obtained a larger root length and stem length better than group A. Besides, previous studies have shown that the EPRC can provide nutrients, such as Ca, Na, K, P, Mg, B, S, and Mo, which promote plant growth and also alleviate the condition of

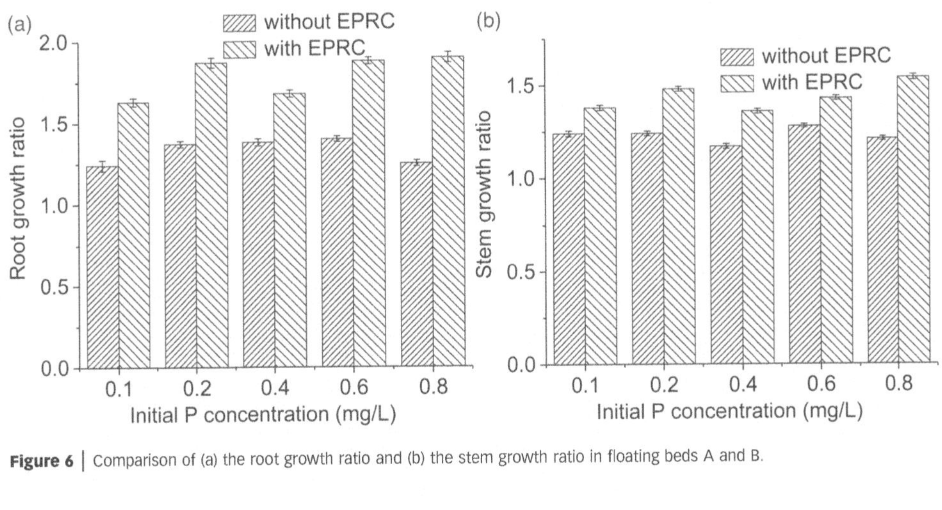

Figure 6 | Comparison of (a) the root growth ratio and (b) the stem growth ratio in floating beds A and B.

nutrient deficiency in low P concentration (Sajwan *et al.* 2003; Kumpiene *et al.* 2007).

Heavy metal leaching test

The concentration of Cu was under the ICP-MS detection limit (0.001 mg/L) (<0.01 mg/L), which met the Class I water quality standard of Environmental Quality Standard for Surface Water (GB 3838-2002). The Zn leaching was 0.0314 mg/L (<0.05 mg/L), which was slightly higher than Cu but also met the Class I water quality standard of Environmental Quality Standard for Surface Water (GB 3838-2002). Although fly ash was reported leaching heavy metals, leaching rates of metals from fly ash were reduced using the cement of which the main components were limestone and clay (Bankowski *et al.* 2004; Liu *et al.* 2011; Li *et al.* 2017).

APPLICATIONS OF CURRENT WORK AND FUTURE RESEARCH

At present, EPRC particles have been used in the P reduction experiment in Taihu Lake, and the results showed that the EPRC has effective P removal efficiency. The test site was a river flowing into Taihu Lake in Yixing City. The width of the river surface was 4 m, the average depth was 0.6 m, and the velocity was 0.5 m/s. During the experiment, the EPRC was placed in nylon mesh bags with a length of 1 m and a diameter of 6 cm. The nylon bags were immersed in river water. The nylon mesh bag filled with the EPRC material weighed about 600 g per bag, totaling 40 strings, and the total dosage of the EPRC was 24 kg. Within 1 week of the initial implementation, the total P concentration in the channel into the lake decreased from 1.2 to less than 0.2 mg/L. It brings two kinds of environmental benefits: it allows the reuse of potentially hazardous waste and increases the removal of pollutants from the wastewater treatment systems. In the future, consideration should be given to increase the hardness of the EPRC. Moreover, the EPRC is a solid material, and only the surface layer and voidage are involved in P removal. Therefore, subsequent studies may consider increasing the specific surface area of the material, such as making it into fibers, so that the material has a higher P removal efficiency than granular materials. In addition, in future research, different plants can be selected for the experiment of composite floating beds, such as plants growing at low temperature, ornamental plants, or plants with high economic value, to provide a reference for a practical engineering application from multiple perspectives. Finally, fly ash, steel slag, and soil with different weight ratios can be considered as the growth matrix of plants in constructed wetlands, and the P removal performance of which can be studied.

CONCLUSION

Findings of the current study revealed that the composite floating bed can significantly enhance the P removal efficiency of the *Ipomoea aquatica* and promote its growth. The presence of the EPRC caused a slight increase in the pH, but the plant released the acidic substance and neutralized a part of the alkalinity. Therefore, the composite floating bed had a negligible effect on the pH of the raw water. In summary, the composite floating bed based on the EPRC can be used as a new way to strengthen P removal in actual water bodies.

REFERENCES

Arcand, M. M. & Schneider, K. D. 2006 Plant- and microbial-based mechanisms to improve the agronomic effectiveness of phosphate rock: a review. *Anais Da Academia Brasileira De Ciencias* **78** (4), 791–807.

Bankowski, P., Zou, L. & Hodges, R. 2004 Using inorganic polymer to reduce leach rates of metals from brown coal fly ash. *Minerals Engineering* **17** (2), 159–166.

Blanco, I., Molle, P., Saenz de Miera, L. E. & Ansola, G. 2016 Basic oxygen furnace steel slag aggregates for phosphorus treatment. Evaluation of its potential use as a substrate in constructed wetlands. *Water Research* **89**, 355–365.

Bowden, L. I., Jarvis, A. P., Younger, P. L. & Johnson, K. L. 2009 Phosphorus removal from waste waters using basic oxygen steel slag. *Environmental Science & Technology* **43** (7), 2476–2481.

Chang, Y. H., Wu, B. Y. & Lai, C. F. 2014 The effect of a green energy landscape fountain on water quality improvement. *Ecological Engineering* **73**, 201–208.

Chang, Y. Y., Cui, H., Huang, M. S. & He, Y. 2017 Artificial floating islands for water quality improvement. *Environmental Reviews* **25** (3), 350–357.

Chen, J., Kong, H., Wu, D., Chen, X., Zhang, D. & Sun, Z. 2007 Phosphate immobilization from aqueous solution by fly ashes in relation to their composition. *Journal of Hazardous Materials* **139** (2), 293–300.

Cichy, B., Kuzdzal, E. & Krzton, H. 2019 Phosphorus recovery from acidic wastewater by hydroxyapatite precipitation. *Journal of Environmental Management* **232**, 421–427.

Ge, Y., Wang, X. Y. & Chang, J. 1999 Comparative studies on the purification ability of plants in different degree eutrophic water. *Acta Scientiae Circumstantiae* **19**, 690–692. (in Chinese).

Hamzah, S., Razali, N. A., Yatim, N. I., Alias, M., Ali, A., Zaini, N. S. & Abuhabib, A. A. M. 2018 Characterisation and performance of thermally treated rice husk as efficient adsorbent for phosphate removal. *Journal of Water Supply Research and Technology - Aqua* **67** (8), 766–778.

Jiang, D., Amano, Y. & Machida, M. 2017 Removal and recovery of phosphate from water by a magnetic Fe_3O_4@ASC adsorbent. *Journal of Environmental Chemical Engineering* **5** (5), 4229–4238.

Khadhraoui, M., Watanabe, T. & Kuroda, M. 2002 The effect of the physical structure of a porous Ca-based sorbent on its phosphorus removal capacity. *Water Research* **36** (15), 3711–3718.

Kumpiene, J., Lagerkvist, A. & Maurice, C. 2007 Stabilization of Pb- and Cu-contaminated soil using coal fly ash and peat. *Environmental Pollution* **145** (1), 365–373.

Lei, Y., Saakes, M., van der Weijden, R. D. & Buisman, C. J. N. 2018 Effects of current density, bicarbonate and humic acid on electrochemical induced calcium phosphate precipitation. *Chemical Engineering Journal* **342**, 350–356.

Li, M., Wu, Y. J., Yu, Z. L., Sheng, G. P. & Yu, H. Q. 2009 Enhanced nitrogen and phosphorus removal from eutrophic lake water by *Ipomoea aquatica* with low-energy ion implantation. *Water Research* **43** (5), 1247–1256.

Li, X. N., Song, H. L., Li, W., Lu, X. W. & Nishimura, O. 2010 An integrated ecological floating-bed employing plant, freshwater clam and biofilm carrier for purification of eutrophic water. *Ecological Engineering* **36** (4), 382–390.

Li, Y., Zhou, Y. B., Zhou, Z. H., Wang, N. S., Zhou, Q. & Wang, F. C. 2014 Effects of sustained-release composite on the oxygen levels and sediment phosphorus fractions of an urban river in Shanghai. *Environmental Technology* **35** (17), 2176–2182.

Li, F., Wu, W., Li, R. & Fu, X. 2016 Adsorption of phosphate by acid-modified fly ash and palygorskite in aqueous solution: experimental and modeling. *Applied Clay Science* **132**, 343–352.

Li, S., Cooke, R. A., Wang, L., Ma, F. & Bhattarai, R. 2017 Characterization of fly ash ceramic pellet for phosphorus removal. *Journal of Environmental Management* **189**, 67–74.

Liu, L., Li, L., Zuo, Y., Huang, Y. & Song, L. 2011 Adsorption of 2-methylisoborneol and geosmin by a low-cost hybrid adsorbent synthesized from fly ash and bentonite. *Journal of Water Supply Research and Technology - Aqua* **60** (8), 478–485.

Liu, Y., Li, L., Zhang, S. J. & Wang, S. H. 2015 Kinetics and thermodynamics of efficient phosphorus removal by a complex material. *Desalination and Water Treatment* **56** (7), 1949–1954.

Liu, J. L., Liu, J. K., Anderson, J. T., Zhang, R. & Zhang, Z. M. 2016a Potential of aquatic macrophytes and artificial floating island for removing contaminants. *Plant Biosystems* **150** (4), 702–709.

Liu, Z. L., Fu, X. Y., Chen, W., Yu, S. & He, X. Y. 2016b Nitrogen and phosphorus removal by five ornamental and wetland plants from upstream of Hunhe river, China. *Fresenius Environmental Bulletin* **25** (5), 1477–1483.

Lu, H. L., Ku, C. R. & Chang, Y. H. 2015 Water quality improvement with artificial floating islands. *Ecological Engineering* **74**, 371–375.

Lyu, Y., Tang, H. L., Li, H. G., Zhang, F. S., Rengel, Z., Whalley, W. R. & Shen, J. B. 2016 Major crop species show differential balance between root morphological and physiological responses to variable phosphorus supply. *Frontiers in Plant Science* **7**, 1939.

Mehner, T. & Benndorf, J. 1995 Eutrophication – a summary of observed effects and possible solutions. *Journal of Water Supply Research and Technology - Aqua* **44**, 35–44.

Oguz, E. 2005 Sorption of phosphate from solid/liquid interface by fly ash. *Colloids and Surfaces A: Physicochemical and Engineering Aspects* **262** (1–3), 113–117.

Sajwan, K. S., Paramasivam, S., Alva, A. K., Adriano, D. C. & Hooda, P. S. 2003 Assessing the feasibility of land application of fly ash, sewage sludge and their mixtures. *Advances in Environmental Research* **8** (1), 77–91.

Santiago Sampaio, V. C., Abdala Neto, E. F., Alves de Lima, A. C., Lima Freitas, I. C. & de Aquino, M. D. 2015 Reduction of natural organic matter by an electrolytic process, conventional coagulation and direct descending filtration in a pilot-scale water treatment system. *Journal of Water Supply Research and Technology - Aqua* **64** (2), 168–175.

Schindler, D. W. 2012 The dilemma of controlling cultural eutrophication of lakes. *Proceedings of the Royal Society B: Biological Sciences* **279** (1746), 4322–4333.

Schindler, D. W., Hecky, R. E., Findlay, D. L., Stainton, M. P., Parker, B. R., Paterson, M. J., Beaty, K. G., Lyng, M. & Kasian, S. E. M. 2008 Eutrophication of lakes cannot be controlled by reducing nitrogen input: results of a 37-year whole-ecosystem experiment. *Proceedings of the National Academy of Sciences of the United States of America* **105** (32), 11254–11258.

State Environmental Protection Administration of the People's Republic of China 2002 *Technical Specifications and Requirements for Monitoring Surface Water and Wastewater*. Beijing, China.

Tian, Z., Wang, C. & Ji, M. 2018 Full-scale dissolved air flotation (DAF) equipment for emergency treatment of eutrophic water. *Water Science and Technology* **77** (7), 1802–1809.

Vidal-Beaudet, L., Rokia, S., Nehls, T. & Schwartz, C. 2018 Aggregation and availability of phosphorus in a Technosol constructed from urban wastes. *Journal of Soils and Sediments* **18** (2), 456–466.

Wang, Y. L., Lysoe, E., Armarego-Marriott, T., Erban, A., Paruch, L., Van Eerde, A., Bock, R. & Liu-Clarke, J. 2018 Transcriptome and metabolome analyses provide insights into root and root-released organic anion responses to phosphorus deficiency in oat. *Journal of Experimental Botany* **69** (15), 3759–3771.

Wang, Y., Wang, W. H., Yan, F. L., Ding, Z., Feng, L. L. & Zhao, J. C. 2019 Effects and mechanisms of calcium peroxide on purification of severely eutrophic water. *Science of the Total Environment* **650**, 2796–2806.

Wu, Q., Hu, Y., Li, S., Peng, S. & Zhao, H. 2016 Microbial mechanisms of using enhanced ecological floating beds for eutrophic water improvement. *Bioresource Technology* **211**, 451–456.

Xing, D. K., Wu, Y. Y., Yu, R., Wu, Y. S., Zhang, C. & Liang, Z. 2016 Photosynthetic capability and Fe, Mn, Cu, and Zn contents in two Moraceae species under different phosphorus levels. *Acta Geochimica* **35** (3), 309–315.

Yang, X. E., Wu, X., Hao, H. L. & He, Z. L. 2008 Mechanisms and assessment of water eutrophication. *Journal of Zhejiang University-Science B* **9** (3), 197–209.

Yeh, N., Yeh, P. & Chang, Y. H. 2015 Artificial floating islands for environmental improvement. *Renewable and Sustainable Energy Reviews* **47**, 616–622.

Zeng, L., Li, W., Zhuang, Y. & Jian, J. 2015 Deep treatment of low concentrated phosphorus by fly ash. *Chinese Journal of Environmental Engineering* **9** (4), 1778–1782.

Zhang, Q., Achal, V., Xu, Y. & Xiang, W. N. 2014 Aquaculture wastewater quality improvement by water spinach (*Ipomoea aquatica* Forsskal) floating bed and ecological benefit assessment in ecological agriculture district. *Aquacultural Engineering* **60**, 48–55.

First received 17 April 2019; accepted in revised form 21 August 2019. Available online 30 October 2019

Endogenous respiration process analysis between aMBR and UV/O_3-aMBR for polluted surface water treatment

Lu Li, Kang Song and Chettiyappan Visvanathan IWA

ABSTRACT

The microbial endogenous respiration process is very important in biological water treatment processes. This study analyzed and compared the endogenous respiration process in an attached growth membrane bioreactor (aMBR) system and a UV/O_3 integrated aMBR system (UV/O_3-aMBR) in treating polluted surface water with COD_{Mn} around 10 mg/L. The endogenous respiration activity of heterotrophic microbes and autotrophic nitrifiers activity in both systems was analyzed and compared. Results show that heterotrophic bacteria and autotrophic nitrifiers enter endogenous respiration at 6 h aeration in an aMBR and 0 h in a UV/O_3-aMBR system. Biomass amount on PVA-gel in aMBR was higher than in UV/O_3-aMBR in terms of specific respiration rates $SOUR_t$, $SOUR_H$, and $SOUR_A$. Substrate remained on PVA-gel in the aMBR system, but no substrate remained on PVA-gel in the UV/O_3-aMBR system. Higher species of microbes, including recoverable and irrecoverable components, existed in the aMBR system as compared to the UV/O_3-aMBR system. The UV/O_3-aMBR system could make full use of the advanced oxidation process (AOP) and biological process, leading to a higher treatment performance, and has the potential to mitigate total energy demand. Thus, the UV/O_3-aMBR system can be used as a new technology for treating polluted surface water with the co-contribution of biological process and AOP treatment.

Key words | attached growth membrane bioreactor (aMBR), endogenous respiration, polluted surface water, respiration map, UV/O_3-aMBR

Lu Li
Kang Song (corresponding author)
State Key Laboratory of Freshwater Ecology and Biotechnology, Institute of Hydrobiology,
Chinese Academy of Sciences,
Wuhan 430072,
China
E-mail: *sk@ihb.ac.cn*

Chettiyappan Visvanathan IWA
School of Environment, Resources and Development,
Asian Institute of Technology,
Pathumthani, 12120,
Thailand

INTRODUCTION

Surface water quality is facing increasing organic contamination due to human activities worldwide (Ma *et al.* 2018; Utaaker *et al.* 2019). Overload of organic matter pollution in surface water leads to challenges during treatment using conventional drinking water treatment plants. These challenges include higher consumption of coagulants and higher disinfection by-product formation potential. The water treatment technology should be able to treat the polluted surface water and reach the drinking water quality standard, which is a big challenge for the world's drinking water supply. Table 1 demonstrates the drinking water national standards of some selected countries and institutions. Improving current wastewater treatment technology or developing high efficiency water treatment technology can be the solution to the polluted surface water issue. A previous study has reported that in an attached growth membrane bioreactor (aMBR) for treating polluted surface water with COD_{Mn} around 10 mg/L, over 50% COD_{Mn} removal was achieved (Li *et al.* 2017). The aMBR system works as an enhanced natural purification process, and polyvinyl alcohol gel (PVA-gel) was used as a bio-carrier to immobilize bacteria for the degradation of organic pollutants from polluted surface water. The aMBR system was operated without adding activated sludge and,

doi: 10.2166/aqua.2019.081

Table 1 | Drinking water quality index of different countries

	Water Quality Convention Index and Limiting Value							
Country/Organization	**NO_3-N (mg/L)**	**NO_2-N (mg/L)**	**TOC (mg/L)**	**Turbidity (NTU)**	**TDS (mg/L)**	**COD_{Mn} (mg/L)**	**NH_3-N (mg/L)**	**pH**
China	10	1	5	1	1,000	5	0.5	6.5–8.5
US-EPA	10	1	–	0.5	500	–	–	6.5–8.5
Japan	10		3	2	500	3	–	5.8–8.6
EC	10		–	1	–	5	0.5	–
WHO	–	–		10	–	–	1.5	6.5–8.5
Australia	3		–	5	600	–	0.5	6.5–8.5

EC, European Commission.

as a result, membrane fouling was largely mitigated, and PVA-gel was used as the main contributor to organic removal in the system. In order to improve the recalcitrant organic compounds' removal, advanced oxidation processes (AOPs) were integrated into the aMBR system as a polishing step (UV/O_3-aMBR), to partially degrade recalcitrant pollutant and improve its biodegradability.

The organic loading rate in the aMBR and UV/O_3-aMBR system was much lower than in conventional wastewater treatment plants, because the COD_{Mn} concentration was only around 10 mg/L. The endogenous respiration of bioprocess with higher organic loading rate, such as wastewater, was reported by earlier studies (Fall *et al.* 2014; Li *et al.* 2014, 2018a; Ordaz *et al.* 2017; Park *et al.* 2017). The bioprocess in the aMBR system is expected to be quite different from conventional activated sludge processes or other biological treatment processes. The AOPs combined with an aMBR system could also lead to changes in substrate utilization and microbe activities. In addition, the bioprocess between aMBR and UV/O_3-aMBR is expected to be quite different and thus needs to be compared and analyzed.

Bioprocess aerobic respiration commonly includes exogenous respiration and endogenous respiration (Hao *et al.* 2010). Endogenous respiration always occurs once no substrate remains for the microbes to use, and the active cell starts utilizing intracellular substances (Gujer *et al.* 1999). Endogenous respiration has been commonly regarded as an evaluation method for active microorganisms, and can be tested quickly and simply with oxygen uptake rate methods (Spanjers & Vanrolleghem 1995). The widely used oxygen uptake rate analysis method is one type of respirometry method, which detects the dissolved oxygen (DO) concentration with a DO meter providing real-time data, to intuitively reflect the substrate utilization by microbes (Blok 1976; Chandran & Smets 2005; Capodici *et al.* 2014; Li *et al.* 2018b).

There has been limited research on the membrane bioreactor (MBR) in treating polluted surface water, let alone the biological process in the MBR in this sector. This study analyzed and compared the endogenous respiration process in aMBR and UV/O_3-aMBR systems in treating polluted surface water by the respirometry method. The microbe activities, including heterotrophic and autotrophic entry time of endogenous respiration, and ability in anti-decay, were also compared and discussed. This study could provide useful information for the understanding and practical application of aMBR and UV/O_3-aMBR processes in treating polluted surface water for drinking purposes in a more environmentally friendly way.

MATERIALS AND METHODS

System setup

A schematic and flowchart of aMBR and UV/O_3-aMBR parallel operation system setup are shown in Figure 1(a) and 1(b), respectively. Polyvinyl alcohol gel (PVA-gel) was used as a bio-carrier with filling ratio of 5% (volume to volume, V/V) in the carrier side of both systems. Hollow fiber PTFE UF membrane with pore size 0.1 μm and surface area 0.1 m^2 was submerged in the membrane side. Hydraulic retention time (HRT) and flow rates controlled in both systems were 2.5 h and 2 L/h, respectively. The aMBR and

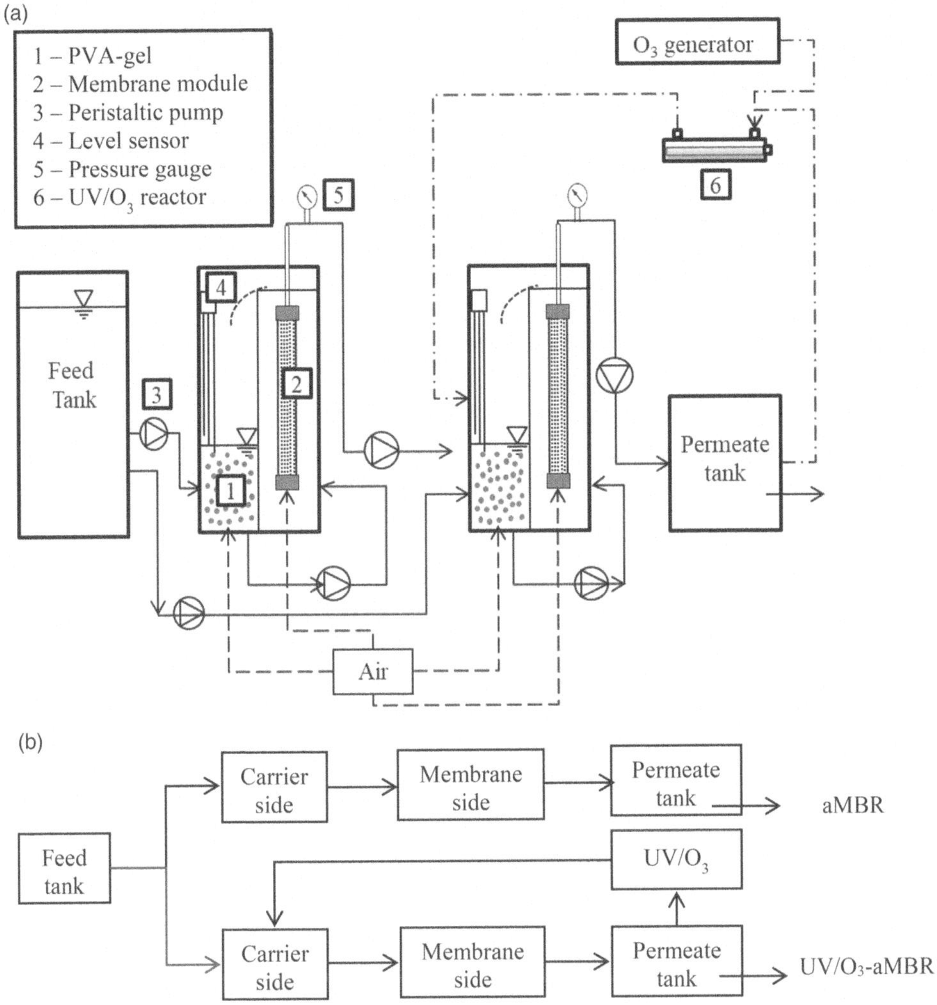

Figure 1 | Schematic (a) and flowchart (b) of aMBR and UV/O_3-aMBR parallel systems.

UV/O_3-aMBR parallel system inlet and permeate pump were maintained by a timer (8 min on and 2 min off). The transmembrane pressure (TMP) of both systems was recorded by a digital pressure gauge. The UV/O_3-aMBR system consisted of an aMBR system with UV/O_3 set in a recirculation stream as a polishing step, as shown in Figure 1(a). aMBR permeate with a recirculation ratio of 60% was pumped into the UV/O_3 reactor to be partially treated and then sent back to the aMBR carrier side. The contact time in the UV/O_3 reactor was based on optimal biodegradable organic carbon (BDOC) results as 7 min. UV light 8 W with a wavelength of 254 nm was used and an O_3 generator (Nippon Ozone Co., Ltd, Japan) with concentration of 1.5 mg/L was used.

Experimental materials

PVA-gel used as a bio-carrier in both systems was taken out for endogenous respiration analysis from the stable operated aMBR and UV/O_3-aMBR system, respectively. The removal performance of basic parameters including COD_{Mn}, DOC, UV_{254}, specific UV absorbance (SUVA), and total nitrogen (TN) was tested before taking out the PVA-gel from both systems. The suspended solids concentration in the carrier side of both systems was less than 100 mg/L and was not taken into consideration for endogenous analysis.

A respirometer was set up with an automatically recording DO meter (HQ40d, HACH, USA) and in a room at constant temperature; the schematic of the system is as shown in Figure 2. During the oxygen uptake rate test, 50 mg/L NH_4Cl was used as the nitrogen source and 100 mg/L CH_3COONa was used as the carbon source. DO values were automatically recorded every 30 seconds for 15 min in all conditions tested. The temperature in all tests was maintained at 25.7 ± 0.2 °C. The pH was 7.79 ± 0.04 in the aMBR system and 7.90 ± 0.03 in the UV/O_3-aMBR system.

Analytical methods

Respirometric analysis calculation

A respirometric method with a DO meter recording the DO concentration changes in the aqueous system was used. Oxygen uptake rate was calculated from the DO values. The PVA-gel bio-carrier amount in each condition was counted after the test, and the specific oxygen uptake rate was calculated accordingly. The respiration rate was composed of exogenous respiration and endogenous respiration, among which, exogenous respiration included autotrophic respiration and heterotrophic respiration as is shown in Equation (1):

$$OUR_T = OUR_{en} + OUR_A + OUR_H \tag{1}$$

Equation (1) divided with OUR_T can produce Equation (2):

$$1 = \frac{OUR_{en}}{OUR_T} + \frac{OUR_A}{OUR_T} + \frac{OUR_H}{OUR_T} \tag{2}$$

where the OUR_T is theoretic total respiration rate, OUR_{en} is real tested endogenous respiration rate, OUR_A is real tested autotrophic respiration rate, OUR_H is real tested heterotrophic respiration rate, and OUR_t is real tested total respiration rate. OUR_{en}/OUR_T represents the ratio of endogenous respiration to the theoretical total respiration rate, while OUR_A/OUR_T represents the ratio of autotrophic respiration to the theoretical total respiration rate, and OUR_H/OUR_T represents the ratio of heterotrophic respiration to the theoretical total respiration rate. $SOUR_T$ is specific total respiration rate, $SOUR_{en}$ is specific endogenous respiration rate, $SOUR_A$ is specific autotrophic respiration rate, and $SOUR_H$ is specific heterotrophic respiration rate.

$$SOUR = OUR/amount\,of\,PVA - gel\,tested\,(pcs) \tag{3}$$

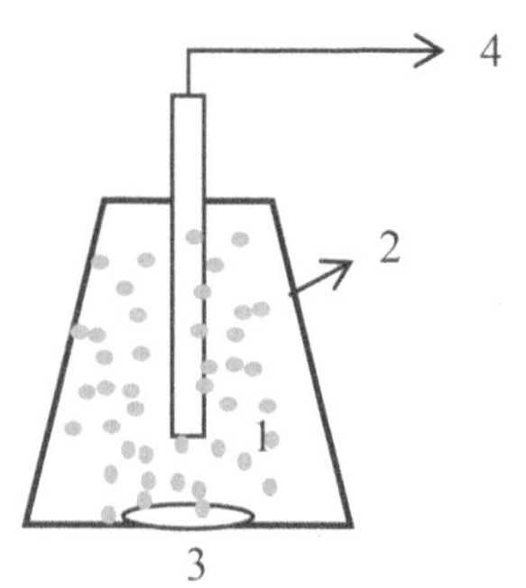

Figure 2 | Schematic of respirometer used in this study.

Respirometric test setup

One hundred and forty milliliters of PVA-gel were sampled from the aMBR and UV/O_3-aMBR systems, of which 20 mL from each system was taken out for testing at an aeration time of 0 h, and the remaining 120 mL PVA-gel was placed in 1.5 L of reverse osmosis (RO) water and aerated in an aeration vessel separately. After that, 20 mL PVA-gel was sampled from the aeration vessel for testing after 6, 12, 24, 48, 96, and 120 h of aeration. The aeration time of 6 to 120 h was selected based on earlier research indicating that activated sludge enters endogenous respiration in 1 h to 5 days (Henze *et al.* 2000; Friedrich & Takács 2013).

The respiration test for one set of aeration time was carried out as follows. (1) 20 mL carrier was put into 400 mL RO water in a respirometer, and OUR_{en} was tested in this situation. (2) The solution in the respirometer was then discharged and the respirometer filled with 400 mL RO water. Ten mL of 2,000 mg/L NH_4Cl was then added to obtain a concentration of 50 mg/L, and OUR_N was then tested. (3) The solution in the respirometer was then discharged and the respirometer filled with 400 mL RO water. Ten mL of 4,000 mg/L CH_3COONa was added to obtain 100 mg/L of a CH_3COONa concentration, after which OUR_c was then tested. (4) The solution in the respirometer was discharged and it was filled with 400 mL RO water, 10 mL of 4,000 mg/L CH_3COONa and 10 mL of 2,000 mg/L NH_4Cl was added to obtain 100 mg/L CH_3COONa concentration and a NH_4Cl concentration of 50 mg/L, then

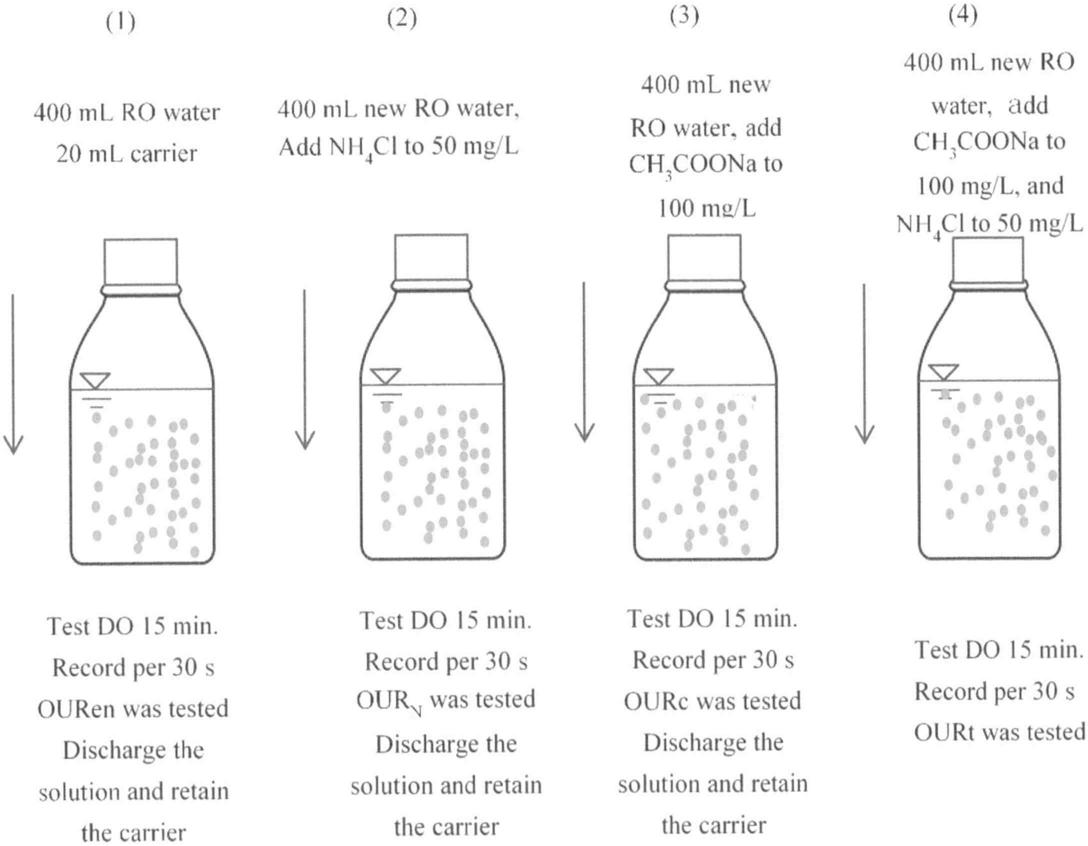

Figure 3 | Respiration map testing procedure for one cycle.

OUR_t was tested. The flowchart of this process is shown in Figure 3.

OUR_{en} was obtained when no substrates were added, OUR_N was obtained when only nitrogen source was added, OUR_c was obtained when only carbon source was added, and OUR_t was obtained when both nitrogen source and carbon source were added:

$$OUR_N = OUR_{en} + OUR_A \tag{4}$$

$$OUR_c = OUR_{en} + OUR_H \tag{5}$$

Due to the low organic loading rate at around 0.096 kg·COD_{Mn}/(m^3 d) in feed, with corresponding low biomass in both systems, the unit of *SOUR* value used in this study was mg/L d pcs, as shown in Figures 4 and 5. To maintain the accuracy of the analysis, *SOUR* was used instead of *OUR*. Even though 20 mL of PVA-gel as carrier was used for testing in all conditions, the accurate amount was counted (as pcs) after each test and used as specific volume parameter to minimize the error.

RESULTS AND DISCUSSION

The removal performance of aMBR and UV/O_3-aMBR systems

The removal performance of the stable operated aMBR and UV/O_3-aMBR system is shown in Table 2. The UV/O_3-aMBR system has higher removal performance, including both COD_{Mn}, DOC, UV_{254}, SUVA, and TN as compared with the aMBR system. With the adding of UV/O_3 as a polish step in the recirculation stream of the aMBR system, the removal performance of recalcitrant organic matters was largely increased, as shown by the UV_{254} results. The SUVA index shows that UV/O_3-aMBR system-treated water has higher hydrophilicity compared with the aMBR system,

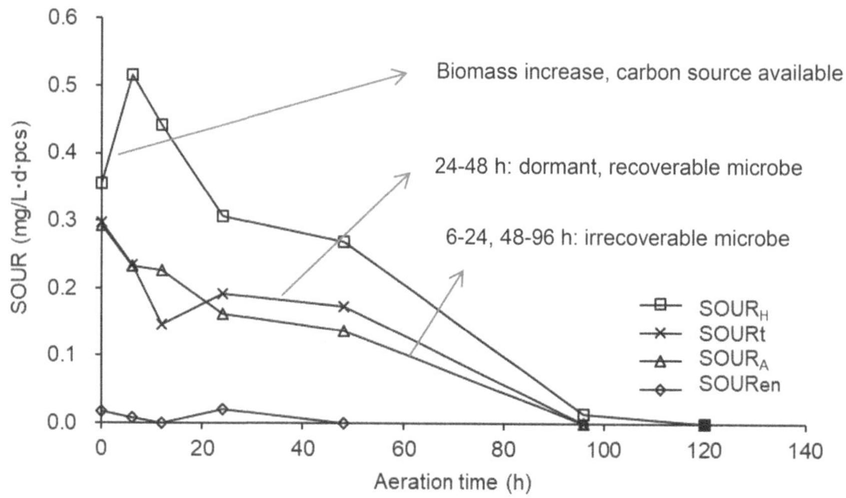

Figure 4 | Specific respiration rate in endogenous respiration processes of carriers from the aMBR system.

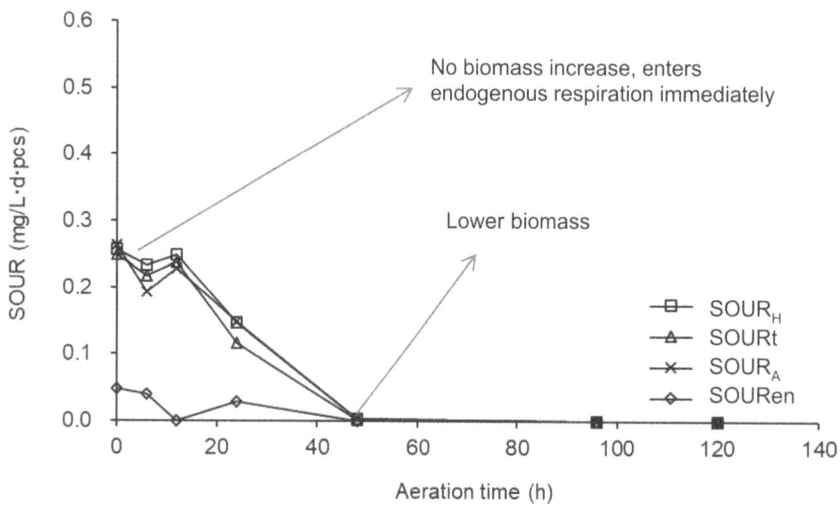

Figure 5 | Specific respiration rate in endogenous respiration processes of carriers from the UV/O_3-aMBR system.

Table 2 | Removal rate of basic parameters in aMBR and UV/O_3-aMBR system

Parameters	Feed	aMBR permeate	UV/O_3-aMBR permeate
COD_{Mn} (mg/L)	10.44 ± 1.88	4.66 ± 0.75	2.91 ± 0.34
DOC (mg/L)	10.33 ± 2.05	6.78 ± 0.87	3.73 ± 0.55
UV_{254} (cm^{-1})	0.20 ± 0.02	0.15 ± 0.02	0.05 ± 0.01
SUVA (L/mg · m)	1.97 ± 0.45	1.93 ± 0.51	1.10 ± 0.21
TN (mg/L)	14.45 ± 3.65	13.26 ± 3.75	8.54 ± 1.21

and this could be the reason why the UV/O_3-aMBR system shows higher organic removal. The contribution of the microbes in both systems was worth analyzing.

The *SOUR* changes in endogenous respiration process

$SOUR_{en}$ value was similar and low in both aMBR and UV/O_3-aMBR systems. Within 0–6 h, the $SOUR_H$ of aMBR was increased sharply, which indicated that the biomass was increasing, and shows the carbon source was available on the PVA-gel. The PVA-gel sampled from the aMBR system had substrate on it, which could be extracellular polymer substrates, organic matter from polluted surface water adsorbed on it, or intracellular storage (Li & Yang 2007). The decrease of $SOUR_N$ ($SOUR_{en} + SOUR_A$) indicated that no nitrogen source remained on the PVA-gel.

This also could suggest that sufficient ammonia oxidation bacteria existed on the PVA-gel in the aMBR system. According to Figure 4, 6 h is a turning point, where $SOUR_H$ starts to decrease after increasing, thus, 6 h can be regarded as the point where heterotrophic bacteria enters endogenous respiration in the aMBR system. As shown in Figure 5, the $SOURt$, $SOUR_H$, and $SOUR_A$ were continuously decreasing, with low increasing turning point observed at 12 h. This shows that, in the UV/O_3-aMBR system, no substrates remained on the PVA-gel surface and the increase observed during 6 h to 12 h may be contributed to by the intracellular storages of the microbes (Ramdani *et al.* 2010). This phenomenon further indicates that the microbes on PVA-gel enter endogenous respiration immediately when taken out from the UV/O_3-aMBR system. The results also suggest that in the UV/O_3-aMBR system, substrate biodegradability was higher than in the aMBR system.

As shown in Figure 4, the decrease of $SOUR_H$ curve showed a rapid decreasing period, 6–24 h and 48–96 h, and a slow decreasing period 24–48 h. This indicated that heterotrophic bacteria can be divided into two components, recoverable and irrecoverable. When there was enough substrate, the respiration rate did not recover at all during 6–24 h and 48–96 h. This indicated that the irrecoverable heterotrophic microbes had decayed. During 24–48 h, the slow decreasing approach to a stable situation indicated that recoverable bacteria component regained activities when enough substrate was added. The $SOUR_A$ curve was decreased with three fast decreasing and two slow decreasing periods and no increasing period was observed. This indicates autotrophic bacteria also included recoverable and irrecoverable components. When substrate is not enough, the recoverable component will start to become dormant, and regain activities once sufficient substrate is provided (Roszak & Colwell 1987). The stable or slow decreasing parts are defined as the dormant period, as shown in Figure 4.

For the aMBR system, $SOUR_t$ shows lower value than $SOUR_H$ and $SOUR_A$ in 0–12 h, and during 0–6 h, it even shows a similar value to $SOUR_A$. This indicated that when there was sufficient carbon source and nitrogen source provided at the same time, autotrophic bacteria respiration dominated and heterotrophic bacteria were probably inhibited in the aMBR system. During 6–12 h, $SOUR_t$ decreased with a similar trend as $SOUR_H$, and $SOUR_A$ decreased quite slowly due to recoverable autotrophic bacteria respiration. $SOUR_t$ was slightly higher than $SOUR_A$ in the 24–96 h period and this might be due to the decrease of effects from autotrophic bacteria inhibition.

The microbe respiration includes endogenous and exogenous respiration, with endogenous respiration showing more stable characteristics as compared with exogenous respiration. Thus, the endogenous respiration to total respiration rate could be used as the microbe activity index (Park *et al.* 2017; Li *et al.* 2019). The $SOUR_{en}/SOUR_t$ in aMBR and UV/O_3-aMBR was increased with increasing aeration time, which indicated that the endogenous respiration will increase when facing adverse environmental conditions such as lacking substrate. The higher endogenous respiration rate implied a higher oxygen consumption required for microbe fundamental metabolism, which also presents a higher operation cost demand of the system. The $SOUR_{en}/SOUR_t$ value in aMBR was lower than in UV/O_3-aMBR, which indicated that the microbe activity was higher in the aMBR system as compared with the advanced oxidation process (AOP) integrated system.

The changes of respiration ratio in endogenous respiration process

$SOUR_H$ quickly decreased by 14.15% from 6 to 12 h, 16.21% from 12 to 24 h, 7.48% from 24 to 48 h, 49.60% from 48 to 96 h, and 2.56% from 96 to 120 h. This indicated that over 90% of the heterotrophic bacteria were irrecoverable in the aMBR system. $SOUR_A$ continuously decreased by 20.52% from 0 to 6 h, 1.71% from 6 to 12 h, 22.09% from 12 to 24 h, 8.68% from 24 to 48 h, and 47% from 48 to 96 h, until it stopped decreasing at 96–120 h. This indicated that some recoverable autotrophic bacteria regained activity from 6 to 12 h and 24 to 48 h, until after 96 h of aeration, all the autotrophic bacteria decayed.

Calculated from the theoretical $SOUR_T$, heterotrophic bacteria occupied around 60% and autotrophic bacteria 40% in the aMBR system. As shown in Figure 4, it can also be seen from the system that the $SOUR_H$ was always higher than $SOUR_A$ throughout the aeration time tested. Also, as shown in Figure 4, there is no absolute stability of $SOUR_H$, $SOUR_A$, and $SOUR_t$ during 120 h aeration and this indicated that the microbes' activity in the aMBR system is tough and active enough. Thus, the addition of any substrates can motivate their respiration processes within 120 h aeration.

The respiration map of bacteria growth on PVA-gel in the UV/O_3-aMBR system is shown in Figure 5. Compared with Figure 4, the values obtained for $SOUR_t$, $SOUR_H$, and $SOUR_A$ were lower than the values obtained from the aMBR system in all samples tested during 120 h aeration. This shows that the biomass amount on PVA-gel in the UV/O_3-aMBR system was much lower than in the aMBR system. During 0–6 h, the $SOUR_t$, $SOUR_H$, and $SOUR_A$ were decreased, which indicated that no extra substrate remained on the PVA-gel, regardless of the presence of carbon source and nitrogen source in any form. Thus, there was no increase in biomass in any form. During 6–12 h, it was observed that there was a slight increase in $SOUR_t$, $SOUR_H$, and $SOUR_A$, which might be due to a part of the microorganisms using the decayed part as substrates. It can also be contributed to by the variation of carriers used, since the difference is quite small, based on the $SOUR$ unit. After 12 h, the $SOUR_t$, $SOUR_H$, and $SOUR_A$ values continued to decrease until 0 at 48 h of aeration and remained at 0 until 120 h. This indicated that both heterotrophic and autotrophic bacteria in the UV/O_3-aMBR system were quite low or there were no recoverable components. Thus, even if sufficient substrates were provided, all $SOUR_t$, $SOUR_H$, and $SOUR_A$ values were decreasing continuously, and there was no steady or dormant period found in its respiration map.

As shown in Figure 5, the $SOUR_t$, $SOUR_H$, and $SOUR_A$ values were quite similar, which indicated that the contribution of the heterotrophic and autotrophic bacteria were nearly equal in the system. However, when the presence of carbon source or nitrogen source is too high, autotrophic bacteria would be the dominant microbes. This is also due to the feed water quality from the parallel aMBR and UV/O_3-aMBR systems with both COD_{Mn} value and NH_4^+-N value of around 10 mg/L. Nitrogen source was sufficient as compared to carbon source in the system, thus the amount of autotrophic bacteria dominated the system compared to the heterotrophic bacteria.

Bacteria were all decayed after 48 h aeration in the UV/O_3-aMBR system and 96 h aeration in the aMBR system as shown in Figures 4 and 5. This also indicated that less biomass and substrate were attached on carriers in UV/O_3-aMBR than aMBR. It is also probable that UV/O_3-aMBR has less species of microbes compared to the aMBR system; attached specific heterotrophic bacteria and autotrophic bacteria provide a high performance in consuming all available substrates, or the integration of the UV/O_3 process had accelerated the biodegradability of organic substrates in the UV/O_3-aMBR system. It is also shown in Figures 4 and 5 that the $SOUR_t$ in the two reactors was nearly the same, and this indicated that, even with higher biomass in the aMBR system, it provides a similar degree of bioactivity as the UV/O_3-aMBR system when sufficient nitrogen source and carbon source were provided.

Practical applications and future perspectives

The membrane-based technology could guarantee the permeate water quality in drinking water treatment. The biological process is an environmentally friendly treatment technology for dealing with organic matter in polluted surface water. With the deterioration of surface water quality, the combination of membrane technology and biological technology-membrane bioreactor system has the potential to be developed as an effective alternative technology in drinking water supply. However, very limited work has been done in illustrating the MBR related system in the treatment of polluted surface water. Polluted surface water was defined as water quality with COD_{Mn} around 10 mg/L and NH_3-N around 3 mg/L. This is much lower as compared with the municipal wastewater, while being higher than the common source water that is used for drinking water supply. The increase of organic matter in surface water is an inevitable trend, which has gradually brought challenges to the conventional coagulation and flocculation drinking water treatment process. The integration of an AOP into the aMBR system as a polishing step to improve the permeate water biodegradability and then to be treated by biological processes could largely improve the recalcitrant organic matter removal. More work has to be focused on the investigation of the MBR technology application in the treatment of polluted surface water.

The biological process has played an important role in the MBR system for treating polluted surface water. The illustration and comparison of the biological process in the aMBR and UV/O_3-aMBR systems could provide detailed information for a future pilot-scale system design. The endogenous respiration map method used in this study is a simple and accurate one. The use of an endogenous respiration map could provide information about microbes in the system, and provide guidance in practical application. It is also recommend to build up a small program to generate a corresponding respiration map simply with the input from the necessary parameters.

CONCLUSIONS

This study investigated the endogenous respiration process in aMBR and UV/O_3-aMBR in treating polluted surface water. Results indicated that microbes enter endogenous respiration at different times in the two systems. With the integration of the UV/O_3 process into the aMBR system, the biomass on the bio-carrier was changed. Higher abundance of microbes including recoverable and irrecoverable components existed in the aMBR system compared to the UV/O_3-aMBR system. The use of AOPs in aMBR for treating polluted surface water could improve recalcitrant removal, while the design of AOP conditions also has to be carefully involved to guarantee the contribution of the biological process.

ACKNOWLEDGEMENTS

The research was supported by Water Pollution Control and Treatment, National Science and Technology Major Project (Grant no. 2018ZX07208001), National Natural Science Foundation of China (Grant no. 41877344), China Postdoctoral Science Foundation (Grant no. 2019M652738). Dr. Kang Song acknowledges the support from 100 Talents Program of Chinese Academy of Sciences (Y82Z08-1-401, Y75Z01-1-401).

REFERENCES

Blok, J. 1976 Measurements of the viable biomass concentration in activated sludge by respirometric techniques. *Water Research* **10** (10), 919–925.

Capodici, M., Di Trapani, D. & Viviani, G. 2014 Co-treatment of landfill leachate in laboratory-scale sequencing batch reactors: analysis of system performance and biomass activity by means of respirometric techniques. *Water Science and Technology* **69** (6), 1267–1274.

Chandran, K. & Smets, B. F. 2005 Optimizing experimental design to estimate ammonia and nitrite oxidation biokinetic parameters from batch respirograms. *Water Research* **39** (20), 4969–4978.

Fall, C., Rogel-Dorantes, J. A., Millan-Lagunas, E. L., Martinez-García, C. G., Silva-Hernández, B. C. & Silva-Trejo, F. S. 2014 Modeling and parameter estimation of two-phase endogenous respirograms and COD measurements during aerobic digestion of biological sludge. *Bioresource Technology* **173**, 291–300.

Friedrich, M. & Takács, I. 2013 A new interpretation of endogenous respiration profiles for the evaluation of the endogenous decay rate of heterotrophic biomass in activated sludge. *Water Research* **47** (15), 5639–5646.

Gujer, W., Henze, M., Mino, T. & Van Loosdrecht, M. 1999 Activated sludge model No. 3. *Water Science and Technology* **39** (1), 183–193.

Hao, X. D., Wang, Q. L., Zhu, J. Y. & Van Loosdrecht, M. C. 2010 Microbiological endogenous processes in biological wastewater treatment systems. *Critical Review of Environmental Science and Technology* **40** (3), 239–265.

Henze, M., Gujer, W., Mino, T. & Loosdrecht, M. V. 2000 *Activated Sludge Models ASM1, ASM2, ASM2D, ASM3*. IWA Publishing, London, UK.

Li, X. Y. & Yang, S. F. 2007 Influence of loosely bound extracellular polymeric substances (EPS) on the flocculation, sedimentation and dewaterability of activated sludge. *Water Research* **41** (5), 1022–1030.

Li, Z. H., Bai, X. L., Zhang, Q., Liu, Y. & He, C. B. 2014 Endogenous respiration process analysis of heterotrophic biomass and autotrophic biomass based on respiration map. *Environmental Science* **35** (9), 3492–3497 (in Chinese).

Li, L., Suwanate, S. & Visvanathan, C. 2017 Performance evaluation of attached growth membrane bioreactor for treating polluted surface water. *Bioresource Technology* **240**, 3–8.

Li, Z. H., Ma, Z. B. & Yu, H. Q. 2018a Respiration adaptation of activated sludge under dissolved oxygen and hypochlorite stressed conditions. *Bioresource Technology* **248**, 171–178.

Li, Z. H., Zhu, Y. M., Yang, C. J., Zhang, T. Y. & Yu, H. Q. 2018b A simple respirogram-based approach for the management of effluent from an activated sludge system. *Bioresource Technology* **261**, 412–419.

Li, Z. H., Hang, Z. Y., Lu, M., Zhang, T. Y. & Yu, H. Q. 2019 Difference of respiration-based approaches for quantifying heterotrophic biomass in activated sludge of biological wastewater treatment plants. *Science of the Total Environment* **664**, 45–52.

Ma, C. Y., Sheu, Y. T., Hsia, K. F., Dong, C. D., Chen, C. W., Huang, Y. C. & Kao, C. M. 2018 Development of water and sediment quality management strategies for an urban river basin: a case study in Taiwan. *Journal of Water Supply: Research and Technology-Aqua* **67** (8), 810–823.

Ordaz, A., Sánchez, M., Rivera, R., Rojas, R. & Zepeda, A. 2017 Respirometric response and microbial succession of nitrifying sludge to m-cresol pulses in a sequencing batch reactor. *Biodegradation* **28** (1), 81–94.

Park, J., Yamashita, N., Wu, G. & Tanaka, H. 2017 Removal of pharmaceuticals and personal care products by ammonia oxidizing bacteria acclimated in a membrane bioreactor: contributions of cometabolism and endogenous respiration. *Science of the Total Environment* **605**, 18–25.

Ramdani, A., Dold, P., Déléris, S., Lamarre, D., Gadbois, A. & Comeau, Y. 2010 Biodegradation of the endogenous residue of activated sludge. *Water Research* **44** (7), 2179–2188.

Roszak, D. B. & Colwell, R. R. 1987 Survival strategies of bacteria in the natural environment. *Microbiological Reviews* **51** (3), 365–379.

Spanjers, H. & Vanrolleghem, P. 1995 Respirometry as a tool for rapid characterization of wastewater and activated sludge. *Water Science and Technology* **31** (2), 105–114.

Utaaker, K. S., Joshi, H., Kumar, A., Chaudhary, S. & Robertson, L. J. 2019 Occurrence of *Cryptosporidium* and *Giardia* in potable water sources in Chandigarh, Northern India. *Journal of Water Supply: Research and Technology-Aqua* **68** (6), 483–494. https://doi.org/10.2166/aqua.2019.157.

First received 1 June 2019; accepted in revised form 24 July 2019. Available online 3 September 2019

Optimization method for joint operation of a double-reservoir-and-double-pumping-station system: a case study of Nanjing, China

Zhihao Gong, Xiaohong Jiang, Jilin Cheng, Yi Gong, Xing Chen and Haomiao Cheng

ABSTRACT

Double-reservoir-and-double-pumping-station systems are commonly used for irrigation water supply in hilly regions of southern China. An optimization model for this water supply system is proposed to minimize water shortage. The model features few coupling constraints, including available water in the system and pumping volume limited by regional water rights. Dynamic programming was adopted to solve the subsystem and aggregation models. The results with the model and that with the standard operation policy were compared; the total water shortage was reduced by 87.7%, total water replenishment from outside was reduced by 2.2%, and total water spill was reduced by 60.6% for a system in Nanjing, China. The method may provide a reference for optimal operation of water supply systems comprising reservoirs and pumping stations.

Key words | decomposition, dynamic programming, joint operation, pumping station, reservoir, water rights

Zhihao Gong
Xiaohong Jiang
Jilin Cheng (corresponding author)
Yi Gong
Xing Chen
School of Hydraulic Energy and Power Engineering, Yangzhou University,
Yangzhou 225009,
China
E-mail: *jlcheng@yzu.edu.cn*

Haomiao Cheng
School of Environmental Science and Engineering, Yangzhou University,
Yangzhou 225127,
China

INTRODUCTION

With the worldwide construction of reservoirs and inter-basin water transfer projects, an optimal joint operation of multi-reservoir systems has recently become a research hot-spot. It has been recognized that benefits derived from the joint operation of a multi-reservoir system may exceed the sum of that from the independent operation of each reservoir in it (Wang *et al.* 2016).

Generally, in an optimization model of reservoir operation, the objective function is to minimize the sum of the squared deviation of actual water supply from the desired target, in an attempt to offer the same deviation at each period as far as possible (Celeste & Billib 2009). Such models are subject to several constraints, such as lower and upper bounds of release and storage, water balance equation, and non-negativity constraints. Furthermore, hydraulic connections between reservoirs should be added into constraints in a multi-reservoir system.

Optimization models include two types of decision variables (water supplies and spills) that increase with the number of reservoirs; increased variables lead to the 'curse of dimensionality' in the solving process (Chen *et al.* 2016). At present, different types of meta-heuristic algorithms featuring high applicability and high efficiency have been applied to solving such problems. The genetic algorithm was applied to optimizing the operation policy of a four-reservoir system (Jothiprakash *et al.* 2011). Particle swarm optimization was adopted for the optimal water resources management of a multi-reservoir system (Nabinejad *et al.* 2017). Gu *et al.* (2017) used a global search algorithm to optimize the joint operation of a system comprising multiple

doi: 10.2166/aqua.2019.094

donor reservoirs and one recipient reservoir. Ehteram *et al.* (2018) suggested the Spider Monkey Algorithm for optimizing operation rules of multiple reservoirs. However, meta-heuristic algorithms may fail to solve models with equality constraints because of their fatal weakness in that they cannot guarantee globally optimal final results, which can be attributed to the random sampling (Birhanu *et al.* 2014; Rani & Srivastava 2015). In addition, the application of meta-heuristic algorithms to reservoir operation problems may lead to spills when the reservoir storage is below its capacity, which is against the normal operation rule.

Dynamic programming based on Bellman's principle (Kennedy 1986) can provide reliable and optimal results. In addition, it has good applicability in multi-stage decision-making processes, regardless of the matter whether the model is continuous or linear. Therefore, it has been widely used in solving optimization problems of reservoir operation. Although the 'curse of dimensionality' (Rani & Moreira 2009; Ahmad *et al.* 2014; Wang *et al.* 2016) is induced under the condition of a large number of reservoirs, appropriate dimension reduction methods, such as the decomposition-aggregation method (Gong & Cheng 2018) and decomposition-coordination method (Mahey *et al.* 2017), can be used to obtain global optimal results by transforming a high-dimensional problem into a series of low-dimensional problems.

The purpose of optimizing the joint operation of a reservoir and a pumping station is to minimize the operation cost, mainly the operation cost of the pumping station. In addition to water supplies and spills of the reservoir, water volume pumped by the pumping station during each period is also a type of decision variable subject to the maximum pumping capacity and pumping volume limited by water rights.

Accordingly, the difficulty of solving the model will increase with increasing numbers of reservoirs and pumping stations in the system. Yu *et al.* (1994) and Pulido-Calvo & Gutiérrez-Estrada (2011) developed a nonlinear optimization model for a reservoir and its replenishment pumping station to minimize the operation cost. Reca *et al.* (2014) constructed a nonlinear model for the joint operation of reservoirs and pumping stations to obtain the minimum operation cost and improved it considering the evaporation of the reservoir. Đurin (2016) derived an optimal operating policy with a regression analysis of the running time mode of pumps, design discharge of pumping station, and storage capacity of the reservoir, according to the deterministic water supply scheme. However, all the above-mentioned works optimized the operation schedule of pumping stations according to seasonal electricity price or time-of-use price and contributed toward saving energy costs. In addition, they did not limit the total water volume pumped by the systems.

In different countries, concepts and connotations of water rights are not the same under different social backgrounds (Molle 2004; Heikkila 2015). However, it especially refers to the water resources usufruct here. Since 2016, the agricultural water price reform has been promoted stage by stage in China (Shen & Wu 2017). As an important premise, the water resources usufruct is defined in the form of the annual total volume of water which is distributed to users. In agriculture, the Chinese government has established an upper bound for the annual amount of irrigation water. Furthermore, the total amount of water that can be obtained from each irrigation gate or pumping station within a year is strictly limited (Sun *et al.* 2016).

This study aimed to develop an optimization method based on the decomposition-aggregation theory for the joint operation of a double-reservoir-and-double-pumping-station system that can provide optimal results following normal operation regulations. For this purpose, an optimization model of this system in Nanjing, China, was established as a case study, and the total available water in the system was coupled into it, considering regional water rights.

DOUBLE-RESERVOIR-AND-DOUBLE-PUMPING-STATION SYSTEM

Research area

Liuhe district of Nanjing, the Jiangsu province is located in eastern China. The hilly regions of this district lie in the subtropical monsoon climate zone. Therefore, local residents build reservoirs and water transfer projects to deal with the uneven temporal and spatial distribution of water resources. The typical irrigation water supply system in this region is the double-reservoir-and-double-pumping-station system shown in Figure 1, which consists of the SH reservoir, HWB reservoir, and their replenishment pumping stations.

1. Reservoirs and pumping stations

The SH reservoir is the second largest reservoir in Liuhe district and its main function is to supply water for irrigation. During the period of operation, the XZ pumping station replenishes it with water from the BL river through a $2\,m \times 2\,m$ square concrete culvert before the water level reaches below the lower boundary limit. Because of the higher topography, the irrigation area of the HWB reservoir is larger although the catchment area is relatively small, and thus, the probability of water shortage is relatively high. During water shortage, the HZ pumping station replenishes the HWB reservoir with water from the SH reservoir through a $1.5\,m \times 1.5\,m$ square concrete culvert. The characteristics of reservoirs and pumping stations are shown in Tables 1 and 2, respectively.

2. Inflows and water demand

Monthly inflows and water demand at the 75% probability of exceedance are shown in Tables 3 and 4, respectively.

3. Evaporation

Evaporation loss is a function of evaporation depth and average free water surface of a reservoir in each period. Evaporation depths were derived from evaporation data collected using the E_{601} evaporator (Table 5), and the data required correction by coefficient k. The average free water surface of each reservoir in a specific period is determined by its surface–volume relationship, which was provided by the Liuhe Water Authority. Finally, the

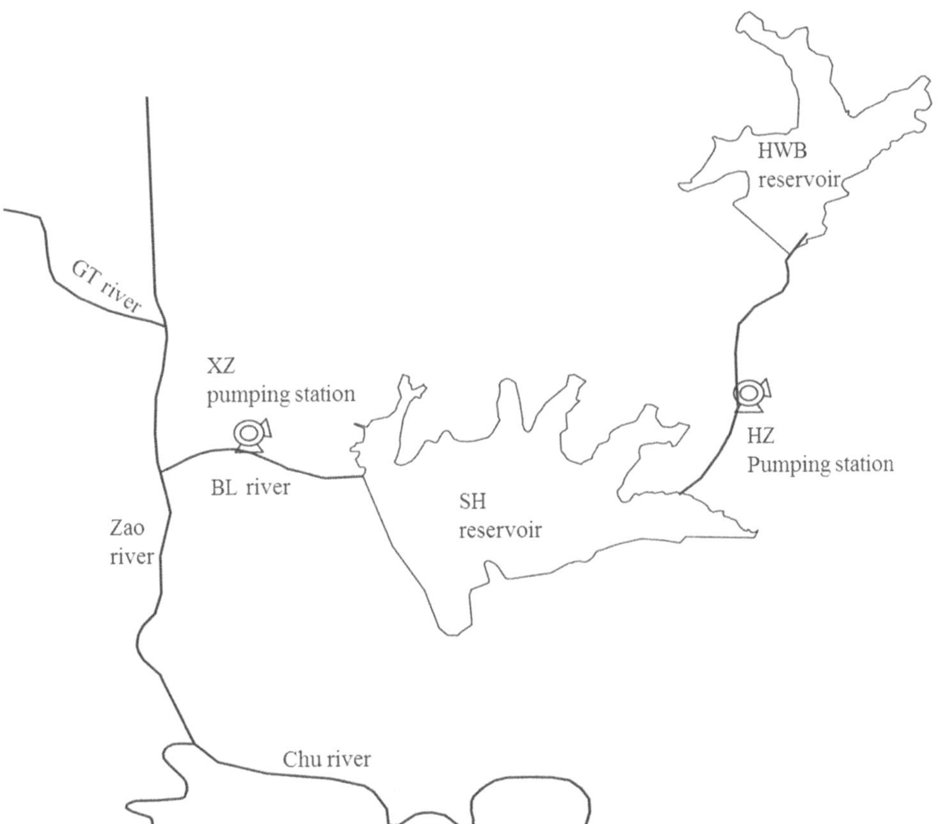

Figure 1 | Location and layout of the system.

Table 1 | Characteristics of reservoirs

Reservoir	Dead storage capacity (10^4 m^3)	Utilizable capacity (10^4 m^3)	Total storage capacity (10^4 m^3)	Limited storage capacity in flood season (10^4 m^3)	Catchment area (km^2)	Irrigation area (hm^2)
SH	600	1,157	2,473	1,000	31.9	1,667
HWB	457	936	1,891	1,393	20.8	2,067

Table 2 | Characteristics of pumping stations

Pumping station	Design discharge (m^3/h)	Design pumping head (m)	Daily operation duration (h)	Water rights (75%) (10^4 m^3)
XZ	10,000	19.4	20	360
HZ	7,500	15.0	20	–

Notes: The annual water rights of the XZ pumping station at 75% probability of exceedance is allocated by the Liuhe Water Authority. The HZ pumping station is an internal pumping station in the system; there is no definite limit on its water rights.

reservoir evaporation for a specific period can be computed using Equation (1):

$$EF_i = 0.1 \times k_i \times E_i \times (\alpha V_i + \beta) \tag{1}$$

where EF_i (10^4 m^3) is the evaporation loss of a reservoir in period i, E_i (mm) is the evaporation depth of E_{601} evaporator in period i; k_i is the correction coefficient for period i; V_i (10^4 m^3) is the average water storage in period i; and α and β are the reservoir coefficients. For the SH reservoir, $\alpha = 1.194 \times 10^{-3}$, $\beta = 2.575$, and for the HWB reservoir, $\alpha = 1.657 \times 10^{-3}$, $\beta = 1.862$.

Generalization of the system

Figure 2 shows a generalized schematic of the double-reservoir-and-double-pumping-station system comprising the SH reservoir, HWB reservoir, and their replenishment pumping stations. The SH reservoir is a donor reservoir, numbered as

Table 3 | Monthly inflows (10^4 m^3)

		Period												
Probability	Reservoir	Oct.	Nov.	Dec.	Jan.	Feb.	Mar.	Apr.	May	Jun.	Jul.	Aug.	Sep.	Total
75%	SH	115	45	11	32	2	28	32	50	445	781	618	320	2,479
	HWB	33	55	1	36	10	9	37	56	218	133	157	20	766

Table 4 | Water demands (10^4 m^3)

		Period												
Probability	Reservoir	Oct.	Nov.	Dec.	Jan.	Feb.	Mar.	Apr.	May	Jun.	Jul.	Aug.	Sep.	Total
75%	SH	23	15	18	17	23	26	30	31	523	323	159	33	1,221
	HWB	59	79	17	16	23	27	31	35	550	559	345	77	1,818

Table 5 | E_i and k_i of each month

Period	Oct.	Nov.	Dec.	Jan.	Feb.	Mar.	Apr.	May	Jun.	Jul.	Aug.	Sep.
E_i (mm)	56	42	38	23	28	41	58	86	98	115	110	71
k_i	1.04	1.12	1.12	1.05	0.92	0.9	0.88	0.92	0.94	0.94	0.98	1.06

Note: The above data are provided by the Liuhe Water Authority.

reservoir 1; the HWB reservoir is a recipient reservoir, numbered as reservoir 2. The XZ pumping station, numbered as pumping station 1, diverts water from the BL river to the SH reservoir; the HZ pumping station, numbered as pumping station 2, diverts water from the SH reservoir to the HWB reservoir.

It has been proved that if there is no appropriate joint operation scheme for such a system, water replenishment, water spills, and water shortages may occur simultaneously. Therefore, an optimization method is necessary for such a system to improve the utilization efficiency of inflows as well as to reduce water spills and shortages, considering limited water rights on the river.

In Figure 2, $YS_{1,i}$, $LS_{1,i}$, $X_{1,i}$, and $PS_{1,i}$ are water demand, inflow, water supply, and water spill in the SH reservoir in period i, respectively; $YS_{2,i}$, $LS_{2,i}$, $X_{2,i}$, and $PS_{2,i}$ are water demand, inflow, water supply, and water spill in the HWB reservoir in period i, respectively; $Y_{1,i}$, and $Y_{2,i}$ are water replenishment by the XZ and HZ pumping stations in period i, respectively.

METHODOLOGY

Optimization model

Objective function

The operation cycle of the reservoir in this study is one year, which is divided into 20 periods: the flood season from June to September is divided by 10 days corresponding to 12 periods from 9 to 20, and the rest of the year is divided into monthly periods corresponding to other 8 periods from 1 to 8. The objective function is to minimize the annual sum of squared water shortage of the system; water shortage is the deviation of the actual water supply from the water demand in each period, and pumping stations do not supply water directly to users. The objective function can be expressed as Equation (2):

$$\min F = \sum_{i=1}^{20} [(X_{1,i} - YS_{1,i})^2 + (X_{2,i} - YS_{2,i})^2] \quad (2)$$

where F is the annual sum of squared water shortage of each reservoir in each period and i is the period number ($i = 1, 2, \ldots, 20$).

Constraints

1. Annual available water in the system

The annual available water in the system includes available water of the two reservoirs and pumping volume limited by the water rights of the river. The constraint can be written as Equation (3):

$$\sum_{i=1}^{20} \sum_{j=1}^{2} X_{j,i} \le \sum_{j=1}^{2} SK_j + BZ_1 \quad (3)$$

where j is the reservoir number ($j = 1,2$); SK_1 (10^4 m^3) and SK_2 (10^4 m^3) are annual available water of the SH and

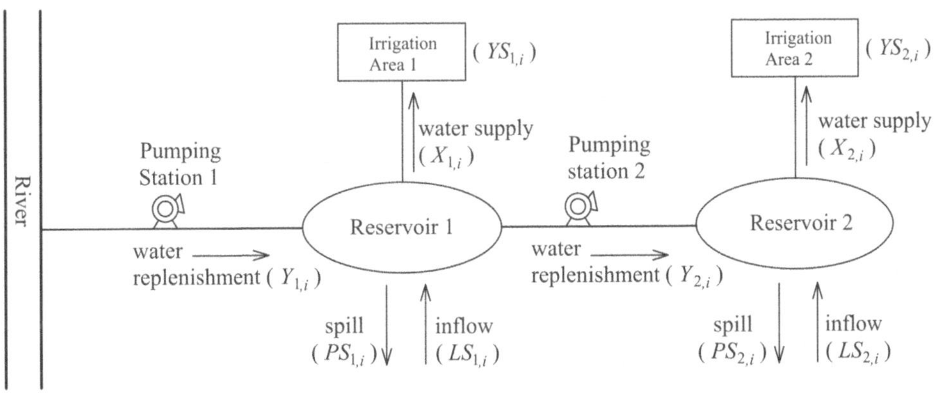

Figure 2 | Generalization of the system.

HWB reservoirs, respectively; and BZ_1 (10^4 m^3) is the maximum annual pumping volume of the XZ pumping station.

2. Maximum annual pumping volume expressed as Equation (4):

$$\sum_{i=1}^{20} Y_{j,i} \leq BZ_j \quad (j = 1, 2) \tag{4}$$

where BZ_2 (10^4 m^3) is the maximum annual pumping volume of the HZ pumping station.

3. Maximum water supply can be expressed as Equation (5):

$$X_{j,i} \leq YS_{j,i} \quad (j = 1, 2) \tag{5}$$

4. Lower and upper bounds of water storage can be expressed as Equation (6):

$$V_{j,i(\min)} \leq V_{j,i} \leq V_{j,i(\max)} \quad (j = 1, 2) \tag{6}$$

According to the water balance principle, for reservoir 1 (SH), the water balance equation can be written as Equation (7):

$$V_{1,i} = V_{1,i-1} + LS_{1,i} + Y_{1,i} - Y_{2,i} - X_{1,i} - PS_{1,i} - EF_{1,i} \tag{7}$$

For reservoir 2 (HWB), the water balance equation can be written as Equation (8):

$$V_{2,i} = V_{2,i-1} + LS_{2,i} + Y_{2,i} - X_{2,i} - PS_{2,i} - EF_{2,i} \tag{8}$$

where $V_{j,i}$ (10^4 m^3) is the water storage of reservoir j in period i; $V_{j,i(\min)}$ and $V_{j,i,(\max)}$ are the lower and upper bounds of water storage

5. Maximum pumping capacity

A pumping station is assumed to pump water at its design operation point. Its maximum pumping capacity in each period is a function of design discharge and maximum operation duration. The constraint can be expressed as Equation (9):

$$Y_{j,i} \leq Q_j \times N_i \times 10^{-4} \quad (j = 1, 2) \tag{9}$$

where Q_1 (m^3/h) and Q_2 (m^3/h) are the design discharges of the XZ and HZ pumping stations and N_i (h) is the maximum operation duration in period i.

6. Initial and boundary conditions

A restriction was imposed to make the final storage $V_{j,20}$ equal to the initial storage $V_{j,0}$. If this limitation is not imposed, the solution would tend to empty the reservoir in the final period.

7. Non-negative constraints

It is also necessary to introduce non-negativity constraints to avoid negative values to the variables, which would be physically impossible.

Solving method

Solving schematic

The above model is a multi-dimensional nonlinear model, which can be divided into multiple stages. There are six types of decision variables, including $X_{1,i}$, $X_{2,i}$, $PS_{1,i}$, $PS_{2,i}$, $Y_{1,i}$, and $Y_{2,i}$, each of which has 20 dimensions. Therefore, it is difficult to solve it directly. As shown in Figure 3, the double-reservoir-and-double-pumping-station system can be decomposed into two subsystems that consist of one reservoir and one pumping station, and hydraulic connections between them are formed through the HZ pumping station. In this manner, the dimensionality of the problem can be reduced.

Solving steps

1. Decomposition of the system

The maximum annual pumping volume of the HZ pumping station, BZ_2, is taken as a coordinating variable, and the model of the double-reservoir-and-double-pumping-station system can be decomposed into two subsystem models that consist of one reservoir and one pumping station as Equations (10)–(13).

HWB reservoir and HZ pumping station (subsystem 2)

$$\min f_2 = \sum_{i=1}^{T} (X_{2,i} - YS_{2,i})^2 \tag{10}$$

$$\sum_{i=1}^{T} X_{2,i} \leq W_2 \tag{11}$$

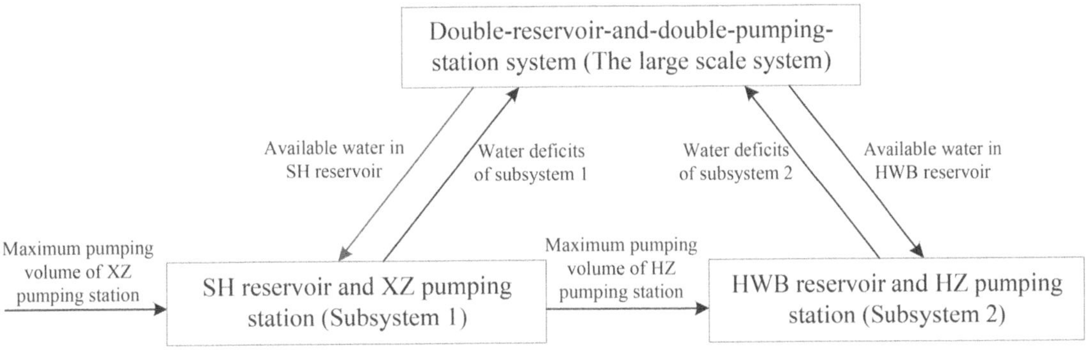

Figure 3 | Decomposition and aggregation of the system.

SH reservoir and XZ pumping station (subsystem 1)

$$\min f_1 = \sum_{i=1}^{T} (X_{1,i} - YS'_{1,i})^2 \tag{12}$$

$$\sum_{i=1}^{T} X_{1,i} \leq W_1 \tag{13}$$

where f_1 is the annual sum of squared water shortage in each period of subsystem 1; f_2 is the annual sum of squared water shortage in each period of subsystem 2; W_1 (10^4 m^3) is the annual available water of subsystem 1; and W_2 (10^4 m^3) is the annual available water of subsystem 2.

The lower and upper bounds of water storage, water balance equation, maximum pumping capacity, maximum annual pumping volume, initial and boundary conditions, and non-negative constraints should also be imposed on the subsystems.

2. Optimization of subsystems

The subsystem models are two nonlinear models with only one coupling constraint. Therefore, one-dimensional dynamic programming (DP) can be used to solve them, taking water supply X_i as a decision variable. The normal operation rule is integrated into recursive procedures of DP (Shi *et al.* 2015; Gong *et al.* 2019) to correct water storage and obtain water spill PS_i and water replenishment Y_i of each period simultaneously; in this manner, the defect of meta-heuristic algorithms, which leads to random occurrences of spills or replenishments against the normal operation rule, can be avoided. The specific procedure is as follows.

- If $V_i < V_{i,\min}$, then the pumping station should replenish the reservoir and the final water storage should be $V_{i,\min}$. The water replenishment and spill in period i are expressed as Equations (14) and (15).

  Water replenishment:

$$Y_i = \min(V_{i,\min} - V_i, Q \times N_i) \tag{14}$$

  Water spill:

$$PS_i = 0 \tag{15}$$

- If $V_i > V_{i,\max}$, then excess water should be drained through the spillway. The water replenishment and spill in period i are expressed as Equations (16) and (17).

  Water replenishment:

$$Y_i = 0 \tag{16}$$

  Water spill:

$$PS_i = V_i - V_{i,\max} \tag{17}$$

- If $V_{i,\min} < V_i < V_{i,\max}$, then the water replenishment and spill in period i should both be zero as Equation (18):

$$Y_i = PS_i = 0 \tag{18}$$

Finally, the corrected water storage in period i is expressed as Equation (19):

$$V'_i = V_i + Y_i - PS_i \tag{19}$$

A flow chart of optimizing the subsystems is shown in Figure 4.

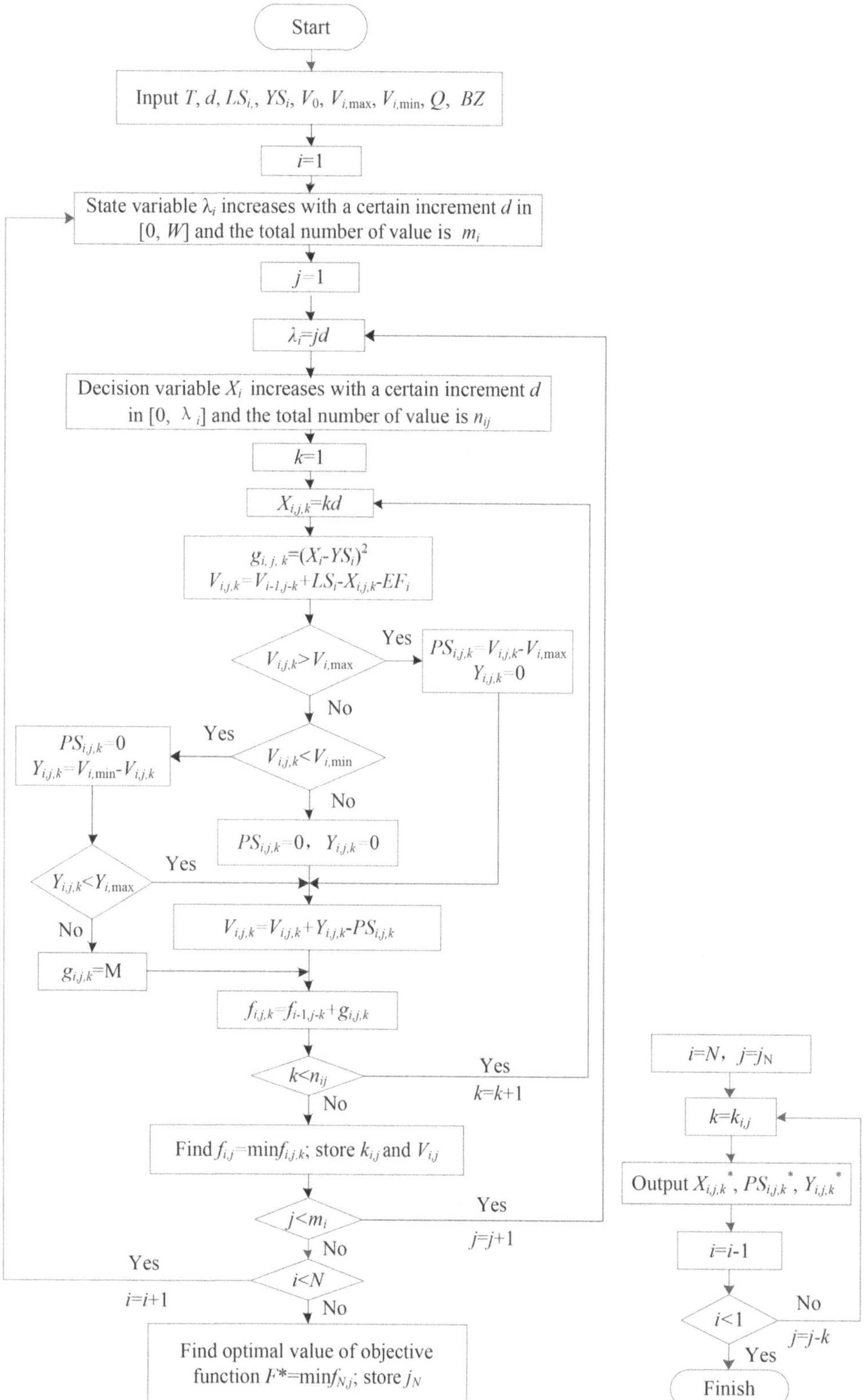

Figure 4 | Flow chart of optimizing subsystems.

Given a list of discrete values of BZ_2 with a certain increment in the feasible region, discrete values of W_1 and W_2 can be determined and substituted into the two subsystem models. DP is used to derive a list of $f_2 \sim W_2$ and $f_1 \sim W_1$ relationships and a list of corresponding solutions, $[X_{1,i}, PS_{1,i}, Y_{1,i}]$ and $[X_{2,i}, PS_{2,i}, Y_{2,i}]$.

3. Aggregation of the system

According to the relationships between the objective function and annual available water of subsystem, $f_2 \sim W_2$ and $f_1 \sim W_1$, an aggregation model can be derived as Equations (20) and (21).

Objective function:

$$\min F = f_1(W_1) + f_2(W_2) \tag{20}$$

Constraints:

$$\sum_{j=1}^{2} W_j \leq \sum_{j=1}^{2} SK_j + BZ_1 \tag{21}$$

The aggregation model can also be solved by DP, with the annual available water of subsystems W_j as decision variables. After optimal results of the aggregation model F^*, W_1^*, and W_2^* are derived, the final optimal operation scheme $[X_{1,i}, PS_{1,i}, Y_{1,i}, X_{2,i}, PS_{2,i}, Y_{2,i}]^*$ of the double-reservoir-and-double-pumping-station system can be obtained by searching optimal results of subsystems corresponding to W_1^* and W_2^*.

RESULTS ANALYSIS

As shown in Figure 5(a), the objective function value F decreases with increasing BZ_2 in the range of [1,100, 1,260] and begins to converge when BZ_2 exceeds 1,220 (10^4 m^3). This means that the benefits of increasing the water transfer volume of the HZ pumping station are no longer remarkable from this point.

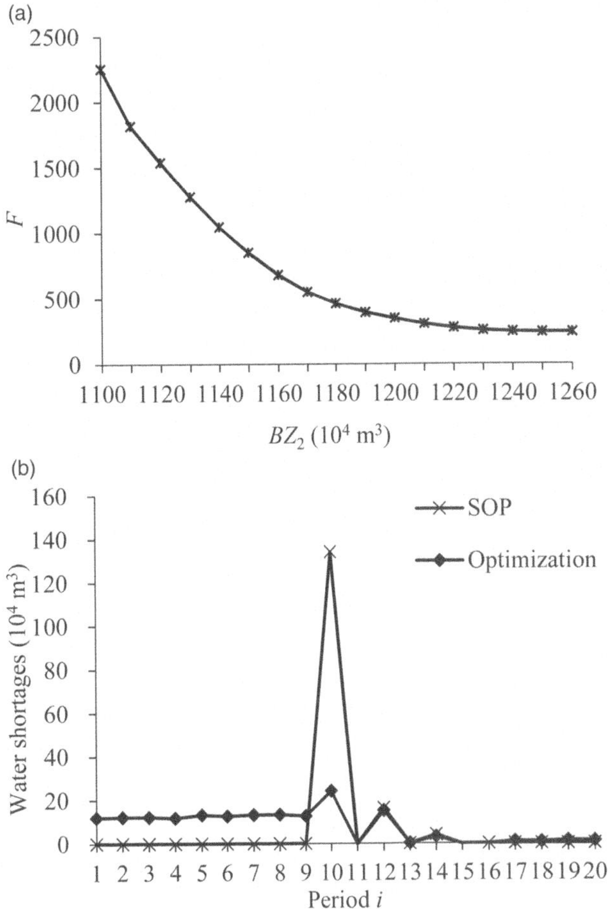

Figure 5 | Optimization results: (a) $F \sim BZ_2$ curve and (b) water shortages of the HWB reservoir.

Finally, the minimum value of F is 241, and the optimal annual pumping volume of the HZ pumping station is 1,245 (10^4 m^3); the corresponding optimal operation results are shown in Table 6. Compared with the results using the

Table 6 | Operation results of the system (10^4 m^3)

Method	Reservoir	Water supply	Water spill	Water replenishment	Water shortage	Evaporation
SOP	SH	1,220	241	360	0	278
	HWB	1,664	0	1,100	154	202
Optimization method	SH	1,220	95	352	0	272
	HWB	1,799	0	1,245	19	211

standard operation policy (SOP), the total water shortage is reduced by 87.7%, total water replenishment from the river is reduced by 2.2%, and total water spill is reduced by 60.6% at the 75% probability of exceedance.

The optimal operation redistributes water resources between two reservoirs (Table 6). Furthermore, the optimal operation not only reduces water shortages of the HWB reservoir but also adjusts its water shortage distribution. As shown in Figure 5(b), the water shortage of the HWB reservoir using SOP reaches up to 1.34 MCM in June, accounting for 87% of the annual water shortage. This shortage can be attributed to the mass water consumption during the steeping period of paddy fields. After optimization, although the frequency of water shortage increases, any severe drought in a period could be avoided. In addition, the optimal operation also changes the evaporation of the system, corresponding to changes in the water storage during each period (Table 6). Nevertheless, the impacts of the changes on the local ecological environment can be ignored because the variation range is within 5%.

Operation curves of reservoirs derived using the optimization method and SOP are shown in Figure 6(a) and 6(b), respectively. As a donor reservoir, water storage of the SH reservoir before the flood season was lower with the optimization than with the SOP, which could help retain more flood water and reduce water spill from 2.41 MCM to 0.95 MCM. In contrast, as a recipient reservoir, the optimal water storage of the SH reservoir before the flood season was higher with the optimization than with the SOP, which could increase available water in the reservoir and avoid water shortages caused by the limited pumping capacity of the HZ pumping station during the peak period of irrigation.

According to the differences in the operation curves of the two reservoirs, operation modes of the two pumping stations in the system (XZ and HZ) were also adjusted. As a replenishment pumping station for the SH reservoir, water pumping patterns using different methods are shown in Figure 6(c). The frequency of water pumping at the XZ station did not change after optimization, but the optimal operation reduced the water pumping volume at the first stage, which was beneficial for relieving the pressure of flood control in the SH reservoir. As a water transfer pumping station for the SH reservoir as well as a replenishment pumping station for the HWB reservoir, different water pumping patterns of HZ station are shown in Figure 6(d).

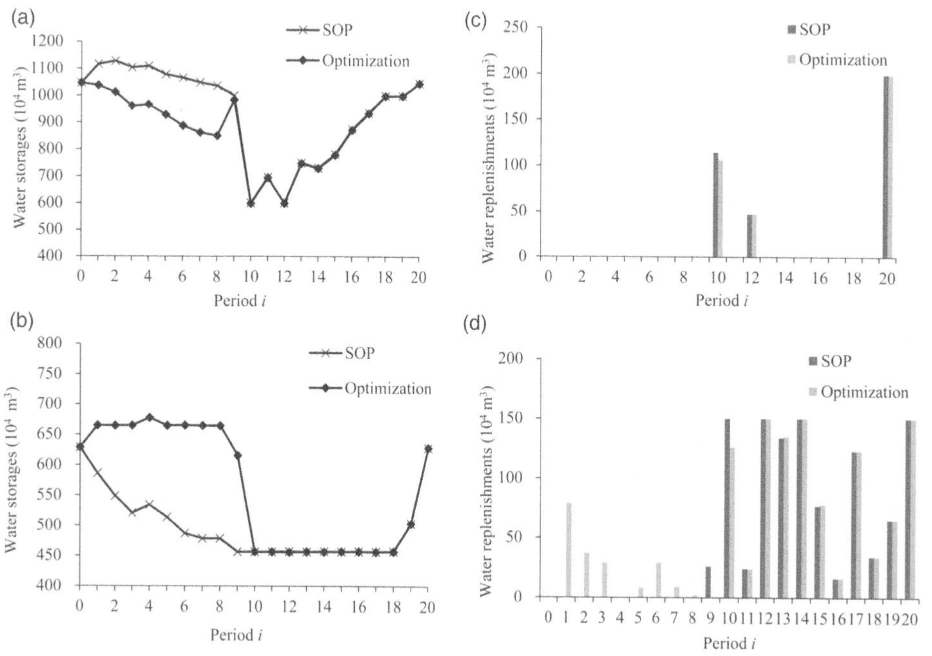

Figure 6 | Operation curves of reservoirs and pumping stations in the system: (a) water storages of the SH reservoir, (b) water storages of the HWB reservoir, (c) water replenishment of the XZ pumping station and (d) water replenishment of the HZ pumping station.

Under the SOP, water was transferred only from June to September, the peak period of irrigation. As a result, the pumping station was fully loaded, but it was still difficult to meet the water demand because of the limited maximum pumping capacity. After optimization, the HZ pumping station transferred 15.5% of the annual water pumping volume before the peak period; this could help the SH reservoir retain more flood water as well as increase available water in the HWB reservoir for irrigation.

In conclusion, the most appropriate approach for the optimal joint operation of the double-reservoir-and-double-pumping-station system is to form a joint operation policy. This would enable full play of the functions of all components of the system, making use of surplus water of the SH reservoir to appropriately compensate water shortages of the HWB reservoir. In this manner, the goal of simultaneously reducing water spills and water shortages can be achieved.

DISCUSSION

At present, meta-heuristic algorithms have become mainstream methods to solve optimal operation models of water resources systems (Bozorgi *et al.* 2017; Mansouri *et al.* 2017). Meta-heuristic algorithms, for instance, genetic algorithm (GA), particle swarm optimization (PSO), and ant colony algorithm (ACO), firstly generate initial solutions and modify them in each iteration according to certain rules until meeting the convergence (Bayat *et al.* 2011; Chang *et al.* 2013; Ehteram *et al.* 2017). However, they cannot guarantee globally optimal results due to the ineluctable random sampling during the iteration process. Comparatively, dynamic programing based on Bellman's principle is adopted to solve the sub models and the aggregation model after the decomposition and aggregation of the system which can ensure the global optimal solution.

In addition, the solving results of meta-heuristic algorithms may depend on the value of some algorithm parameters (Allawi *et al.* 2018; Ehteram *et al.* 2018), such as crossover rate and mutation rate in GA and inertia weight and acceleration factor in PSO. Therefore, it is necessary to carry out sensitivity analysis on these parameters for the final results. However, there is no need to input any algorithm parameter when using DP which can make the solving procedure more simple and reliable (Kennedy 1986; Peng *et al.* 2018).

These findings provide a reference for the optimal operation of similar systems that consist of reservoirs and pumping stations in hilly regions of southern China, northern Vietnam, Myanmar, and Laos. The mean annual rainfall is over 1,000 mm in these regions, but 70–90% of that occurs in the flood season due to the monsoon climate, resulting in seasonal water shortages (Wang *et al.* 2017; Quinn *et al.* 2018). What is worse, in recent years, agricultural irrigation, as a traditional water user, has been greatly squeezed because of the rapid development of industry in south China and southeast Asia (Zhang *et al.* 2018). However, it is difficult for the original empirical operation method to handle these challenges. Under the limited agricultural water rights, the optimization model proposed in this paper can effectively adapt to the changes of water resources and alleviate seasonal water shortages by inflow regulation. In future research, the model should consider the stochastic nature of both inflows and water demands for real-time system optimization.

CONCLUSION

An optimization model of a double-reservoir-and-double-pumping-station system was developed in this study to minimize water shortages. One such system in Nanjing, China, was selected for a case study, and the constraint of total available water in the system was coupled into it, considering regional water rights. The system was decomposed into two subsystems based on the decomposition-aggregation theory, and DP was adopted to solve the sub models and the aggregation model. A joint operation rule of the system is integrated into the optimization model to obtain an appropriate operation scheme of both reservoirs and pumping stations.

ACKNOWLEDGEMENTS

This work was supported by the National Key Research and Development Program of China [grant number

2017YFC0403205] and the Postgraduate Research & Practice Innovation Program of Jiangsu Province [grant number KYCX19_2105]. The authors wish to acknowledge the managers of Liuhe Water Authority, Nanjing, China, for their support of providing reservoir and irrigation data.

REFERENCES

Ahmad, A., El-Shafie, A., Razali, S. F. M. & Mohamad, Z. S. 2014 Reservoir optimization in water resources: a review. *Water Resources Management* **28** (11), 3391–3405.

Allawi, M. F., Jaafar, O., Ehteram, M., Hamzah, F. M. & El-Shafie, A. 2018 Synchronizing artificial intelligence models for operating the dam and reservoir system. *Water Resources Management* **32** (10), 3373–3389.

Bayat, B., Mousavi, S. J. & Namin, M. M. 2011 Optimization–simulation for short-term reservoir operation under flooding conditions. *Journal of Water Supply: Research and Technology-AQUA* **60** (7), 434–447.

Birhanu, K., Alamirew, T., Olumana Dinka, M., Ayalew, S. & Aklog, D. 2014 Optimizing reservoir operation policy using chance constraint nonlinear programming for Koga irrigation dam, Ethiopia. *Water Resources Management* **28** (14), 4957–4970.

Bozorgi, A., Bozorg-Haddad, O., Rajabi, M., Latifi, M. & Chu, X. 2017 Applications of the anarchic society optimization (ASO) algorithm for optimizing operations of single and continuous multi-reservoir systems. *Journal of Water Supply: Research and Technology-AQUA* **66** (7), 556–573.

Celeste, A. B. & Billib, M. 2009 The role of spill and evaporation in reservoir optimization models. *Water Resources Management* **24** (4), 617–628.

Chang, J., Bai, T., Huang, Q. & Yang, D. 2013 Optimization of water resources utilization by PSO-GA. *Water Resources Management* **27** (10), 3525–3540.

Chen, D., Leon, A. S., Gibson, N. L. & Hosseini, P. 2016 Dimension reduction of decision variables for multireservoir operation: a spectral optimization model. *Water Resources Research* **52** (1), 36–51.

Đurin, B. 2016 Some aspects of the operation work of pump station and water reservoir. *Periodica Polytechnica Civil Engineering* **60** (3), 345–353.

Ehteram, M., Karami, H., Mousavi, S., Farzin, S. & Kisi, O. 2017 Evaluation of contemporary evolutionary algorithms for optimization in reservoir operation and water supply. *Journal of Water Supply: Research and Technology-AQUA* **67** (1), 54–67.

Ehteram, M., Karami, H. & Farzin, S. 2018 Reducing irrigation deficiencies based optimizing model for multi-reservoir systems utilizing spider monkey algorithm. *Water Resources Management* **32** (7), 2315–2334.

Gong, Y. & Cheng, J. L. 2018 Optimization of cascade pumping stations' operations based on head decomposition-dynamic programming aggregation method considering water level requirements. *Journal of Water Resources Planning and Management* **144** (7), 253–264.

Gong, Z. H., Cheng, J. L., Yang, S. T., Chang, B. C. & Jiang, Y. 2019 山湖水库"一库两站"系统联合运行优化调度方法研究 (Research on the approach of water resources optimal allocation to the combined operation of Shanhu reservoir and its two pumping stations). *Journal of Drainage and Irrigation Machinery Engineering* **37** (2), 124–129 (in Chinese).

Gu, W., Shao, D., Tan, X., Shu, C. & Wu, Z. 2017 Simulation and optimization of multi-reservoir operation in inter-basin water transfer system. *Water Resources Management* **31** (11), 3401–3412.

Heikkila, T. 2015 Coordination in water resource management: the impact of water rights institutions. *Water Policy* **5** (4), 331–348.

Jothiprakash, V., Shanthi, G. & Arunkumar, R. 2011 Development of operational policy for a multi-reservoir system in India using genetic algorithm. *Water Resources Management* **25** (10), 2405–2423.

Kennedy, J. O. S. 1986 *Dynamic Programming: Applications to Agriculture and Natural Resources*, 1st edn. Elsevier, New York, USA.

Mahey, P., Koko, J. & Lenoir, A. 2017 Decomposition methods for a spatial model for long-term energy pricing problem. *Mathematical Methods of Operations Research* **85** (1), 137–153.

Mansouri, A., Aminnejad, B. & Ahmadi, H. 2017 Introducing modified version of penguins search optimization algorithm (PeSOA) and its application in optimal operation of reservoir systems. *Water Science and Technology: Water Supply* **18** (4), 1484–1496.

Molle, F. 2004 Defining water rights: by prescription or negotiation. *Water Policy* **6** (3), 207–227.

Nabinejad, S., Jamshid Mousavi, S. & Kim, J. H. 2017 Sustainable basin-scale water allocation with hydrologic state-dependent multi-reservoir operation rules. *Water Resources Management* **31** (11), 3507–3526.

Peng, Y., Zhang, X., Xu, W., Shi, Y. & Zhang, Z. 2018 An optimal algorithm for cascaded reservoir operation by combining the grey forecasting model with DDDP. *Water Science and Technology: Water Supply* **18** (1), 142–150.

Pulido-Calvo, I. & Gutiérrez-Estrada, J. C. 2011 Selection and operation of pumping stations of water distribution systems. *Environmental Research Journal* **5** (3), 1–20.

Quinn, J. D., Reed, P. M., Giuliani, M., Castelletti, A., Oyler, J. W. & Nicholas, R. E. 2018 Exploring how changing monsoonal dynamics and human pressures challenge multireservoir management for flood protection, hydropower production, and agricultural water supply. *Water Resources Research* **54**, 4638–4662.

Rani, D. & Moreira, M. M. 2009 Simulation-optimization modeling: a survey and potential application in reservoir systems operation. *Water Resources Management* **24** (6), 1107–1138.

Rani, D. & Srivastava, D. K. 2015 Optimal operation of Mula reservoir with combined use of dynamic programming and genetic algorithm. *Sustainable Water Resources Management* **2** (1), 1–12.

Reca, J., García-Manzano, A. & Martínez, J. 2014 Optimal pumping scheduling for complex irrigation water distribution systems. *Journal of Water Resources Planning and Management* **140** (5), 630–637.

Reca, J., García-Manzano, A. & Martínez, J. 2015 Optimal pumping scheduling model considering reservoir evaporation. *Agricultural Water Management* **148**, 250–257.

Shen, D. & Wu, J. 2017 State of the art review: water pricing reform in China. *International Journal of Water Resources Development* **33** (2), 198–232.

Shi, Z. T., Cheng, J. L., Yang, S. T., Chang, B. C. & Jiang, Y. 2015 基于 DPSA算法的"单库-单站" 水资源优化调度方法研究 (Research on methods of optimal water resources allocation for single reservoir single pumping station system using DPSA). *Journal of Irrigation and Drainage* **34** (2), 37–40 (in Chinese).

Sun, T., Wang, J., Huang, Q. & Li, Y. 2016 Assessment of water rights and irrigation pricing reforms in Heihe river basin in China. *Water* **8**, 333. doi:10.3390/w8080333.

Wang, L. K., Yang, C. T. & Wang, M. H. S. 2016 *Advances in Water Resources Management*. Springer, New York.

Wang, Z., Zhou, X., Liu, Y., Zhou, D., Zhang, H. & Sun, W. 2017 Precipitable water vapor characterization in the coastal regions of China based on ground-based GPS. *Advances in Space Research* **60** (11), 2368–2378.

Yu, G., Powell, R. S. & Sterling, M. J. H. 1994 Optimized pumping scheduling in water distribution systems. *Journal of Optimization and Application* **83** (3), 463–488.

Zhang, W., Mao, H., Yin, H. & Guo, X. 2018 A pricing model for water rights trading between agricultural and industrial water users in China. *Journal of Water Supply: Research and Technology-AQUA* **67** (4), 347–356.

First received 22 June 2019; accepted in revised form 20 September 2019. Available online 17 October 2019

Economic and environmental impact analysis of ammoniacal nitrogen removal from landfill leachate using sequencing batch reactor: a case study from Czech Republic

R. Badri Narayan, B. I. Zargham, Audrey Ngambia and Arlieza R. Riyanto

ABSTRACT

Lany landfill, located in the Czech Republic, generates around 10 m^3/d of leachate, heavily polluted with ammoniacal nitrogen, heavy metals and salts, which needs to be treated onsite in order to minimize their effect on the ecosystem and on human health. A sequencing batch reactor (SBR) was designed for ammoniacal nitrogen removal from the leachate and the economic feasibility and environmental impact of the designed SBR was assessed. From the cost-benefit analysis, capital expenditure (CAPEX) of 33,500 €/year and operational expenditure (OPEX) of 13,521 €/year were estimated. A shadow price concept tool was used to calculate the environmental benefit as 21,000 €/year. The net present value of the project was evaluated to be 19,528 € with an internal rate of return of 21.6%. For environmental assessment, triple bottom line (TBL) analysis on the existing practice of discharging the leachate to a nearby wastewater treatment plant and on-site treatment using SBR was performed. The total score for existing practice was calculated to be 55.1% while for on-site treatment it was 59.6%. Based on the results it was concluded that on-site treatment is both economically and environmentally feasible. A mitigation plan was also prepared for the impacts identified in the environmental assessment.

Key words | cost-benefit analysis, reverse osmosis, sequencing batch reactor, shadow price, triple bottom line

R. Badri Narayan (corresponding author)
B. I. Zargham
Audrey Ngambia
Arlieza R. Riyanto
Department of Water Technology and Environmental Engineering,
University of Chemistry and Technology,
Technická 5, Dejvice, 166 28, Praha 6,
Czech Republic
E-mail: rbn21096@gmail.com

INTRODUCTION

Landfill leachate is defined as an aqueous liquid generated as a consequence of the percolation of precipitation (rain, snow melt) and infiltration of groundwater through wastes, biochemical processes in waste's cells and the inherent water content of wastes themselves (Joshi & Gogate 2019). Leachate is a complex wastewater comprising of a large number of different substances or compounds, mainly organic matter, ammoniacal nitrogen, heavy metals and other dissolved solids. It is estimated that from about a 907 tonne of municipal solid waste, approximately 0.05–0.2 tonnes of leachate can be generated during the whole operational lifetime of the landfill (Kurniawan & Lo 2009; Souza *et al.* 2014). However, the volume of the leachate generated depends on the composition of the waste, age and size of the landfill, the compaction of waste in the landfill, geology of the landfill site and the weather conditions. The components of the leachate are highly polluting substances or compounds with high potential to cause adverse effects on human health and the environment if they not properly managed (Iskander *et al.* 2018). It is therefore essential to properly manage the leachate generated from the landfill and to treat it in order to

doi: 10.2166/aqua.2019.084

minimize the effect of highly polluting contaminants on the ecosystem and on human health.

The landfill considered in this case study is located at a municipality in the Czech Republic named Lany. Lany landfill receives approximately 8,000 tonne/yr of mixed municipal solid waste from its municipality. The landfill has been operating since 1993 and produces approximately 10 m^3/d of leachate which is highly polluted with organic load, ammoniacal nitrogen, heavy metals, and salts. The landfill currently sends the leachate for treatment to a nearby small municipal wastewater treatment plant (WWTP), which is considered to be an expensive option. However, it is also expected that in the near future, legislation will become more stringent, which may prohibit the treatment of leachate in that particular WWTP. As a result, the landfill installed a reverse osmosis (RO) plant to treat the leachate onsite. However, the existing RO plant seems to have a low ammoniacal nitrogen removal efficiency. Hence, leachate pre-treatment for ammoniacal nitrogen removal before the RO unit is indispensable. This would make sure that the treated leachate will comply with the legislative discharge limit of ammoniacal nitrogen into surface waters. Therefore, the authors have restricted the scope of nutrient removal in this study only with respect to ammoniacal nitrogen, as the other nutrients are considered to be removed efficiently by the RO plant onsite.

Physical–chemical treatment methods are mainly used either as pre-treatment or post-treatment methods to treat landfill leachate. They are employed along with a biological treatment process in order to enhance its efficiency since they have the potential to remove bio-refractory and undesirable compounds such as heavy metals, humic acid, fulvic acid, adsorbable organic halogens (AOXs), polychlorinated biphenyls (PCBs), etc. Different processes such as air stripping, chemical precipitation, and membrane technology have been studied for the removal of ammoniacal nitrogen from leachate (Abbas *et al.* 2009). However, biological treatment was given preference because of the major disadvantages of the physico-chemical processes as mentioned below.

- In air stripping, there is a problem of generation of contaminated gases which requires further treatment with sulphuric and hydrochloric acid, hence making this option unfeasible (De *et al.* 2019).
- Ammoniacal nitrogen can also be removed by precipitating leachate with magnesium ammonium phosphate (MAP). The problem associated with chemical precipitation is that it requires a large amount of chemicals as it takes place at a particular pH range (8.5–9.0) (Chu *et al.* 2018), thus resulting in large sludge production.
- The highest removal efficiency of ammoniacal nitrogen was observed by membrane technology but the drawbacks for this technology are that membranes are susceptible to fouling, it requires treatment of membrane concentrate (Truong *et al.* 2018) and the operation of membrane reactor is very complex, thus requiring a skilled operator.

Sequencing batch reactor (SBR) is a recognized biological treatment strategy with abundant reports in the literature of its successful application to landfill leachate treatment (Gao *et al.* 2015; Torretta *et al.* 2016; Yong *et al.* 2018). It offers various benefits such as suitability to treat low amounts of influent, versatility and robustness to adapt to the pre-visible variations in leachate characteristics over time. If parameters in influent change, process conditions can be modified accordingly, i.e. fill volume, filling rate, aerobic reaction time, anoxic reaction time, sludge age, sludge concentration, carbon source type and amount, among others (Pelaz *et al.* 2018). Considering all the above advantages of SBR, it was hence chosen as the feasible alternative to treat the leachate in this work.

Preliminary studies on the economics and environmental impacts of a project are important to obtain an insight on the feasibility of its application on a larger scale. For economic assessment, cost-benefit analysis (CBA) with environmental impact valuation is a result of the incorporation of environmental criteria in the decision-making process. The EU Water Framework Directive (WFD) provides a very important role to economic analysis and requires that CBA should be carried out with a goal of involving all the benefits, including those whose value is not determined by the market but have a high value because they uniquely contribute to improving societal welfare (Helming & Reinhard 2009). The CBA is a tool which is used to compare the economic feasibility of the implementation of different projects. CBA revolves around a simple concept, that a project should only be commissioned if the sum of all the benefits is greater than the sum of aggregate costs (Molinos-Senante *et al.* 2011). As regards the water resources context, it is known that wastewater treatment has significant environmental benefits. The environmental benefits of wastewater treatment refer to the elimination or removal of pollutants such as suspended solids, chemical oxygen demand (COD), nitrogen, phosphorus, etc., from wastewater. However, most of these environmental benefits are not monetarily quantified because they do not have any market value. They are termed as positive externalities in economics. In spite of this, the monetary evaluation of these benefits is quintessential to prove the economic feasibility of wastewater treatment projects using CBA (Molinos-Senante *et al.* 2010).

In this view, the economic assessment of these externalities in the field of water treatment and reuse is gaining particular importance. Godfrey *et al.* (2009) worked on the application of CBA on a greywater reuse facility in India. Likewise, Seguí *et al.* (2009) showed that a travel cost method could be used to evaluate the environmental benefits resulting from wastewater reuse in a wetland restoration project. Chen & Wang (2009) used a net benefit value model for cost-benefit evaluation of reuse projects in a residential area of China. Molinos-Senante *et al.* (2011) studied the CBA of water reuse projects in wastewater treatment plants in Spain.

Project implementation can have several environmental impacts. The concept of sustainability assessment has mainly resulted from the works carried out by technicians of Environmental Impact Assessment (EIA), and Strategic Environmental Assessment (SEA), which in turn has been manipulated by techniques related to policy making (Pope *et al.* 2004). It was suggested by the EIA and SEA practitioners that the environmental assessment could contribute to sustainability by extending its scope to include social and economic considerations along with environmental ones. This refers to a three pillar method or triple bottom line (TBL) concept. The TBL approach considers the three pillars of sustainability, namely, social, economic and environmental aspects, with equal importance in decision making principle (Di Maria & Sisani 2019). The main highlight of the TBL approach is that it emphasises the fact that material gains are not sufficient measures or preservers of well-being of human society and the environment. Although TBL is a sustainable concept, it has certain drawbacks as it promotes potentially competing interests between the three pillars of sustainability, rather than the bondage and interrelationship between them (Plakas *et al.* 2016). Hence, the task of integration becomes cumbersome and endorsing trade-offs results at the expense of the environment. Environmental quality deterioration can occur during different project phases, thus assessment of potential impacts is required. Through environmental impact identification, the feasibility of this project could be assessed.

This work is a case study on a landfill site in Lany municipality of the Czech Republic. The specific objectives of this work are: (i) to design SBR process for removal of ammoniacal nitrogen from the leachate; (ii) to perform economic assessment on the proposed design of SBR using CBA; (iii) to evaluate the economic performance of the project by estimating cash flow, net present value and internal return rate; (iv) to compare and assess the impact of existing practice and on-site treatment on various aspects (environmental, economic, and community); and (v) to prepare a mitigation plan for the impacts observed during the environmental assessment.

MATERIALS AND METHODS

Process design

The process has been designed to treat the landfill leachate having the characteristics shown in Table 1. The proposed process was designed based on the previous works conducted by Neczaj *et al.* (2005) for the treatment of landfill leachate having characteristics similar to those mentioned in this work. Process conditions were set for this design at 10% leachate dilution in the reactor, 4,000 mgMLSS/L, 0.8 mgVSS/mgSS, 10 days sludge age. With these conditions, preliminary calculations were performed to develop the conceptual design for this process. A scheme for SBR treatment process of landfill leachate is presented in Figure 1.

With a leachate production of 10 m^3/d, the required total volume in the reactor was calculated to be 50 m^3 as shown in Table S1 (Supplementary Material). In order to handle two cycles per day, 5 m^3 of leachate at a dilution of

Table 1 | Average composition of the leachate to be treated

Parameter	Concentration (mg/L)
pH	7.6
BOD	150
COD	2,786
$N\text{-}NH_4$	800
TIC	780
TSS	100
TDS	13,000
SO_4^{2-}	70
NO_3^-	6
Cl^-	2,200
Ba	1.8
Ca	110
Cr	1.5
Cu	0.48
Fe	3
K	1,200
Mg	190
Na	2,050
Pb	0.03
Zn	3.7
Conductivity (microS/cm)	14,000

10% can be treated in each cycle. The number of cycles to be carried out per day was considered to be two in order to avoid the need for a night shift labour. The required dimensions for the SBR tank were estimated to be around 3.5 m in diameter and 6 m in height (Kong *et al.* 2009). For a robust design, the cycle time was estimated at 8 h, divided as: 0.5 h aerobic fill, 4 h aerobic reaction, 2 h anoxic reaction, 0.5 h settle time and 0.5 h draw time (Neczaj *et al.* 2005). Considering the aforementioned conditions, required nitrification and denitrification rates were evaluated for this particular situation. The detailed process calculations are presented in Table S1. Nevertheless, if parameters change, some process conditions can be modified accordingly. In case of concern about nitrifiers' retention in the reactor given the low organic load available for floc forming microorganisms in the aerobic phase, bio-carriers could be added to favour biofilm growth. Given the low organic load present in the leachate, the addition of an external carbon source was required. Methanol was chosen because of its high availability, high COD, low yield and price. Considering the typical COD requirement for denitrification as 4.5 $mgCOD/mgN\text{-}NO_3$ (von Sperling 2007), the TOD of methanol as 1.5 $mgO_2/mgCH_3OH$, and the estimated amount of $N\text{-}NO_3$ to be removed, the amount of methanol required was estimated using Equation (1). Methanol consumed per day was estimated to be about 28 L/d.

$$\text{Estimated methanol required for denitrification} = \frac{\text{amount of nitrates to be denitrified} * 4.5\text{ mg COD/mgN} - \text{nitrate}}{1.5\text{ mg } O_2/\text{mg methanol}} \quad (1)$$

The optimum pH range for the nitrification–denitrification process to occur is 7.0–8.3 (Raper *et al.* 2018). Nitrification consumes alkalinity which can lead to a drop in the pH level if the buffer capacity of the wastewater is not enough.

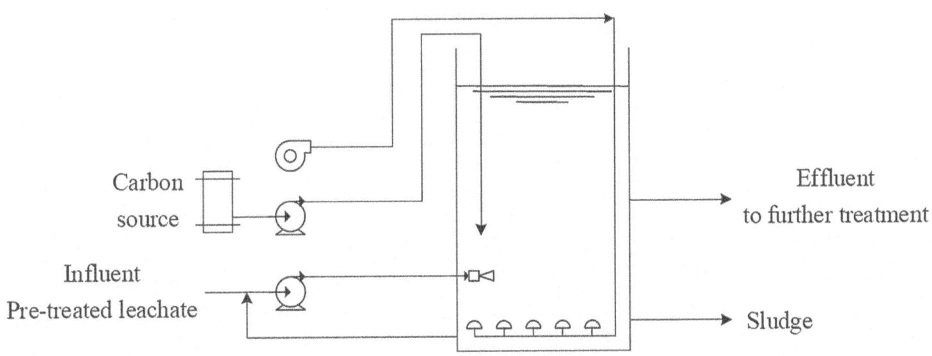

Figure 1 | Schematic of biological leachate treatment using SBR.

The buffer capacity of this leachate could hence be adjusted as required by adding a suitable form of alkalinity such as $CaCO_3$ (Raper *et al.* 2018). Thus, pH can be brought back to an optimal level for the denitrification process to take place. The tank was filled by using a centrifugal pump. Flow capacity was estimated at 15 m^3/h and pump head was estimated at 7 m.

Aeration will be required to provide oxygen and agitation during the aerobic phase. Considering typical requirements, 1 $mgO_2/mgBOD$ and 4.3 $mgO_2/mgNH_3$ (von Sperling 2007), and typical values for aeration parameters air requirement was estimated to be 100 Nm^3/h at a pressure of around 600 mbar, which is achievable by means of a regenerative blower. Agitation would be required during the anoxic phase. This could be achieved using a mechanical suspended mixer or submersible mixer. To keep costs low in this case, it was achieved by recirculating mixed liquor through a jet nozzle using the same pump used for filling the reactor. A smaller pump was required to add the carbon source for denitrification and some additional electricity was used by the control system and eventual ancillary equipment. Electricity consumption was estimated as 1 kW during half of the cycle to maintain a conservative approach. Sludge will have to be dewatered to 50% solids content using a filter press and then be sent for incineration.

Several reasonable assumptions were made in this conceptual design and the results were compared with typical ranges to check the biological feasibility of the process. Nevertheless, the limitations of this conceptual design are acknowledged and the validation of process parameters and assumptions would be required as an early stage of the project execution. This will require tests that are out of the scope of this work.

Economic assessment

The following assumptions/considerations were made while performing the economic assessment:

- The whole estimation was based on the consideration of an existing RO pilot plant operating in the landfill.
- Environmental benefits were based on ammoniacal nitrogen removal (Molinos-Senante *et al.* 2010).
- Capital costs were estimated from catalogues of wastewater treatment plants.
- No new land was required to construct the SBR tank as the landfill already has enough space.
- Labour cost was not accounted for in the calculation of operation expenditure as there are already operators involved in the existing RO plant and this process will not require additional labour.
- The total period of operation of SBR per year was taken as 350 days, considering 15 days of maintenance.
- The time horizon chosen for this assessment was 10 years to obtain a short conservative evaluation.
- The inflation rate of 3% and the discount rate of 10% was assumed to be constant throughout the period of 10 years (Molinos-Senante *et al.* 2012).

To perform this analysis, the net profit at each year was estimated in a simplified way by estimating capital costs, operation costs and environmental benefits of removing nitrogen from the effluent. Capital costs were split into eight sub items (tank, blower and pumps, diffusers, piping, electricity and control, engineering, construction and sludge dewatering). Operation costs were divided into six different categories (energy, staff, reagents, waste management, maintenance, and others) and estimated using the amounts required according to the conceptual design and market prices obtained from reliable sources. The annualised equivalent cost was calculated to illustrate and quickly compare against the benefits. For the environmental benefit, the shadow price of removing nitrogen from the effluent was taken as 8 €/kg of nitrogen (Molinos-Senante *et al.* 2010). To evaluate the economic performance of the proposed process, the net present value and internal return rate were calculated from Equations (2)–(4):

$$NP = \Sigma B_i - \Sigma C_i \tag{2}$$

where NP is the net profit; B_i is the value of the benefit item i and C_i is the value of the cost item i.

$$NPV = \sum_{t=0}^{T} \frac{NP_t}{(1+r)^t} \tag{3}$$

where NPV is the net present value; NP_t is the net profit at time t; r is the discount rate and t is the time horizon of the project.

$$IRR = NPV = \sum_{t=0}^{T} \frac{NP_t}{(1+r)^t} = 0 \quad (4)$$

where *IRR* is the internal return rate.

Cost-benefit estimates for the first year were increased according to expected inflation and the net profit at each year was discounted into the present value by means of a proper discount rate. This is in agreement with the methodology proposed by Molinos-Senante *et al.* (2012). In a detailed assessment for execution, this methodology could be refined and additional details could be included, but for the preliminary feasibility assessment within the scope of this work the current approach was deemed satisfactory.

Environmental assessment

An EIA was performed for two options: (i) hauling of leachate to the nearest municipal WWTP; and (ii) on-site treatment of leachate by the proposed technology (SBR). The method used for impact assessment of the project was TBL evaluation methodology. TBL is a framework of sustainability to measure the performance and success of a business with three criteria: people, planet, profit. It balances the criteria of environmental protection, minimizing community impacts, and minimizing economic impacts (costs) (Di Maria & Sisani 2019). The scoring system was carried out based on how the alternatives under investigation impact various criteria and the weight of each criterion was assigned according to its importance (Alhaddi 2015). In this project, environmental and community impacts were weighted equally at 30% and economic impact is weighted higher at 40%. The scoring, which indicates how effective the requirements have been achieved, as shown in Table S2 (Supplementary Material), was based on a TBL assessment. The scoring was performed on a scale of 1–3 following these guidelines (Plakas *et al.* 2016):

- Score 1, the lowest level, indicates that the option proposed has achieved the least criteria goals, causing major negative impacts and requires significant mitigation.
- Score 2, the mid-level, indicates the option proposed does not achieve all the criteria, causing moderate negative impacts and requiring some mitigation.
- Score 3, the highest level, indicates the option proposed have met all the criteria, giving maximum overall benefit and no mitigation is required.

Table S2 indicates how different phases of the project may affect various aspects such as the environment, community, and economics. Two levels of scoring were used in this assessment, '+' represents phases with minor impacts on the aspects and '++' indicates phases with major impacts (Pope *et al.* 2004).

With regard to these options, several assumptions were considered:

- The community residential area is located in a radius of more than 10 km from the landfill.
- The scope of this evaluation is limited only to the landfill and the activity associated with it.
- The landfill is already in operation, and environmental studies like flora and fauna biodiversity has already been conducted.
- The landfill has plenty of available space, thus no purchase of extra land for construction of a treatment plant is required.
- All the constructions discussed in this work were related to SBR treatment plants only and not for the existing RO plant.

RESULTS AND DISCUSSION

Economic assessment

Using the method presented in the previous section, the estimated capital costs and operational costs were calculated and are presented in Tables 2 and 3 respectively. From Table 2, it can be clearly seen that the SBR tank is the most expensive commodity as it is the significant part of the process and since a great many processes like biological reaction and sedimentation take place in the same tank, it is more expensive than the conventional stirred tank reactors for biological nutrient removal. Capital costs were estimated from catalogues of wastewater treatment plants. Costs estimated from actual purchase orders were increased by 20% to account for inflation and geographical differences, and rounded up to provide a conservative estimation. Complete

Table 2 | Estimated capital costs

Items	Value (€)
Tank	13,700
Blower and pumps	3,000
Diffusers	300
Piping	1,500
Electricity and control	2,000
Engineering	3,000
Construction	5,000
Sludge dewatering	5,000
Total	33,500

Table 3 | Estimated operational costs

Items	Value (€)
Energy	1,789
Staff	0
Reagents	3,938
Waste management	2,100
Maintenance	2,345
Others	3,350
Total	13,521

calculation report for costs estimation is presented in Table S3 (Supplementary Material).

Energy and reagents consumption, and amount of sludge generated were calculated according to conceptual design as shown in Table S1. The total amount of electricity required per cycle of operation was calculated to be 36.5 kWh. This accounts for the electricity consumed mainly by blowers and pumps as shown in Table S3. The unit electricity price for industries in the Czech Republic was found to be 0.07 €/kWh (Statista 2018). Hence the total cost for energy consumption per year of operation came to be around 1,789 €. The spot price of methanol was estimated to be 0.5 €/kg, from the contract price offered by Methanex Corporation (2019). The cost of disposal for dewatered sludge was found to be 150 €/ton from the current Plzen city waste incinerator rates (ZEVO Plzeň 2019). Maintenance costs were estimated with a heuristic criteria of 7% capital expenditure (CAPEX) according to the authors' experience and is consistent with typical values suggested by practitioners (Wendland 2005). The amount of ammoniacal nitrogen removed per cycle as a result of nitrification and denitrification processes was calculated to be 3.75 kg as shown in Table S1. For the calculation of benefits of nitrogen removal, a shadow price of 8 €/kg of nitrogen removed was used and the total benefits gained per year as a result of nitrogen removal was estimated to be 21,000 € (Molinos-Senante *et al.* 2010).

Inflation was set to 3% since it is the upper limit target for the Czech National Bank (2019) and they have historically succeeded in achieving the lower target. The discount rate was set to 10% yearly, being slightly higher than the Czech National Bank interest rate of 0.75% monthly (Czech National Bank 2018). With these parameters, economic performance of the project was estimated as shown in Table 4 and Figure 2. Total equivalent annualised cost

Table 4 | Cash flow over a period of 10 years

	Year										
Items	0	1	2	3	4	5	6	7	8	9	10
CAPEX (€/year)	−33,500	0	0	0	0	0	0	0	0	0	0
OPEX (€/year)	0	−13,927	−14,344	−14,775	−15,218	−15,675	−16,145	−16,629	−17,128	−17,642	−18,171
Benefits (€/year)	0	21,630	22,279	22,947	23,636	24,345	25,075	25,827	26,602	27,400	28,222
Cash flow (€/year)	−33,500	7,703	7,934	8,173	8,418	8,670	8,930	9,198	9,474	9,758	10,051
Discounted cash flow (€/year)	−33,500	7,003	6,557	6,140	5,749	5,384	5,041	4,720	4,420	4,139	3,875
Cumulative discounted cash flow (€/year)	−33,500	−26,497	−19,940	−13,799	−8,050	−2,666	2,374	7,095	11,514	15,653	19,528

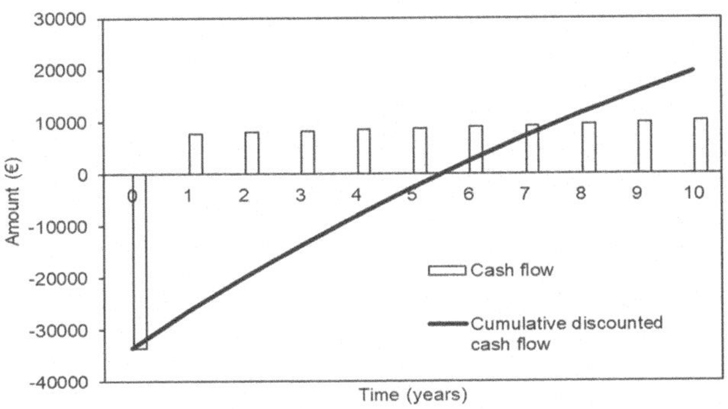

Figure 2 | Economic performance of the design over a period of 10 years.

is 18,973 €/year, which is already a good indicator of good economic performance since it is lower than the estimated benefit. The net present value (NPV) was found to be 19,526 € and the internal return rate (IRR) was estimated to be 21.6%. The payback time was found to be around 5.5 years from Figure 2 and it seems to be a short period of time for a project of this scale.

Environmental assessment

Impacts on the environment and community for both alternatives have been discussed separately in Tables 5 and 6 respectively.

Economic impacts

Option 1: Existing practice: No capital cost is involved and all the cost involved is linked to operation which is the hauling and treatment of leachate at WWTP.

Option 2: On-site treatment: Huge capital investment will be required for the construction of the treatment plant. Regarding the operational cost, it is less i.e. ~4 € as compared to the hauling of leachate to municipal WWTP that costs 8–10 €. Scoring for all the impacts mentioned above are included in Table S2.

Mitigation and monitoring measures

Noise: Noise is one of the main factors that can adversely affect human health if it is more than the permissible limit. It can cause different health related problems such as hearing loss, cardiovascular diseases, reduced productivity or focusing, social handicaps and many more (Münzel *et al.* 2018). It can also affect the calmness and peace of the local community surrounding the landfill facility. Noise at the landfill is mainly produced by the movement of heavy-duty vehicles such as compactors, trucks, etc. So, noise generated within the landfill facility will not only affect the labourers or personnel working there but the surrounding communities will also be effected. In order to mitigate these problems, several measures are proposed (Omar *et al.* 2012):

- an adequate buffer zone between the landfill site and local community should be provided;
- trees such as palm oil can be planted around the facility to absorb the noise;
- keeping the compactors and trucks in good condition to minimize vibrations and noise produced by them;
- providing workers with PPEs (ear plugs) to reduce the exposure to noise;
- the arrangement of cycle schedule for mobilization of materials and equipment;
- instalment of a sound barrier and vibration isolator;
- use of materials that can absorb noise such as foam or fiberglass.

Monitoring of noise level is required to be carried out regularly in the project area using a sound level meter in different representative points and times. The measurements are analysed and compared with the permissible level.

Odour: Another important environmental impact is the emission of odorous compounds from landfill and nuisance created by it. These compounds are produced as a result of decomposition of organic matter, by microorganisms, which is dumped in landfill. According to one study (Mitiani *et al.* 2008), more than 300 trace compounds were detected in the landfill gas. Among these compounds, unpleasant odour is mainly caused by sulphur-containing compounds such as sulphides and mercaptans. Apart from the decomposition of solid waste, odorous emissions are also produced during the landfill leachate treatment such as volatile fatty acids. Similarly, like noise, odour can also create an issue for nearby communities. Thus proper measures need to be taken in order to mitigate this odour problem (Mitiani *et al.* 2008):

- locate the site far from the local communities;

Table 5 | (a) Environmental impacts of hauling leachate to municipal WWTP and (b) community impacts of hauling leachate to municipal WWTP

	Description
(a) Environmental impacts	
Surface water quality	This option does not affect the surface water quality directly because landfill is not associated with the discharge to surface water but it can have some indirect impacts. For instance, if the leachate contains high concentrations of heavy metals and/or ammoniacal nitrogen, it can affect the performance of the municipal WWTP which can lead to a deterioration of surface water. Thus, a score of 2 was assigned
Greenhouse gases	Greenhouse gases are emitted only during the hauling of leachate which will affect the environment
Chemical use	Since no chemicals are used in this option, it does not have any impact
Treatment plant performance	Since leachate is being mixed with the municipal WWTP, so the plant has to treat wastewater which has high concentrations of heavy metals and ammoniacal nitrogen which can affect its performance
Spills	There can be moderate risks of spilling during the loading, hauling and unloading of leachate. These spills can affect soil, surface water and ground water quality
Flora and fauna	The biodiversity of flora and fauna is not affected since the habitat of the biotic component is not being interfered
Less burden on downstream treatment	Since the leachate is only being transported into the municipal WWTP with no preliminary treatment, it will cause burden to the downstream treatment
(b) Community impacts	
Aesthetics	The hauling process would not have any impacts on the aesthetics since the landfill is isolated
Health and safety	The transportation of the leachate may affect the community nearby and also indirectly impact on the surface water quality that has been mentioned above. Since there is not much labour involved, therefore the magnitude of safety is not the main concern
Job opportunity	No labour is involved in this practice
Odour	During transportation and loading/unloading of leachate some volatile gases can create a fouling smell
Noise	The transportation of leachate using vehicles may create noise
Construction duration	No construction involved
Society perception	Since the leachate is mixed with municipal wastewater, society may have a negative impression, especially on non-governmental organizations (NGOs) in this particular field

- design waste delivery routing and schedule in such a way that receptors have minimum exposure to odorous compounds;
- placing an adequate cover on waste in active cells of landfill every day;
- construction of gas abstraction wells and treatment plant for treating these polluted gases;
- plantation of Golden pothos (*Epipremnum aureum*) (Oyabu *et al.* 2003);
- proper sealing of source (SBR) from where it is originating;
- use of odour neutralizer such as metazene.

Measurement of odour is a very complex process but the chemical or compound responsible for creating that odour can be quantified using different gas detectors or odour monitoring software (Laor *et al.* 2014).

Health and safety: Waste or garbage, if not properly dumped in the landfill site, serves as a major breeding ground for vectors (disease carrying organisms) such as rodents, flies, pest and insects. Moreover, pathogenic bacteria and viruses are also present in this waste. These pathogens cause different diseases to site workers when they are exposed. Diseases include malaria, diarrhoea, cholera, skin and respiratory allergies, Dengue fever, jaundice and many more. Some of these diseases are transferable, so if one person is affected there is a great chance that the staff or workers working with that person can also be affected. Apart from these pathogens, landfill gas is also responsible for causing many

Table 6 | (a) Environmental impacts of onsite leachate treatment and (b) community impacts of onsite leachate treatment

	Description
(a) Environmental impacts	
Surface water quality	All parameters (pollutants) have been brought under the limits by onsite treatment of leachate. Hence it does not impose any threat to surface water quality
Greenhouse gases	Methane is produced during the decomposition of organic matter which is a potential greenhouse gas
Chemical use	Methanol is used during the biological treatment which is toxic and flammable
Treatment plant performance	Coupling a biological (SBR) with a physical (RO) process enhances the overall removal efficiency of pollutants
Spills	There is no risk of spilling
Flora and fauna	The biodiversity of flora and fauna is slightly affected since the habitat of the biotic component is affected during construction
Less burden on downstream treatment	Since the on-site treatment plant will treat the leachate so there will be less burden on the downstream treatment plant
Gaseous emissions	Ammonia is produced during anaerobic decomposition, which is toxic. Moreover, there will be generation of particulate matter (dust) during the construction phase
(b) Community impacts	
Aesthetics	The construction of the treatment plant would have minor impacts aesthetically
Health and safety	Since there is more labour involved, therefore their safety is important
Job opportunity	Operators will be required for the operation of SBR and RO thus creating new employment opportunities
Odour	Ammonia and hydrogen sulphide are produced during anaerobic decomposition which creates nuisance
Noise	Noise will be produced during the construction and operation phase of onsite treatment plant
Construction duration	It will take approximately six months for the treatment plant to be constructed
Society perception	Since the leachate is mixed with municipal wastewater, society may have a negative impression, especially on NGOs in this particular field. However, society perception may be influenced with the effort of company to perform Corporate Social Responsibility (CSR)
Solid waste	Sludge will be produced which requires proper handling or disposal
Land use	The treatment plant construction will occupy land which may be used for other purposes such as expansion of office buildings, parking lot, etc.

health issues such as birth problems, cancer etc. So, the health and safety of the public and workers can be protected by several mitigation measures (Maheshwari *et al.* 2015):

- the use of personal protection equipment (PPE) such as mask, earmuff, etc.;
- proper training for workers on SBR operation and waste and chemical handling;
- periodic workshop on health and safety;
- by following good hygiene practices;
- ensuring safe and proper handling of waste;
- creating awareness in workers and public regarding the healthcare to improve the life standards.

The health of workers can be monitored by regular medical check-ups and regular supervision or checking of PPE.

Solid waste: toxic sludge is produced as a result of operation of the SBR. The sludge cannot be dumped directly into landfill because it contains active biomass and heavy metals, and thus is toxic in nature. So, in order to dispose of this toxic sludge, it has to be treated. The following methods can be adapted for the treatment of sludge:

- solidification and usage of sludge as a raw material for concrete (Jamshidi *et al.* 2011);
- sludge can be transported to the incineration plant where it can be used as fuel (Stasta *et al.* 2006);

- utilizing solar energy for on-site drying of sludge beds (Salihoglu *et al.* 2007).

PRACTICAL APPLICATIONS AND FUTURE RESEARCH

Due to the Landfill Directive (99/31/EC), many developed countries in Europe have shifted to other sustainable ways of disposing solid waste instead of landfilling. The number of landfill disposal sites in most of the European countries has been decreasing over the past few years, due to the increase in environmental concern and awareness among governments. Thus, the results obtained from this study would be impertinent with regard to many developed countries in Europe. However, for countries where disposing solid waste in landfill is still prevalent, this research could be fruitful. Many developing countries such as Indonesia, Vietnam, Pakistan, India, Bangladesh, etc., are not equipped with proper landfill leachate treatment facilities and are still looking for cost effective and eco-friendly options. For these countries, this work could be applied as a baseline study for the development of on-site treatment plants. In Cipayung, Indonesia, stabilization ponds such as anaerobic, facultative, and maturation ponds are being used for landfill leachate treatment. The treatment plant's effluent quality still exceeds the standard limits of discharge (Noerfitriyani *et al.* 2018). To achieve the required effluent leachate quality that meets the discharge limits, SBR could be an option to be implemented. Moreover, for developed countries like USA or UK, where financial budget is not a limiting factor, SBR can be coupled with a RO system in order to reclaim the water for reuse inside landfill premises itself.

Based on the above analysis conducted, the following points could be investigated in future for broadening the scope of this study:

- Efficiency evaluation of SBR integrated treatment trains for ammoniacal nitrogen removal and salinity in landfill leachates.
- Economic and environmental assessment of alternative landfill leachate treatment technologies like membrane separation, coagulation, precipitation, etc.
- Study on the effects of other nutrients present in the leachate on ammoniacal nitrogen removal.
- Economic feasibility of landfill leachate management: impact of landfill specific site conditions, local regulations and leachate quality.
- This study is relevant for cold climatic countries. However, a similar study can be carried out for hot climatic countries in order to assess the effect of temperature on the performance of SBR.
- Various sludge management and treatment methods can also be compared and studied in the future.

CONCLUSIONS

The SBR was designed to remove ammoniacal nitrogen from Lany landfill leachate and has now been found to be both economically and environmentally feasible. The economic performance was assessed using cost-benefit analysis, estimating the environmental benefit through a shadow price concept tool, and satisfactory results were obtained. Shadow price methodology provided the implication of monetary value to the unaccountable positive externalities, thereby emphasizing significant importance to environmental benefits in addition to monetary benefits, which is very important in wastewater treatment studies due to growing concern over pollution and more stringent standards in the near future. Profit is estimated to be positive since total annualised equivalent cost is lower than benefit. The net present value is 19,526 € and the internal return rate is 21.6%. The TBL method was used for environmental impact analysis due to its sustainable approach of taking all the three main pillars, namely society, economics and environment, into account. Assessment of environmental, economic, and community impact on the existing practice (total score 55.1%) and on-site treatment (total score 59.6%) was compared. On-site treatment provides a lower impact in various aspects and it can be concluded that it is feasible to be implemented economically and environmentally. A mitigation plan was also prepared for the impacts identified, especially for noise, odour, health and safety aspects.

ACKNOWLEDGEMENTS

The authors extend their gratitude to the European Commission for providing funding support (project number: 2017-1957/001-001-EMJMD) to pursue the Erasmus Mundus master's degree program – International Master of Science in Environmental Technology and Engineering (IMETE) at University of Chemistry and Technology, Prague. The authors also thank Ekologie for providing necessary data regarding the landfill and leachate characteristics.

SUPPLEMENTARY MATERIAL

The Supplementary Material for this paper is available online at https://dx.doi.org/10.2166/aqua.2019.084.

REFERENCES

Abbas, A. A., Jingsong, G., Ping, L. Z., Ya, P. Y. & Al-Rekabi, W. S. 2009 Review on landfill leachate treatments. *J. Appl. Sci. Res.* **5** (5), 534–545.

Alhaddi, H. 2015 Triple bottom line and sustainability: a literature review. *Bus. Manage. Stud.* **1** (2), 6–10.

Chen, R. & Wang, X. C. 2009 Cost–benefit evaluation of a decentralized water system for wastewater reuse and environmental protection. *Water Sci. Technol.* **59** (8), 1515–1522.

Chu, D., Ye, Z. L., Chen, S. & Xiong, X. 2018 Comparative study of heavy metal residues in struvite products recovered from swine wastewater using fluidised bed and stirred reactors. *Water Sci. Technol.* **78** (8), 1642–1651.

Czech National Bank 2018 *CNB Discount Rate Changed Over Time.* Available from: www.cnb.cz/en/faq/how_has_the_cnb_discount_rate_changed_over_time.html (accessed 18 March 2019).

Czech National Bank 2019 *Inflation Report I/2019.* Available from: www.cnb.cz/miranda2/export/sites/www.cnb.cz/en/monetary_policy/inflation_reports/2019/2019_I/download/ir_I_2019.pdf (accessed 18 March 2019).

De, S., Hazra, T. & Dutta, A. 2019 Treatment of landfill leachate by integrated sequence of air stripping, coagulation–flocculation and adsorption. *Environ. Dev. Sustain.* **21** (2), 657–677.

Di Maria, F. & Sisani, F. 2019 A sustainability assessment for use on land or wastewater treatment of the digestate from bio-waste. *Waste Manage.* **87**, 741–750.

Gao, J., Oloibiri, V., Chys, M., Audenaert, W., Decostere, B., He, Y., Van Langenhove, H., Demeestere, K. & Van Hulle, S. W. 2015 The present status of landfill leachate treatment and its development trend from a technological point of view. *Rev. Environ. Sci. Biotechnol.* **14** (1), 93–122.

Godfrey, S., Labhasetwar, P. & Wate, S. 2009 Greywater reuse in residential schools in Madhya Pradesh, India-a case study of cost-benefit analysis. *Resour. Conserv. Recycl.* **53** (5), 287–293.

Helming, J. & Reinhard, S. 2009 Modelling the economic consequences of the EU water framework directive for Dutch agriculture. *J. Environ. Manage.* **91** (1), 114–123.

Iskander, S. M., Zhao, R., Pathak, A., Gupta, A., Pruden, A., Novak, J. T. & He, Z. 2018 A review of landfill leachate induced ultraviolet quenching substances: sources, characteristics, and treatment. *Water Res.* **145**, 297–311.

Jamshidi, A., Mehrdadi, N. & Jamshidi, M. 2011 Application of sewage dry sludge as fine aggregate in concrete. *J. Environ. Stud.* **37** (59), 4–6.

Joshi, S. M. & Gogate, P. R. 2019 Treatment of landfill leachate using different configurations of ultrasonic reactors combined with advanced oxidation processes. *Sep.. Purif. Technol.* **211**, 10–18.

Kong, Y., Liu, Y. Q., Tay, J. H., Wong, F. S. & Zhu, J. 2009 Aerobic granulation in sequencing batch reactors with different reactor height/diameter ratios. *Enzyme Microb. Technol.* **45** (5), 379–383.

Kurniawan, T. A. & Lo, W. H. 2009 Removal of refractory compounds from stabilized landfill leachate using an integrated H_2O_2 oxidation and granular activated carbon (GAC) adsorption treatment. *Water Res.* **43** (16), 4079–4091.

Laor, Y., Parker, D. & Pagé, T. 2014 Measurement, prediction, and monitoring of odors in the environment: a critical review. *Reviews in Chemical Engineering* **30** (2), 139–166.

Maheshwari, R., Gupta, S. & Das, K. 2015 Impact of landfill waste on health: an overview. *J. Environ. Sci. Toxicol. Food Technol.* **1** (4), 17–23.

Methanex Corporation 2019 *Methanex Posts Regional Contract Methanol Prices for North America, Europe and Asia.* Available from: https://www.methanex.com/our-business/pricing (accessed 18 March 2019).

Mitiani, Y., Shoji, Y. & Kuriyama, K. 2008 Estimating economic values of vegetation restoration with choice experiments: a case study of an endangered species in Lake Kasumigaura, Japan. *Landscape Ecol. Eng.* **4**, 103–113.

Molinos-Senante, M., Garrido-Baserba, M., Reif, R., Hernández-Sancho, F. & Poch, M. 2012 Assessment of wastewater treatment plant design for small communities: environmental and economic aspects. *Sci. Total Environ.* **427**, 11–18.

Molinos-Senante, M., Hernández-Sancho, F. & Sala-Garrido, R. 2010 Economic feasibility study for wastewater treatment: a cost-benefit analysis. *Sci. Total Environ.* **408**, 4396–4402.

Molinos-Senante, M., Hernández-Sancho, F. & Sala-Garrido, R. 2011 Cost–benefit analysis of water-reuse projects for environmental purposes: a case study for Spanish wastewater treatment plants. *J. Environ. Manage.* **92** (12), 3091–3097.

Münzel, T., Schmidt, F. P., Steven, S., Herzog, J., Daiber, A. & Sorensen, M. 2018 Environmental noise and the cardiovascular system. *J. Am. Coll. Cardiol.* **71** (6), 688–697.

Neczaj, E., Okoniewska, E. & Kacprzak, M. 2005 Treatment of landfill leachate by sequencing batch reactor. *Desalination* **185** (1–3), 357–362.

Noerfitriyani, E., Hartono, D. M., Moersidik, S. S. & Gusniani, I. 2018 Leachate characterization and performance evaluation of leachate treatment plant in Cipayung landfill, Indonesia. *In IOP Conference Series: Earth Environ. Sci.* **106** (1), 012086.

Omar, D., Karuppanan, S. & AyuniShafiea, F. 2012 Environmental health impact assessment of a sanitary landfill in an urban setting. *Procedia Soc. Behav. Sci.* **68**, 146–155.

Oyabu, T., Sawada, A., Onodera, T., Takenaka, K. & Wolverton, B. 2003 Characteristics of potted plants for removing offensive odors. *Sens. Actuators B Chem.* **89** (1–2), 131–136.

Pelaz, L., Gómez, A., Letona, A., Garralón, G. & Fdz-Polanco, M. 2018 Sequencing batch reactor process for the removal of nitrogen from anaerobically treated domestic wastewater. *Water Sci. Technol.* **77** (6), 1581–1590.

Plakas, K. V., Georgiadis, A. A. & Karabelas, A. J. 2016 Sustainability assessment of tertiary wastewater treatment technologies: a multi-criteria analysis. *Water Sci. Technol.* **73** (7), 1532–1540.

Pope, J., Annandale, D. & Morrison-Saunders, A. 2004 Conceptualising sustainability assessment. *Environ. Impact Assess. Rev.* **24** (6), 595–616.

Raper, E., Fisher, R., Anderson, D. R., Stephenson, T. & Soares, A. 2018 Alkalinity and external carbon requirements for denitrification-nitrification of coke wastewater. *Environ. Technol.* **39** (17), 2266–2277.

Salihoglu, N. K., Pinarli, V. & Salihoglu, G. 2007 Solar drying in sludge management in Turkey. *Renew. Energy* **32** (10), 1661–1675.

Seguí, L., Alfranca, O. & Garcia, J. 2009 Techno-economical evaluation of water reuse for wetland restoration: a case study in a natural park in Catalonia, Northeastern Spain. *Desalination* **246** (1–3), 179–189.

Souza, M. A. B. B., Oliveira, M. B., Araújo, A. D. S. F. & de Castro, J. A. 2014 Analyze of the density and viscosity of landfill leachate in different temperatures. *Am. J. Environ. Eng.* **4** (4), 71–74.

Stasta, P., Boran, J., Bebar, L., Stehlik, P. & Oral, J. 2006 Thermal processing of sewage sludge. *Appl. Therm. Eng.* **26** (13), 1420–1426.

Statista 2018 *Industrial Prices for Electricity in Czechia 2000–2017.* Available from: www.statista.com/statistics/596267/electricity-industry-price-czech-republic/ (accessed 18 March 2019).

Torretta, V., Ferronato, N., Katsoyiannis, I., Tolkou, A. & Airoldi, M. 2016 Novel and conventional technologies for landfill leachates treatment: a review. *Sustainability* **9** (1), 9–47.

Truong, H. T. B., Nguyen, T., Thi, P. & Bui, H. M. 2018 Integration of aerobic granular sludge and membrane filtration for tapioca processing wastewater treatment: fouling mechanism and granular stability. *J. Water Supply Res. Technol. Aqua* **67** (8), 846–857.

von Sperling, M. 2007 *Activated Sludge and Aerobic Biofilm Reactors.* IWA Publishing, London, UK.

Wendland, A. 2005 *Operation Costs of Wastewater Treatment Plants, Hamburg.* Available from: https://cgi.tu-harburg.de/~awwweb/wbt/emwater/documents/slides_c2.pdf (accessed 18 March 2019).

Yong, Z. J., Bashir, M. J., Ng, C. A., Sethupathi, S. & Lim, J. W. 2018 A sequential treatment of intermediate tropical landfill leachate using a sequencing batch reactor (SBR) and coagulation. *J. Environ. Manage.* **205**, 244–252.

ZEVO Plzeň 2019 *Disposal Pricelist.* Available from: www.zevoplzen.cz/cenik (accessed 18 March 2019).

First received 12 June 2019; accepted in revised form 5 September 2019. Available online 29 November 2019

Remediation of groundwater contamination from an old, non-functional landfill in Hořkovec open cast mine, Czech Republic

Adyl Anvarov, Adelaida Pelaez Angel, Beatriz Felices Rando and Jimena Lazaro Gil

ABSTRACT

The main aim of this work was to assess different suitable strategies for the remediation of groundwater contaminated with the leachate from an old, not functional landfill located next to Hořkovec open cast mine, in the Czech Republic. The leachate consisted of mainly chlorinated aliphatic compounds and aromatic volatile compounds. The site, that had already been treated, was observed to show rebounding effects after the first remediation treatment. This article analyses the possibilities of using different types of remediation technologies that include *in-situ* chemical oxidation (ISCO) with different oxidants (potassium permanganate, sodium permanganate and sodium persulfate), as well as *in-situ* bioremediation (ISB), and the combination of both types of treatment. From the results of the analysis it was concluded that the best option for this case was to carry out a pre-treatment of the area by ISCO with sodium persulfate as the oxidant agent and then a further biological treatment.

Key words | bioremediation, BTEX, groundwater remediation, *in-situ* chemical oxidation, landfill leachate, petroleum hydrocarbons

Adyl Anvarov (corresponding author)
Adelaida Pelaez Angel
Beatriz Felices Rando
Jimena Lazaro Gil
Department of Water Technology and Environmental Engineering,
University of Chemistry and Technology,
Technická 5, Dejvice, 166 28 Praha 6,
Czech Republic
E-mail: *adylanvar@gmail.com*

INTRODUCTION

Contamination of groundwater resources as a result of inappropriate construction and management of municipal and industrial waste landfills is a common problem worldwide. In the Czech Republic, the number of active landfills (for industrial, municipal and inert waste storage) increased drastically after the breakup of Czechoslovakia. However, the number of active landfills decreased over time as a result of the development of new legislation. In detail, the number of active landfills decreased from a total of 10,000 in 1990 to 380 in 1996 (Fagin 2001). Moreover, the entry of the Czech Republic into the European Union in 2004 entailed the entry into force of new environmental laws and regulation. Consequently, a total number of 178 waste disposal sites are currently under operation in the country, while the rest have been closed and are under post-closure operations (Martinát *et al.* 2016). Within the post-closure operations, the Council Directive 1999/31/EC of 26 April 1999 states the requirement to perform a monitoring plan for at least 25 years in order to ensure the absence of adverse effects on the environment and people. Therefore, research into suitable soil and water remediation techniques is essential to avoid the negative impact of the landfills into the environment.

The studied landfill was established on January 22nd, 1968 and it was used as a storage of wastes from paint and varnish industries, as well as pharmaceutical and personal care products. The groundwater in the contaminated site was isolated from the leachate coming from the

doi: 10.2166/aqua.2019.198

aquifer and received further treatment in 1990. However, the monitoring of the treated area showed a rebounding effect, and therefore, an increment in contaminants' concentration during the last years. The concentration of the groundwater contaminants, characterized by a monitoring well, is presented in Table 1. The most abundant compounds found were chlorinated aliphatic compounds (CACs) and aromatic volatile compounds – benzene, toluene, ethylbenzene and xylenes (BTEX).

Table 1 | Concentration of detected contaminants in the monitoring well; Czech Republic standards and pollution indicator concentration

Compound	Czech Republic standards [μg/L]	Pollution indicator [μg/L]	Monitoring well [μg/L]
Benzene	1	0.39	597
Toluene	0.2	860	187
Ethylbenzene	0,2	1.3	64.3
o-xylene	0.2	–	54.5
1,2-dichlorethane	3	0.15	1,960
Chloroform	2.5	0.19	19,400
Tetrachlorethene	10	9.7	21.1
Trichlorethene	10	0.44	664
cis-1,2-dichlorethene	0.1	28	73.2
Vinylchloride	0.5	0.015	<100
Dichlormethane	0.1	9.9	1,620

Regarding the distribution of the contaminants within the aquifer, aromatic volatile compounds and chlorinated hydrocarbons can be present in four different states (Figure 1): dissolved in water, adsorbed on solid particles, vapour and free phase (when solubility limit of pollutant in water is exceeded). The free phase, known as non-aqueous phase liquid (NAPL), can be divided into two different subdivisions according to their specific gravity – specific weight higher than water (dense – DNAPLs) and specific weight lower than water (light – LNAPLs). The vast majority of petroleum hydrocarbons belongs to LNAPLs (Bouchard 1989).

Numerous methods and techniques have been developed for groundwater remediation during the last decades. However, the differences in the nature of the contamination, the water properties, and the aquifer soil properties do not allow the use of identical remediation approaches at different sites.

Wasay *et al.* (1999) studied the efficiency of using granular activated carbon, ferric chloride (coagulation agent) and granular activated alumina for heavy metals' removal from a contaminated groundwater reservoir. Results showed that the maximum cation separation of Cd (II), Cu (II), Cr (III), Mn (II), Pb (II) and Zn (II) was achieved by granular

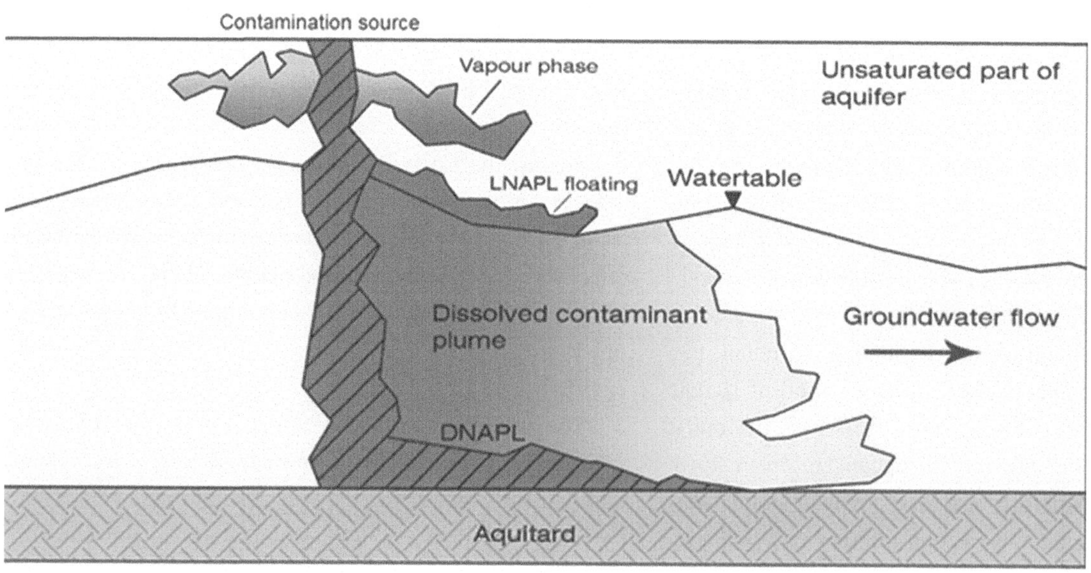

Figure 1 | Petroleum hydrocarbons' distribution through unsaturated and saturated part of the aquifer profile.

activated carbon at neutral pH. By using this technique, heavy metals' removal percentages between 80% and 96% were obtained. Liang & Liu (2008) investigated the possibilities for treating leachate from a municipal solid waste landfill by means of anammox microorganisms. This technique showed a removal capacity of 97% for ammonia nitrogen, 87% for total nitrogen and 89% for chemical oxygen demand (COD). Moreover, Li *et al.* (2009) studied the performance of a field-scale aged refuse biofilter for treating a landfill leachate. Results showed that the continuous operation of the biofilter can decrease the amount of ammonia and COD content in the leachate. The average COD concentration dropped from 5,478 mg/L to 267 mg/L and the mean ammonia concentration from 811 mg/L to 6 mg/L. Liang *et al.* (2008) researched the possibility of treating groundwater and soil slurries contaminated with gasoline and diesel fuels by oxidative degradation. To this aim, activated persulfate was used as the main oxidizing agent. The treatment efficiency was considered to be dependent on the activation method (thermal or chemical), the oxidizing agent dose and the presence of chelating agents. Moreover, membrane technologies have also been studied for separating wastewater from petroleum hydrocarbons. In this paper, an analysis of the advantages, disadvantages and limitations of different remediation technologies is presented. In view of their suitability for the treatment of the studied leachate, a combination of different technologies is proposed. Additionally, a financial assessment and an environmental impact assessment of the proposed treatment are also included.

MATERIALS AND METHODS

Different methods for *in-situ* chemical oxidation (ISCO) and *in-situ* bioremediation (ISB) were reviewed in order to assess their suitability for contamination removal.

In-situ chemical oxidation

ISCO is a technology that consists of the introduction of chemical oxidants into the groundwater subsurface with the aim of destroying organic pollutants. ISCO is feasible when the pollutants can be mineralized to carbon dioxide and water or when they can be oxidized into a less harmful sub-product that can be treated afterwards. The main limitation of ISCO technologies is to deliver the oxidant agent in the sense that it can be physically in contact with the target contaminant. Another important limitation is that the oxidant agent is consumed by subsurface chemical reactions that are not related to the degradation of the target pollutants. This loss of oxidant is called natural oxidant demand (NOD) (Haselow *et al.* 2003).

The suitability of using potassium permanganate, sodium permanganate, and sodium persulfate as oxidants was assessed with regard to the characteristics of the contaminants and the treatment area. In this sense, several factors need to be considered when selecting the most appropriate oxidant: (a) delivery and distribution of the oxidant; (b) rate and extent of NOD; (c) rate and extent of reaction with the contaminant; and (d) geochemical effects from the oxidation.

Potassium and sodium permanganate are the most common forms of permanganate (MnO_4^-). Both of them are used for ISCO of chlorinated solvents, as they produce MnO_2 and CO_2 when reacting with organic compounds, as well as other intermediate organic products (Yin & Allen 1999). They are non-specific oxidizers with a low oxidation potential and high NOD that can be used within a wide range of pH values and they do not require a catalyst (Environmental Protection Agency). The main advantages of using permanganate as an oxidant are: (i) no exothermic reactions; (ii) catalysts are not required; (iii) pH control is not required; and (iv) bioremediation can accompany oxidation because neither the oxidants nor the by-products are toxic for the microbial population (Kluck & Achari 2004). However, some drawbacks must be taken into account as well: (i) manganese remains dissolved in the area, as well as unreacted permanganate, that leaves a purple colour in groundwater (Interstate Technology Regulatory Council 2005); and (ii) MnO_2 can precipitate, occupying pore space, reducing soil permeability and creating a coating that would reduce the contact surface between the oxidant and the NAPL (Environmental Security Technology Certification Program 1999). However, there is not an exact approach to quantify to what extent this drawback affects contaminant oxidation (Interstate Technology Regulatory Council 2005).

However, permanganate can have an adverse effect on bioremediation if used in large amounts (Major *et al.* 2002). Figure 2 shows the percentages of removal efficiency of bioremediation and different oxidation techniques when combined (Sutton *et al.* 2011). It is observed that the best results are obtained for sodium persulfate (SPS) and permanganate, and that the percentage of bioremediation efficiency for the last one is noticeably reduced.

Regarding SPS, it is considered as an emerging oxidant for ISCO technology for the treatment of organic pollutants in groundwater, soils and sediments (Brown *et al.* 2002). Persulfate is a strong oxidant with relatively high stability under normal subsurface conditions, thus being able to reach the contaminated zone. In comparison with permanganate, persulfate has lower NOD, so less quantity of persulfate is needed in order to treat pollutants, which is an important factor in terms of cost-effectiveness and sustainability (Petri *et al.* 2011).

Regarding its reaction chemistry, it can react by direct electron transfer (direct oxidation) or by its decomposition into free radicals that can degrade target organic compounds. Persulfate decomposition into free radicals requires previous activation by: heat, chelated or non-chelated transition metals (especially iron), hydrogen peroxide, and alkaline pH conditions. Activated persulfate

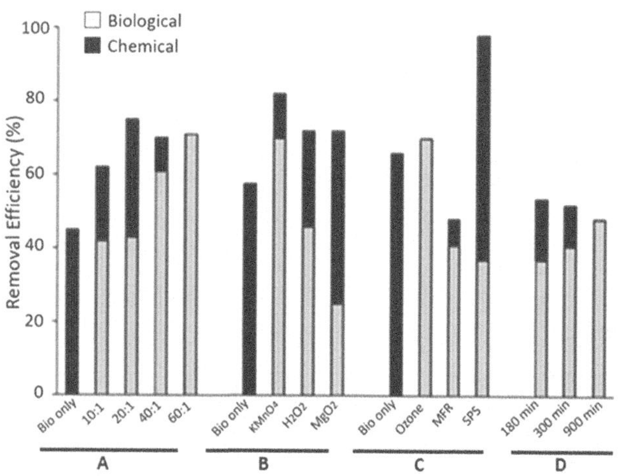

Figure 2 | Summary of removal efficiency for combined chemical and biological remediation in instances where a variety of *in-situ* chemical oxidation treatments were tested (sodium persulfate and permanganate were used as an oxidation agent). Labels indicate what follows. A: H_2O_2/Fe ratio used for treatment; B: type of chemical oxidant used prior to incubation for 134 days; C: type of chemical oxidant, including modified Fenton's reagent and iron-activated SPS, and data are for the first 20 days of experimentation; and D: duration of ozone treatment (Sutton *et al.* 2011).

reaction is more rapid and less selective than direct oxidation. However, persulfate is consumed more rapidly and it could be less productive (Petri *et al.* 2011).

Different types of persulfate activation were reviewed with the aim of assessing the most suitable one. In the case of heat activation, it may result in very aggressive oxidizing conditions; reaction mechanisms may shift with increasing temperature, affecting the overall reaction efficiency. Installation of a parallel heating system to heat the aquifer matrix to the desired temperature implies high operational costs. Regarding its effectiveness against DNAPL contamination, the increasing temperature may produce volatilization and dissolution of contaminants, enhancing further treatment of dissolved DNAPL and contaminant desorption. Another degradation process that also occurs at high temperatures is hydrolysis. However, chlorinated ethenes are generally resistant to hydrolysis (Jeffers *et al.* 1989).

Oxidant activation by special metal compounds is another option to generate strong radicals. Several studies show that iron-chelated complexes improve the effectiveness of treatment compared to non-chelated iron systems (Block *et al.* 2004). The reason is that the oxidant/iron/chelator molar ratio appears to have a strong impact on the overall effectiveness. Another advantage of chelated activation in comparison to the non-chelated one is that the first one occurs at neutral pH conditions, while the last one requires acid pH (Petri *et al.* 2011).

Finally, in terms of its potential efficiency, several studies have demonstrated that persulfate is able to degrade all CAC present in this study area (Waldemer *et al.* 2007). Block *et al.* (2004) developed an investigation about the degradation of these compounds in aqueous systems under a variety of persulfate activation methods: Fe (II) and Fe (III), chelated Fe (III), hydrogen peroxide, alkaline pH, and without activation. The results showed that in the case of activation methods, the concentrations of chloroethene were substantially lower in comparison to the inactivated ones, proving to be more effective (Petri *et al.* 2011).

Determining dosing requirements

A sufficient amount of oxidant injection into the targeted treatment area is one of the most important factors in the

ISCO process. In order to calculate the oxidant dosing requirement, the sum of three different factors has to be performed: (1) the stoichiometric demand, (2) the reaction with reduced species in the soil matrix (NOD) or soil oxidant demand and (3) the loss due to decomposition.

In the case of persulfate, the only component of NOD that reacts with this compound is reduced metals, as it does not react with natural organic material (NOM) (Brown *et al.* 2004). In this sense, the reaction with reduced metals is rapid and near stoichiometric, and the reaction with NOM is very slow and indistinguishable from decomposition, which is the most important non-beneficial reaction. For this reason, persulfate dose is a function of the stoichiometric demand and of the quantity required to compensate the amount lost by decomposition (Brown 2010).

The stoichiometric demand is the stoichiometric requirement times the mass of the contaminant, and is calculated using the equivalent weight of persulfate and the equivalent weight of the contaminant being oxidized (Equation (1)). The equivalent weight is the molecular weight divided by the number of electrons transferred (usual valence).

$$\text{Stoichimetric requirement} = \frac{\text{Equivalent weight oxidant}}{\text{Equivalent weight contaminant}} \tag{1}$$

Persulfate, as peroxygen compound, will decompose to oxygen and sulfate. The decomposition test carried out by Brown (2010) suggests that persulfate losses continue to occur over time for all concentrations of persulfate, and decomposition increases with increasing levels of added persulfate and with experiment duration time. A pilot test can be developed in order to calculate the amount of persulfate to be decomposed. Moreover, several titration procedures by potassium permanganate can estimate the amount of oxidant lost due to the decomposition.

In order to calculate the decomposition of persulfate by titration, the following procedure is used:

- Preparation of 50% (wt-wt) slurries of persulfate and soil from the site. Persulfate concentrations should be 50, 100 and 150 gl/L (5, 10 and 15%).
- Take a 5 mL aliquot of the liquid, add 25 mL of 25% of sulfuric acid, and stir for 2 minutes.
- Add dried (105 °C for 24 hours) ferrous sulfate heptahydrate ($FeSO_4 \times H_2O$). Add 0.75 g for the 5%, 1.4 g for the 10% and 2.0 g for 15%. Stir for 10 minutes.
- Titrate to a persistent pink colour with 0.1 N KMnO4.
- Calculate the concentration of persulfate by Equation (2):

$$Na_2S_2O_8\left(\frac{g}{L}\right) = \left[\frac{((\text{Weight } FeSO_4 \times H_2O)/278.01) - ((\text{mL } KMnO_4/1000) \times N\ KMnO_4)}{\text{mL sample}/1000}\right] \times 119 \tag{2}$$

This procedure needs to be done every 2–5 days for a total of 10–20 days of time. The length of time between samples depends on the rate of decomposition of the persulfate.

In-situ bioremediation (ISB)

ISB is a biological treatment employing microorganisms such as bacteria and fungi to transform organic pollutants into non-toxic compounds or mineralize them to inorganic compounds like carbon dioxide, methane, water and inorganic salts (Farhadian *et al.* 2008; Wolicka *et al.* 2009; Perelo 2010). In this sense, considering the literature review performed, Table 2 presents the most suitable bacteria strains capable of degrading the pollutants present in the aquifer.

ISB has two major advantages: the first is the reduction of cost while the second one is that it can lead to a complete mineralization of the pollutants (Farhadian *et al.* 2008; Perelo 2010). However, it is important to emphasize that bioremediation is not always suitable; this depends on such factors as the pollutant biodegradability and the geophysical conditions of the aquifer to be treated. That includes soil structure and stratification, pH, temperature, hydraulic conductivity, presence of nutrients, etc. (National Research Council 1993). Especially important for ensuring microbial growth is the presence of elements such as nitrogen, phosphorus and potassium, which can be provided as additional nutrients (Henry 2010).

Table 2 | Bacteria used for biodegradation of target contaminants (Aulenta *et al.* 2006; Farhadian *et al.* 2008; Pérez-de-Mora *et al.* 2014)

Pollutant	Microorganisms
BTEX	*Rhodococcus rhodochrous, Rhodococcus* sp., *Pseudomonas* sp., *Pseudomonas putida, Pseudomonas fluorescens, Cladophialophora* sp., *Achromobacter xylosoxidans*
BTX	*Geobacteraceae, Rhodococcus pyridinovorans*
BTE, TCE	*Pseudomonas putida*
Toluene	*Ralstonia picketii, Burkholderia cepacia, Pseudomonas putida, Blastochloris sulfoviridis, Geobacter* sp., *Thauera aromatica, Desulfobacula toluolica, Dechloromonas* sp.
Benzene	*Pseudomonas aeruginosa, Dechloromonas* sp.
Ethylbenzene	*Azoarcus* sp. strain EB1
Aromatic compounds and chloroaliphatics	*Pseudomonas putida*
PCE, TCE, DCE, VC	*Dehalococcoides ethenogenes, Dehalococcoides* sp.
PCE, TCE	*Desulfitobacterium* sp., *Clostridium bifermentans, Dehalobacter restrictus, Enterobacter agglomerans, Desulfuromonas chloroethenica, Desulfuromonas michiganensis, Dehalospirillum multivorans, Sulfurospirillum halorespirans, Desulfomonile tiedjei*

PCE, perchloroethene; TCE, trichloroethene; DCE, dichloroethene; VC, vinylchloride.

Rapid impact assessment matrix

The rapid impact assessment matrix (RIAM) method was developed by Pastakia & Madsen (1996) and consists of evaluating each scenario within the following categories of impacts: (i) physical and chemical impacts, (ii) ecological and biological impacts, (iii) sociological and cultural impacts and (iv) economical and operational impacts. These categories of impacts were subdivided into specific impacts that were further evaluated one by one in terms of their severity. The evaluation was done by considering a previously defined criteria. Two main groups were distinguished: (A) criteria related to the condition, which can affect individually the score obtained; (B) criteria related to the situation that should not affect individually the score obtained. These groups are composed of the following criteria: importance (A1), magnitude (A2), permanence (B1), reversibility (B3) and cumulation (B3). Thus, a score was assigned to each impact and for each of the four different criteria related to the performance and scope of the project.

The aim was to categorize all the identified impacts as positive or negative. To achieve this, the scores for the criteria (group B) were summed together, which helps to account for all factors and create an overall, joint result. The sum of the B group criteria was multiplied by the product of the group A criteria to obtain a final environmental score (ES) for each of the impacts (Equation (3)):

$$ES = (A1 \times A2) \times (B1 + B2 + B3) \quad (3)$$

In group A, the importance (A1) refers to the degree of significance of a particular impact, taking into consideration the spatial boundaries of the project. The score ranges from +4 (important to national/international interests) to −4 (no importance). The magnitude (A2) is stated as a measure of the scale of benefit/disbenefit of an impact; it can take values in the range of +3 (major positive benefit) to −3 (major disbenefit or change) (Figure 3).

Regarding group B, permanence (B1) defines if an impact is temporary or permanent, describing the temporal status of the condition. The values range from 1 (no change/not applicable) to 3 (permanent). Reversibility (B2) refers to the possibility of reconstruction of the factor affected by the project, i.e., the possibility of returning to the initial conditions prior to the action by natural means. The values range from 1 (no change/not applicable) to 3 (irreversible). Finally, cumulation is a measure of whether the impact will have a single direct impact or whether cumulative effects will take place over time, or a synergistic effect with other impacts. The value ranges from 1 (no change/not applicable) to 3 (cumulative/synergistic) (Figure 3). The ES

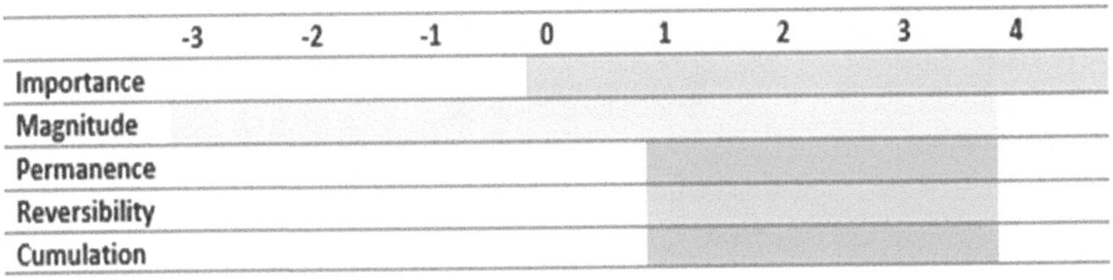

Figure 3 | Range of environmental impact values assigned to each factor.

values can be further classified according to the bands shown in Table 3.

Financial analysis

The cost for each remediation technique is highly dependent on the specific conditions of each site. Therefore, adequate approximation of the costs were used based on the values proposed in the literature (Harkness & Farnum 2010).

Additionally, the cost categories were also divided according to the protocol given by Harkness & Farnum (2010). According to the authors, the cost information of each remediation technique can be classified into four general categories (design, capital, operation and maintenance; and monitoring) and in turn each category is divided into other sub-categories as shown in Figure 4.

The total investment required for the proposed scenario was obtained by calculating the net present value (NPV), according to Equation (4), where NC_t is the net cost at time t, r is the discount rate and T is the time horizon of the project. In this study, the value for r was taken as 4.5%, the average discount rate for the Czech Republic in the past 24 years (Trading Economics 2019).

$$\mathrm{NPV} = \sum_{t=0}^{T} \frac{NC_t}{(1+r)^t} \qquad (4)$$

RESULTS AND DISCUSSION

In-situ chemical oxidation (ISCO)

After the analysis of the properties of the site pollutants and the different possible technologies for their remediation, as well as the experience of scientists in the field (Table 4), two scenarios were proposed: (1) ISB combined with ISCO using sodium persulfate or (2) ISB combined with

Table 3 | Range bands used for ES score of the environmental impact assessment

RIAM environmental score (ES)	Range value (RV) (Alphabetical)	Description of range band
108 to 72	E	Major positive change/impact
71 to 36	D	Significant positive change/impact
35 to 19	C	Moderate positive change/impact
10 to 18	B	Positive change/impact
1 to 9	A	Slight positive change/impact
0	N	No change/status quo/not applicable
−1 to −9	−A	Slight negative change/impact
−10 to −18	−B	Negative change/impact
−19 to −35	−C	Moderate negative change/impact
−36 to −71	−D	Significant negative change/impact
−72 to −108	−E	Major negative change/impact

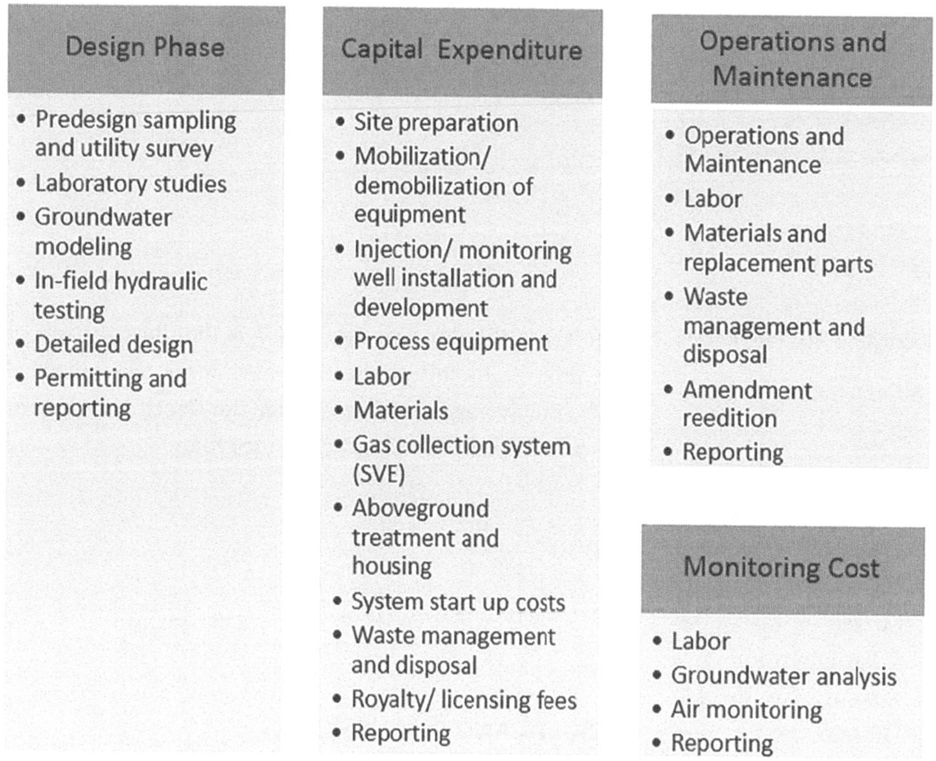

Figure 4 | Categories and sub-categories for cost analysis (adapted from Harkness & Farnum 2010).

ISCO using permanganate. ISB was the preferred method due to all its advantages: it is a natural process that provides the possibility of a complete mineralization of the pollutants, and the cost of the treatment and equipment needed are very low. However, the high concentration and toxicity of the pollutants found in the site requires the use of additional chemical treatment of the contaminated area. This additional treatment includes the previously exposed ISCO methods prior to ISB by means of the oxidants already presented: potassium/sodium permanganate or sodium persulfate. Since the intention is to use this technology together with bioremediation, the preferred oxidant is sodium persulfate because, as shown in Figure 2, the efficiency of biological treatment when using sodium persulfate is much higher than using permanganate, and the global removal efficiency using biological treatment combined ISCO with sodium persulfate appears to be the highest. The success of the treatment relies on two factors: (a) the site characterization and (b) the ability to deliver an adequate amount of oxidant (Henry 2010).

Pilot test and injection plan

In order to characterize the site and the contaminants present a pilot test is necessary. In this sense, the pilot test is an important phase in every type of remediation study in order to understand how the targeted site will respond to the specified remediation strategy. In addition, it provides information to assess the geological conditions and geochemical properties of the aquifer (Brown 2010). Pilot tests are carried out at the site, under *in-situ* conditions, by monitoring wells and soil borings, so they provide an accurate insight about what to expect during full-scale application. They also provide insight into design parameters such as achievable radius of influence (ROI) and field injection rates for full-scale applications. In addition, information about oxidant dosage and its distribution within the aquifer, NOD, and NADP and metals' presence can also be obtained. Finally, initial phase characterization provides information regarding geological conditions and geochemical properties of the aquifer (Table 5) (Brown 2010).

Table 4 | Scientific experience on the treatment of leachate plume from different types of landfill

Author	Target contaminants	Methodology	Results
Chellam & Clifford (2002)	Leachate from uranium mining and milling sites (molybdenum, selenium, uranium and radium)	Coagulation with iron ions (Fe^{3+}) at different pH values (low, neutral and high)	Efficiency of coagulation process directly depends on the type of chemical compound. For instance, radium was successfully removed at both high and low pH, while molybdenum was removed only at low pH
Zhong *et al.* (2017)	Recalcitrant chemicals present in leachate	Physico-chemical operations (coagulation-flocculation, chemical precipitation, filtrations, adsorption and ion exchange)	It was assessed that different methods have both advantages and disadvantages. For instance, coagulation and precipitation have a high sludge production rate that must undergo further treatment; Different type of filtrations (microfiltration, ultrafiltration and reverse osmosis) typically costly and required pre-treatment operation; Ion exchange and adsorption has high efficiency but typically used as a polishing step in the end of treatment train
Jun *et al.* (2009)	Leachate from industrial landfill containing mostly ammonium, heavy metals and organic contaminants	Sequence of permeable reactive barrier filled by zeolites and granular zero valent iron as a reactive media	Reactor used in the pilot-scale operation showed that it has potential to decompose highly recalcitrant chemical compounds and convert them to easily biodegradable substances. As a result of the research it was investigated that BOD to COD ratio rapidly increased. Reactive media were acting as a reducing agent and consequently performed redox reactions
Christensen *et al.* (2000)	Leachate plume from an old non-functional landfill (petroleum hydrocarbons and chlorinated solvents)	Natural attenuation	Results showed that application of cost-effective strategy (natural attenuation) can provide mass reduction of the target contaminants. However, it is suggested to use such approach only in sites where conventional treatment (pump and treat, physico-chemical operations, etc.) were already performed

Regarding the injection of the required oxidant, an injection plan has to be previously designed. This design should ensure an appropriate contact between the ISCO reagents and the target contaminant. Due to the fact that the distribution of the reagent depends on the characteristics of the site, the injection plan needs to be based on the pilot test results and model development. In this sense, hydraulic data are crucial, and can be obtained from pump tests, dipole test and conventional slug tests conducted in specific areas of the contaminant plume (Naval Facilities Engineering Command-NAVFAC 2013).

Attending to the characteristics of the studied area, the 'direct injection method' is selected as the preferred one, as it is less equipment intensive and quicker in comparison to other methods such as recirculation. It consists of the direct injection of the reagent into the subsurface in a determined volume of water from an external source, consequently displacing groundwater according to the volume of reagent injected.

In order to design the layout and spacing of the injection wells, a key design parameter is the ROI around a treatment well. The ROI determines the lateral area of the aquifer that can be impacted by the treatment well over the duration of the remediation treatment. In order to accurately calculate it, it is necessary to develop a groundwater model, using the design pumping rate to

Table 5 | Site characterization parameters and design implications (based on pilot test)

Parameters	Description	Design impact
Nature and extent of contamination	Horizontal and vertical distribution of contaminants Presence of NAPLs	Horizontal or vertical locations to introduce oxidants Impact oxidant demand
Human and ecological health risks	Determination of contaminants and their ecotoxicology	Influence on treatment goals
Fate and transport of contaminants	Movement of contaminants within the aquifer (flow rate)	Location of injection wells Concentration of oxidants and their introduction
Hydrogeology	Lithology (heterogeneities, grain size, permeability) Gradients Confined or unconfined aquifer Saturated thickness Hydraulic conductivities Darcy velocity Mineralogy Groundwater flow	Approach to introduce oxidants into aquifer Metal mobilization (mineralogy) Low conductivity regions might need additional injections Radius of influence NOD (persulfate reacts with reduced metals)
Hydrogeochemistry	Distribution coefficients (Kd) pH Buffering capacity Dissolved oxygen (DO) Oxidation reduction potential	Approach to introduce oxidants into aquifer
Plume size		Oxidant dosification Well layout
Depth to groundwater		Well design

determine the expected hydraulic gradients and imparted groundwater flow field.

Environmental impact assessment (EIA) and financial analysis

Regarding the EIA, the ES values obtained from the RIAM matrix were classified according to Table 6. No values belonging to E or −E categories were found, meaning that there will not be any extremely strong impact from any of the factors (neither positive nor negative). Table 6 shows all the factors included in the categories D and −D which are defined as significant positive and negative impacts, respectively, while Table 7 shows the most significant impacts. Finally, a bubble diagram (Figure 5) was used as an alternative technique to represent environmental impact magnitude.

For the purposes of financial analysis, the treatment train proposed (ISCO + ISB) is expected to last four years with a monitoring plan for 30 years. Table 8 shows the site characteristic values obtained from the literature, own calculations, preliminary design and other reliable sources. Table 9 summarizes the cost for the treatment train proposed, where it was assumed that a crew of four people will be hired to perform the injection of the treatments and this strategy will be done annually (up to four years) over 3 weeks. Thus, practically for the first two years, persulfate for ISCO treatment will be injected, while during the last two years, lactic acid will be injected as substrate for the microorganism in order to stimulate ISB. It was also assumed that the wells installed will be suitable for both ISCO and ISB, where the injection plan was also assumed to be through a manifold using a dosing pump connected to the river that passes nearby. As additional analysis, monitoring was ascertained as the category with the highest cost. Indeed, this category consumes a great deal of time, monitoring instruments and labour. Operation and maintenance is the second most expensive category.

Table 6 | Summary of RIAM results

		ES	RB	A1 Importance (0 → 4)	A2 Magnitude (−3 → 3)	B1 Perfomance (1 → 3)	B2 Reversibility (1 → 3)	B3 Cumulation (1 → 3)
Physical and chemical								
Environmental quality	Increase of gases emissons level	−64	− **D**	4	−2	2	3	3
	Increase of particular matter emissons level	−64	− **D**	4	−2	2	3	3
Hydrogeology	Groundwater quality alternation	63	**D**	3	3	2	2	3
	Surface water quality alternation	63	**D**	3	3	2	2	3
Ecological and biological								
Flora and fauna	Loss of biodiversity	−48	− **D**	3	−2	2	3	3
	Affectation to protected species	−48	− **D**	3	−2	2	3	3
Economical and operational impacts								
Demographic system	Direct or indirect employment creation	40	**D**	4	2			
	Public health	48	**D**	2	3	2	3	3
Sociological and cultural impacts								
Cultural and historical heritage	Impact on the value of the protected area	42	**D**	3	2	2	2	3

Table 7 | Most significant positive (+D) and negative (−D) impacts

Significant negative impacts (−D)	Significant positive impacts (+D)
Increase of gaseous emissions levels	Alteration of groundwater quality
Increase of particulate matter levels	Alteration of surface water quality
Risk of loss of biodiversity	Direct or indirect employment
Effect on the protected species	Public health
	Impact on the value of the protected area

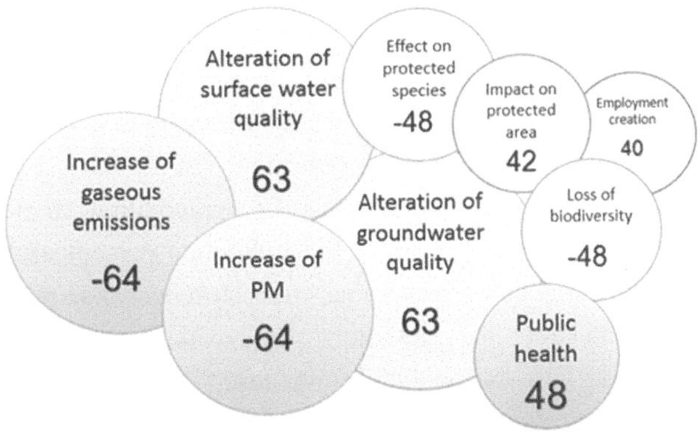

Figure 5 | Bubble diagram showing the magnitude of impacts from the project.

CONCLUSIONS

The composition of the contaminants present in the leachate from the landfill was analysed and, according to it, the suitability of ISCO (with either $KMnO_4$, $NaMnO_4$ or SPS as the main oxidants) and ISB was evaluated as a remediation treatment, showing that the best option encountered was to perform a pre-treatment by ISCO with SPS and a following treatment using bioremediation. Since the variety of contaminants present in the leachate is very complex, difficulties were faced during the literature review due to the lack of similar case studies.

Table 8 | Summary of site parameters and values used to calculated the treatment costs

Parameter	Value
Length of the landfill (m)	125
Width of landfill (m)	75
Area of the landfill (m^2)	9,375
Injection rate (L/min)	15
Natural oxidant demand-NOD (persulfate) (g/kg solids)	3
Substrate adsorption (g/kg solids)	1.5
Well ROI (m)[a]	12
Length between wells (m)	24
Number of wells needed	21

[a]Value from Krembs *et al.* (2011).

Table 9 | Summary of costs for the treatment done by ISCO + ISB

Summary of costs (€)	
Total design	71,100
Total capital	75,690
Total operation and maintenance (4 yr NPV)	110,800
Total monitoring (30 yr NPV)	376,218
Overall total (30 yr NPV)	**633,808**

Moreover, the geology and geochemistry of the area is another important factor while evaluating the suitability of different technologies, such as the presence of chelating agents, etc. In this sense, the performance of a pilot test is highly required before designing the remediation treatment in order to characterize the pollutants present and the characteristics of the treated area. Moreover, key parameters needed for designing the oxidant injection method can also be obtained. Pilot tests also provide an insight about the future performance and efficiency of the selected remediation technology.

Finally, performing a financial assessment is an important part of assessing the economical feasibility of the proposed remediation project. Moreover, an environmental impact assessment also needs to be carried out in order to identify the negative impacts of the remediation project. In this sense, mitigation activities can be assessed in order to reduce these impacts.

REFERENCES

Aulenta, F., Majone, M. & Tandoi, V. 2006 Enhanced anaerobic bioremediation of chlorinated solvents: environmental factors influencing microbial activity and their relevance under field conditions. *Journal of Chemical Technology & Biotechnology* **81** (9), 1463–1474.

Block, P. A., Brown, R. A. & Robinson, D. 2004 Novel activation technologies for sodium persulfate in situ chemical oxidation. In: *Proceedings of the Fourth International Conference on the Remediation of Chlorinated and Recalcitrant Compounds*, Monterey, CA, USA.

Bouchard, D. C. 1989 Contaminant transport in the subsurface: sorption equilibrium and the role of nonaqueous phase liquids. In: *Intermedia Pollutant Transport* (D. T. Allen, Y. Cohen & I. R. Kaplan, eds). Springer, Boston, MA, USA, pp. 189–211.

Brown, R. A. 2010 Chemical oxidation and reduction for chlorinated solvent remediation. In: *In Situ Remediation of Chlorinated Solvent Plumes* (H. Stroo & C. Ward, eds). Springer, New York, USA, pp. 481–535.

Brown, R., Robinson, D., Sethi, D. & Block, P. 2002 Second generation persulfate ISCO. In: *Second International Conference on Oxidation and Reduction Technologies for In-Situ Treatment of Soil and Groundwater*, Toronto, Ontario, Canada.

Brown, R. A., Robinson, D., Skladany, G. & Loeper, J. 2004 Response to naturally occurring organic material: permanganate versus persulfate. In: *Proceedings of the Fourth International Conference of Remediation of Chlorinated and Recalcitrant Compounds*, Monterey, CA, USA.

Chellam, S. & Clifford, D. A. 2002 Physical–chemical treatment of groundwater contaminated by leachate from surface disposal of uranium tailings. *Journal of Environmental Engineering* **128** (10), 942–952.

Christensen, T. H., Bjerg, P. L. & Kjeldsen, P. 2000 Natural attenuation: a feasible approach to remediation of groundwater pollution at landfills? *Groundwater Monitoring & Remediation* **20** (1), 69–77.

Environmental Security Technology Certification Program. 1999 *Technology Status Review: In Situ Oxidation*. ESTCP, Arlington, VA, USA.

Fagin, A. 2001 Environmental capacity building in the Czech Republic. *Environment and Planning A* **33** (4), 589–606.

Farhadian, M., Vachelard, C., Duchez, D. & Larroche, C. 2008 *In situ* bioremediation of monoaromatic pollutants in groundwater: a review. *Bioresource Technology* **99** (13), 5296–5308.

Harkness, M. & Farnum, R. 2010 Cost analyses for remedial options. In: *In Situ Remediation of Chlorinated Solvent Plumes* (H. F. Stroo & C. H. Ward, eds). Springer, New York, USA, pp. 677–697.

Haselow, J. S., Siegrist, R. L., Crimi, M. & Jarosch, T. 2003 Estimating the total oxidant demand for in situ chemical oxidation design. *Remediation Journal* **13** (4), 5–16.

Henry, B. M. 2010 Biostimulation for anaerobic bioremediation of chlorinated solvents. *In Situ Remediation of Chlorinated Solvent Plumes* (H. Stroo & C. Ward, eds). Springer, New York, USA, pp. 357–423.

Interstate Technology and Regulatory Council. 2005 *Technical and Regulatory Guidance for in Situ Chemical Oxidation of Contaminated Soil and Groundwater*, 2nd edn. ISCO-2, Washington, DC, USA.

Jeffers, P. M., Ward, L. M., Woytowitch, L. M. & Wolfe, N. L. 1989 Homogeneous hydrolysis rate constants for selected chlorinated methanes, ethanes, ethenes, and propanes. *Environmental Science & Technology* **23** (8), 965–969.

Jun, D., Yongsheng, Z., Weihong, Z. & Mei, H. 2009 Laboratory study on sequenced permeable reactive barrier remediation for landfill leachate-contaminated groundwater. *Journal of Hazardous Materials* **161** (1), 224–230.

Kluck, C. & Achari, G. 2004 Chemical oxidation techniques for in situ remediation of hydrocarbon impacted soils. *Environmental Engineering*, 1–8.

Krembs, F. J., Clayton, W. S. & Marley, M. C. 2011 Evaluation of ISCO field applications and performance in situ chemical oxidation for groundwater remediation. In: *In Situ Chemical Oxidation for Groundwater Remediation, SERDP ESTCP Environmental Remediation Technology* (R. L. Siegrist, M. Crimi & T. J. Simpkin, eds). Springer Publishers, New York, USA, pp. 319–354.

Li, H., Zhao, Y., Shi, L. & Gu, Y. 2009 Three-stage aged refuse biofilter for the treatment of landfill leachate. *Journal of Environmental Sciences* **21** (1), 70–75.

Liang, Z. & Liu, J. 2008 Landfill leachate treatment with a novel process: anaerobic ammonium oxidation (Anammox) combined with soil infiltration system. *Journal of Hazardous Materials* **151** (1), 202–212.

Liang, C., Huang, C. F. & Chen, Y. J. 2008 Potential for activated persulfate degradation of BTEX contamination. *Water Research* **42** (15), 4091–4100.

Major, D. W., McMaster, M. L., Cox, E. E., Edwards, E. A., Dworatzek, S. M., Hendrickson, E. R. & Buonamici, L. W. 2002 Field demonstration of successful bioaugmentation to achieve dechlorination of tetrachloroethene to ethene. *Environmental Science & Technology* **36** (23), 5106–5116.

Martinát, S., Navrátil, J., Dvořák, P., Van der Horst, D., Klusáček, P., Kunc, J. & Frantál, B. 2016 Where AD plants wildly grow: the spatio-temporal diffusion of agricultural biogas production in the Czech Republic. *Renewable Energy* **95**, 85–97.

National Research Council. 1993 *In Situ Bioremediation: When Does it Work?* National Academies Press, Washington, DC, USA.

Pastakia, C. & Madsen, K. N. 1996 A rapid assessment matrix for use in water related projects. In: *Stockholm Water Conference*, Stockholm, Sweden, pp. 12–15.

Perelo, L. W. 2010 In situ and bioremediation of organic pollutants in aquatic sediments. *Journal of Hazardous Materials* **177** (1–3), 81–89.

Pérez-de-Mora, A., Zila, A., McMaster, M. L. & Edwards, E. A. 2014 Bioremediation of chlorinated ethenes in fractured bedrock and associated changes in dechlorinating and nondechlorinating microbial populations. *Environmental Science & Technology* **48** (10), 5770–5779.

Petri, B. G., Watts, R. J., Tsitonaki, A., Crimi, M., Thomson, N. R. & Teel, A. L. 2011 Fundamentals of ISCO using persulfate. In: *In Situ Chemical Oxidation for Groundwater Remediation* (R. L. Siegrist, M. Crimi & T. Simpkin, eds). Springer, New York, USA, pp. 147–191.

Sutton, N. B., Grotenhuis, J. T. C., Langenhoff, A. A. M. & Rijnaarts, H. H. M. 2011 Efforts to improve coupled in situ chemical oxidation with bioremediation: a review of optimization strategies. *Journal of Soils and Sediments* **11** (1), 129–140.

Trading Economics 2019 TRADING ECONOMICS. https://tradingeconomics.com/ (accessed September 2019).

Waldemer, R. H., Tratnyek, P. G., Johnson, R. L. & Nurmi, J. T. 2007 Oxidation of chlorinated ethenes by heat-activated persulfate: kinetics and products. *Environmental Science & Technology* **41** (3), 1010–1015.

Wasay, S. A., Barrington, S. & Tokunaga, S. 1999 Efficiency of GAC for treatment of leachate from soil washing process. *Water, Air, and Soil Pollution* **116** (3–4), 449–460.

Wolicka, D., Suszek, A., Borkowski, A. & Bielecka, A. 2009 Application of aerobic microorganisms in bioremediation in situ of soil contaminated by petroleum products. *Bioresource Technology* **100** (13), 3221–3227.

Yin, Y. & Allen, H. E. 1999 *In Situ Chemical Treatment*. Technology evaluation report, GWRTAC E-series report. www.gwrtac.org.

Zhong, H., Tian, Y., Yang, Q., Brusseau, M. L., Yang, L. & Zeng, G. 2017 Degradation of landfill leachate compounds by persulfate for groundwater remediation. *Chemical Engineering Journal* **307**, 399–407.

First received 29 June 2019; accepted in revised form 20 September 2019. Available online 28 November 2019

IWA Publishing's authorised EU representative for General Product Safety Regulations is Diane D'Arras, 15 rue Duret, 75116 Paris, France, e-mail: safety@iwap.co.uk.

Printed and bound by CPI Group (UK) Ltd, Croydon, CR0 4YY
12/05/2026
02108881-0001